Dynamic Random Walks

Theory and Applications

Dynamic Random Walks

Theory and Applications

by

NADINE GUILLOTIN-PLANTARD
Institut C. Jordan
Université Claude Bernard
Lyon, France

RENÉ SCHOTT
IECN and LORIA
Université Henri Poincaré
Nancy, France

ELSEVIER
Amsterdam – Boston – Heidelberg – London – New York – Oxford
Paris – San Diego – San Francisco – Singapore – Sydney – Tokyo

ELSEVIER B.V.
Radarweg 29
P.O. Box 211, 1000 AE Amsterdam
The Netherlands

ELSEVIER Inc.
525 B Street, Suite 1900
San Diego, CA 92101-4495
USA

ELSEVIER Ltd
The Boulevard, Langford Lane
Kidlington, Oxford OX5 1GB
UK

ELSEVIER Ltd
84 Theobalds Road
London WC1X 8RR
UK

© 2006 Elsevier B.V. All rights reserved.

This work is protected under copyright by Elsevier B.V., and the following terms and conditions apply to its use:

Photocopying
Single photocopies of single chapters may be made for personal use as allowed by national copyright laws. Permission of the Publisher and payment of a fee is required for all other photocopying, including multiple or systematic copying, copying for advertising or promotional purposes, resale, and all forms of document delivery. Special rates are available for educational institutions that wish to make photocopies for non-profit educational classroom use.

Permissions may be sought directly from Elsevier's Rights Department in Oxford, UK: phone (+44) 1865 843830, fax (+44) 1865 853333, e-mail: permissions@elsevier.com. Requests may also be completed on-line via the Elsevier homepage (http://www.elsevier.com/locate/permissions).

In the USA, users may clear permissions and make payments through the Copyright Clearance Center, Inc., 222 Rosewood Drive, Danvers, MA 01923, USA; phone: (+1) (978) 7508400, fax: (+1) (978) 7504744, and in the UK through the Copyright Licensing Agency Rapid Clearance Service (CLARCS), 90 Tottenham Court Road, London W1P 0LP, UK; phone: (+44) 20 7631 5555; fax: (+44) 20 7631 5500. Other countries may have a local reprographic rights agency for payments.

Derivative Works
Tables of contents may be reproduced for internal circulation, but permission of the Publisher is required for external resale or distribution of such material. Permission of the Publisher is required for all other derivative works, including compilations and translations.

Electronic Storage or Usage
Permission of the Publisher is required to store or use electronically any material contained in this work, including any chapter or part of a chapter.

Except as outlined above, no part of this work may be reproduced, stored in a retrieval system or transmitted in any form or by any means, electronic, mechanical, photocopying, recording or otherwise, without prior written permission of the Publisher.
Address permissions requests to: Elsevier's Rights Department, at the fax and e-mail addresses noted above.

Notice
No responsibility is assumed by the Publisher for any injury and/or damage to persons or property as a matter of products liability, negligence or otherwise, or from any use or operation of any methods, products, instructions or ideas contained in the material herein. Because of rapid advances in the medical sciences, in particular, independent verification of diagnoses and drug dosages should be made.

First edition 2006

Library of Congress Cataloging in Publication Data
A catalog record is available from the Library of Congress.

British Library Cataloguing in Publication Data
A catalogue record is available from the British Library.

ISBN-13: 978-0-444-52735-6
ISBN-10: 0-444-52735-4

∞ The paper used in this publication meets the requirements of ANSI/NISO Z39.48-1992 (Permanence of Paper).
Printed in The Netherlands.

to Anne, Bertrand,
Marjorie, Maxence,
Micheline and Nelly.

Contents

Preface   xi

Part I   THEORETICAL ASPECTS

1. PRELIMINARIES ON DYNAMIC RANDOM WALKS   3
   1. Introduction   3
   2. Definitions   5
   3. A riemannian dynamic random walk   6
   4. Examples   7

2. LIMIT THEOREMS FOR DYNAMIC RANDOM WALKS   11
   1. A strong law of large numbers   11
   2. A central limit theorem   12
   3. A local limit theorem   14
   4. A Strassen's Functional Law of the Iterated Logarithm   20
   5. A functional large deviation principle   23

3. RECURRENCE AND TRANSIENCE   33
   1. Introduction   33
   2. The one-dimensional case   33
   3. The higher-dimensional case   35

4. DYNAMIC RANDOM WALKS IN A RANDOM SCENERY   39
   1. The recurrent case   41
   2. The transient case   59
   3. A particular dynamical system: The rotation on the torus   66

5. ERGODIC THEOREMS   71
   1. Introduction   71
   2. Principal Results   72

3.	Proof of Theorems 5.4 and 5.5	76
4.	Proof of Theorem 5.3	77
5.	Proof of Theorem 5.6	81

6. DYNAMIC RANDOM WALKS ON HEISENBERG GROUPS — 83
1.	Introduction	83
2.	Generalities on Heisenberg groups	84
3.	Limit theorems	85

7. DYNAMIC QUANTUM BERNOULLI RANDOM WALKS — 99
1.	Introduction	99
2.	Quantum probabilistic notions	99
3.	Quantum Bernoulli random walks	100
4.	The dual of $SU(2)$	102
5.	Quantum Bernoulli random walks as random walks on the dual of $SU(2)$	105
6.	Dynamic random walks on the dual of $SU(2)$	105

Part II   APPLICATIONS

8. DISTRIBUTED ALGORITHMS WITH DYNAMICAL RANDOM TRANSITIONS — 119
1.	Colliding stacks	119
2.	The banker algorithm	128

9. DATA STRUCTURES WITH DYNAMICAL RANDOM TRANSITIONS — 143
1.	Introduction	143
2.	Preliminaries	143
3.	Dynamic linear lists	146
4.	Dynamic priority queues	154
5.	Dynamic dictionaries	155
6.	An example: Linear lists and rotation on the torus	158

10. TRANSIENT RANDOM WALKS ON DYNAMICALLY ORIENTED LATTICES — 167
1.	Introduction	167
2.	Model and results	168
3.	Proofs	172
4.	Examples	186

11. ASSET PRICING IN DYNAMIC (B,S)-MARKETS — 191
1.	Introduction	191

2.	Absence of Arbitrage of Dynamic (B,S)-Markets	194
3.	Completeness of Dynamic (B,S)-Markets	206
4.	Fair Pricing and Hedging Strategies in Complete Dynamic Markets	210
5.	γ-Pricing and γ-Hedging	217
6.	Asymptotic Behavior of Binary (B,S)-markets	221

Appendices		230
A– Ergodic theory		231
1.	Some definitions and basic theorems	231
2.	Examples of dynamical systems	233
B– Some Results on Diophantine Approximations		235
C– Skorohod metric		241
D– Fourier series		243
E– Hilbert spaces, representations, *-algebras, von Neumann algebras		245
1.	Hilbert spaces	245
2.	Lie algebras and representations	246
3.	*-algebras and von Neumann algebras	247

References	249
Index	265

Preface

The theory of random walks has a long history which goes back (at least) to the start of the last century. Feller's books [49, 50] contain preliminary material on the topic. Spitzer's book [177] is a friendly introduction to the domain. Self-avoiding random walks have been intensively investigated in connection with problems in physics and percolation theory. In the 70's the study of random walks on non-commutative algebraic structures (Lie groups, homogeneous spaces, hypergroups, ...) has become an important area of research [79, 81]. Meanwhile random walk tools have found many applications in computer science including algorithm analysis [132] as well as in other fields. More recently [68, 69, 70] dynamic random walks have been introduced and investigated. The main purpose of this book is to report on the progress realized in the emerging domain of dynamic random walks. We provide a smooth introduction to this area and present some applications in computer science and option pricing in financial markets. Most of the material is scattered throughout available literature, however, we have nowhere found all of this material collected in accessible form. The theory of dynamic random walks as exposed in this book has been developed by N. Guillotin-Plantard [68, 69, 70]. The applications in computer science presented in this book are mostly based on joint works between N. Guillotin-Plantard and R. Schott. More applications of random walks in computer science are described in [132]. The model of dynamic financial market based on (binomial) dynamic random walks presented in this book may be considered to be a small step towards option pricing when decisions are time dependent random variables.

Part I is devoted to theoretical aspects of dynamic random walks: Chapter 1 contains some preliminaries and basic definitions. Limit theorems (law of large numbers, central limit theorem, local limit theorem, law of the iterated logarithm, large deviation principle as well as results about recurrence and transience which are consequences of three first mentioned limit theorems) are given in Chapter 2 and 3. Some aspects of dynamic random walks in a random scenery are exposed in Chapter 4 (the reader may consult the bibliography for complements). Chapter 5 is devoted to ergodic theorems, here we use a slightly different dynamic random-walk model. Dynamic random walks on Heisenberg groups are investigated in Chapter 6. Chapter 7 is devoted to dynamic quantum Bernoulli random walks.

Part *II* contains several applications of the results stated in Part *I*: Chapter 8 revisits some distributed algorithms whose analysis has been started in the 70's by D.E. Knuth then continued by A. Yao, P. Flajolet, G. Louchard, R. Maier, and the second author of the present book. A similar study is done in Chapter 9 for dynamic data structures whose combinatorial analysis goes also back to D.E. Knuth and was also continued by some of the above mentioned researchers in collaboration with J. Françon, C. Kenyon-Mathieu, C. Puel and J. Vuillemin. Chapter 10 presents results obtained by N. Guillotin-Plantard and A. Le Ny on random walks on dynamically oriented lattices. Chapter 11 reports on work in progress on the use of the binomial dynamic random-walk model in financial mathematics. Few appendices will help refreshing memories (if necessary!) and provide adapted places for some intricate technical results.

Finally we would mention that M. Lin, B. Rubshtein and A.L. Wittmann [116] have developed a more abstract theory of random walks with dynamic random transitions on locally compact groups.

This book is intended for mathematicians, computer scientists and all researchers interested by recent developments in probability theory and their applications. Each chapter contains didactical material as well as advanced technical sections.

The authors are particularly grateful to B. Pinçon for his help on performing simulations and solving a differential equation, to D. Petritis for his scientific advises, to F. Comets, F. Delarue, C. Dombry, A. Le Ny and H. Schurz for allowing us to use material which is joint work with the present authors and to our colleagues from IECN, LORIA and Institut C. Jordan for stimulating discussions.

I
THEORETICAL ASPECTS

Chapter 1

PRELIMINARIES ON DYNAMIC RANDOM WALKS

1. INTRODUCTION

Consider a particle which moves randomly along the x-axis, coming to rest only at the points $x = \ldots, -2, -1, 0, 1, 2, \ldots$ with integral coordinates. Suppose the particle's motion is such that once at a point i, it jumps at the next step to either the point $i+1$ or the point $i-1$ with probabilities p and $q = 1 - p$, respectively. Let $S(n)$ be the particle's position after n steps. Then the sequence $S(0) \to S(1) \to S(2) \to \ldots$ is a Markov chain with transition probabilities

$$p_{ij} = \begin{cases} p & \text{if } j = i+1 \\ q & \text{if } j = i-1 \\ 0 & \text{otherwise} \end{cases}$$

$S(n)$ is a one-dimensional random walk which is often used in many areas including physics, computer science, information theory, mathematical finance, games theory, ... For example physicists use such a random-walk model as a crude approximation to one-dimensional diffusion or Brownian motion: A physical particle is exposed to a great number of molecular collisions which impart to it a random motion. The case $p > q$ corresponds to a drift to the right when shocks from the left are more probable, when $p = q = \frac{1}{2}$, the random walk is called symmetric.

The classical ruin problem uses a similar model: Consider the gambler who wins or loses a dollar with probability p and q, respectively. Let his capital be z and let him play against an adversary with initial capital $a - z$, so that the combined capital is a. The game continues until the gambler's capital either is reduced to zero or has increased to a, that is, until one of the two players is ruined. One is interested in the probability distribution of the duration of the game.

4 DYNAMIC RANDOM WALKS

The banker algorithm (whose analysis is presented in the second part of this book, see also [120, 121, 124, 132]) provides another interesting application of similar random-walk models: Consider (for simplicity) two customers C_1 and C_2 sharing a given resource m (money). There are fixed upper bounds m_1 and m_2 on how much of the resource each of the customers is allowed to use at any time. The banker decides to affect to the customer C_i, $i = 1, 2$ the required resource units only if the remaining units are sufficient in order to fulfill the requirements of C_j, $j = 1, 2$; $j \neq i$. The natural formulation of this algorithm is in terms of random walks in a rectangle with a broken corner, i.e.,

$$\{(x_1, x_2) : 0 \leq x_1 \leq m_1, 0 \leq x_2 \leq m_2, x_1 + x_2 \leq m\}$$

where the last constraint generates the broken corner. The random walk $Y_m(.)$ has four reflecting barriers (R) and one absorbing barrier (A) as shown in Figure 1.1. The distribution of steps (steps of unit length) ΔY is given by:

$$P(\Delta Y = (1,0)) = p_1, \ P(\Delta Y = (-1,0)) = q_1, \ P(\Delta Y = (0,1)) = p_2,$$
$$P(\Delta Y = (0,-1)) = q_2,$$

$p_1 + q_1 + p_2 + q_2 = 1$, with the boundary conditions of Figure 1.1. The

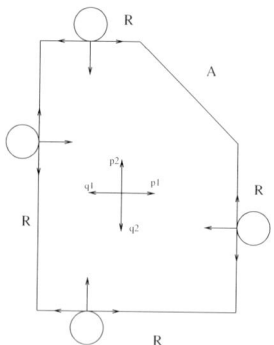

Figure 1.1.   Banker algorithm

hitting place and the hitting time of the absorbing boundary are the parameters of interest in computer science.

We believe that the reader is aware of similar examples where such random-walk models fit in.

The above introduced random-walk model would be more realistic if p_i (and therefore q_i), $i = 1, 2$, would be time and space dependent. Random walks with probabilities varying from place to place have been intensively studied (see [49, 50] for example).

The main purpose of this book is to introduce and investigate a dynamic random-walk model which is closer to real-world applications.

2. DEFINITIONS

Let $S = (E, \mathcal{A}, \mu, T)$ be a dynamical system where $(E, \mathcal{A}, \mu)$ is a probability space and T is a transformation defined on E. Let $d \geq 1$ and $f_1, \ldots, f_d$ be functions defined on E with values in $[0, \frac{1}{d}]$. Let $(X_i)_{i \geq 1}$ be a sequence of independent random vectors with values in $\mathbb{Z}^d$. Let $x \in E$ and $(e_j)_{1 \leq j \leq d}$ be the unit coordinate vectors of $\mathbb{Z}^d$. For every $i \geq 1$, the law of the random vector X_i is given by

$$\mathbb{P}(X_i = z) = \begin{cases} f_j(T^i x) & \text{if } z = e_j \\ \frac{1}{d} - f_j(T^i x) & \text{if } z = -e_j \\ 0 & \text{otherwise} \end{cases}$$

We write

$$S_0 = 0, \quad S_n = \sum_{i=1}^{n} X_i \text{ for } n \geq 1$$

for the $\mathbb{Z}^d$-random walk generated by the family $(X_i)_{i \geq 1}$. The random sequence $(S_n)_{n \geq 0}$ is called a dynamic $\mathbb{Z}^d$-random walk.

It is worth remarking that if the functions f_j are constant then we have the classical random walks but if these functions are not all constant, $(S_n)_{n \in \mathbb{N}}$ is a non-homogeneous Markov chain. The dynamic random walks were introduced and studied by the first author in [68], [69], [70]. They are of some relevance in the statistical mechanics of quasiperiodic systems in the presence of external spatial disorder (see [51], [83] and [104]).

Let $\mathcal{C}_1(S)$ denote the class of functions $f \in L^1(E, \mu)$ satisfying the following condition (H_1): there exists a set $E_0 \subseteq E$ with probability 1 such that for every $x \in E_0$,

$$\left| \sum_{i=1}^{n} \left(f(T^i x) - \int_E f(x) d\mu(x) \right) \right| = o\left(\frac{\sqrt{n}}{\log n} \right)$$

Let $\mathcal{C}_2(S)$ denote the class of functions $f \in L^1(E, \mu)$ satisfying the following condition (H_2):

$$\sup_{x \in E} \left| \sum_{i=1}^{n} \left(f(T^i x) - \int_E f(x) d\mu(x) \right) \right| = o(\sqrt{n})$$

Let $C_3(S)$ denote the class of functions $f \in L^1(E,\mu)$ satisfying the following condition (H_3):

$$\sup_{x\in E}\left|\sum_{i=1}^{n}\left(f(T^ix)-\int_E f(x)d\mu(x)\right)\right|=o\left(\frac{\sqrt{n}}{\log n}\right).$$

We assume that for $1 \leq i \leq d$, $\int_E f_i(x)d\mu(x) = \frac{1}{2d}$ and denote by $A = (a_{ij})_{1\leq i,j\leq d}$ the matrix with coefficients

$$a_{jj} = \frac{1}{d^2}\int_E (d+1-4d^2 f_j^2(x))d\mu(x)$$

$$a_{ij} = a_{ji} = \frac{1}{d^2}\int_E (1-4d^2 f_i(x)f_j(x))d\mu(x).$$

The matrix A is in fact the limit of the covariance matrix of $\frac{S_n}{\sqrt{n}}$ so A is nonnegative definite.

3. A RIEMANNIAN DYNAMIC RANDOM WALK

Let $f_1,\ldots,f_d$ be one-periodic functions defined on $\mathbb{R}$ with values in $[0,\frac{1}{d}]$ and $(X_{i,n})_{1\leq i\leq n}$ be a sequence of independent random vectors with values in $\mathbb{Z}^d$ with distribution

$$\mathbb{P}(X_{i,n}=z)=\begin{cases} f_j(x+\frac{i}{n}) & \text{if } z=e_j \\ \frac{1}{d}-f_j(x+\frac{i}{n}) & \text{if } z=-e_j \\ 0 & \text{otherwise} \end{cases} \quad (1.1)$$

We write

$$S_0 = 0, \quad S_n = \sum_{i=1}^{n} X_{i,n} \text{ for } n \geq 1$$

this n-dynamic random walk. This random walk is more difficult to study than the previous one due to the presence of n in the transition probabilities which creates much more temporal inhomogeneity.

PROPOSITION 1 *Let f be a one-periodic function which can be expanded into a Fourier series $f(x) = \sum_{h\in\mathbb{Z}} c_h e^{2\pi ihx}$.*
When there exists $\beta > 1$ such that $|c_n| + |c_{-n}| = \mathcal{O}(n^{-\beta})$, then

$$\sup_{x\in[0,1]}\left|\sum_{i=1}^{n}(f(x+\frac{i}{n})-\int_{[0,1]} f(t)dt\right|=\mathcal{O}(n^{1-\beta}).$$

The proof of the proposition is straightforward. When the functions $f_1,\ldots,f_d$ can be expanded into a Fourier series

$$f_j(x) = \sum_{h\in\mathbb{Z}} c_h^{(j)} e^{2\pi ihx}$$

where the coefficients $(c_h^{(j)})_{h \in \mathbb{Z}}$ satisfy: $c_0^{(j)} = \frac{1}{2d}$ and there exists $\beta_j > 1$ such that
$$|c_n^{(j)}| + |c_{-n}^{(j)}| = \mathcal{O}(n^{-\beta_j}).$$

Remark:
Thanks to the above uniform inequality all limit theorems which we are going to prove in the next chapters for the dynamic random walk defined in 2. can be adapted for the dynamic random walk given by (1.1).

4. EXAMPLES

4.1 THE DYNAMIC $\mathbb{Z}^D$-RANDOM WALK GENERATED BY AN IRRATIONAL ROTATION ON THE TORUS

Let S be the dynamical system $(\mathbb{T}^r, \mathcal{B}(\mathbb{T}^r), \lambda, T_\alpha)$ where T_α is the rotation on the r-dimensional torus $\mathbb{T}^r$ associated with the r-dimensional irrational vector $\alpha = (\alpha_1, \ldots, \alpha_r)$ defined by $x \to x + \alpha \mod 1$, and λ is the Lebesgue measure on $\mathbb{T}^r$. Two cases are considered:

1. $r = 1$
   We denote a_n the n-th partial quotient of α, i.e.
   $$\alpha = [\alpha] + \cfrac{1}{a_1 + \cfrac{1}{a_2 + \ldots}}.$$

   From the following proposition, the classes $\mathcal{C}_i(S), i = 1, \ldots, 3$ can be chosen as the set of functions of bounded variation defined on $\mathbb{T}^1$ with values in $\mathbb{R}$.

   PROPOSITION 2 *Let f be a function with bounded variation $V(f)$. For every irrational α such that the inequality $a_m < m^{1+\epsilon}$, where $\epsilon > 0$, is satisfied eventually for all m,*
   $$\sup_{x \in \mathbb{T}^1} \left| \sum_{l=1}^n (f(T_\alpha^l x) - \int_{\mathbb{T}^1} f(t)dt) \right| = \mathcal{O}(\log^{2+\epsilon} n).$$

   The proof of this result can be found in [68] or Appendix B. It is based on Denjoy-Koksma's inequality and diophantine approximations.

2. $r \geq 1$
   The following proposition also permits us to choose large classes $\mathcal{C}_i(S), i = 1, \ldots, 3$ for the dynamical system $S = (\mathbb{T}^r, \mathcal{B}(\mathbb{T}^r), \lambda, T_\alpha)$.

PROPOSITION 3 *Let f be a function with bounded variation in the sense of Hardy and Krause, and α an irrational vector of type η, then*

$$\sup_{x \in \mathbb{T}^r} \left| \sum_{l=1}^{n} (f(T_\alpha^l x) - \int_{\mathbb{T}^r} f(t)dt) \right| = \begin{cases} \mathcal{O}(\log^{r+1} n) & \text{if } \eta = 1 \\ \mathcal{O}(n^{1 - \frac{1}{((\eta-1)r+1)}} \log n) & \text{if } \eta > 1. \end{cases}$$

To prove this result, we used the Hlawka-Zaremba theorem and determined the asymptotic behavior of the discrepancy of the sequence $(x + k\alpha)_{k \in \mathbb{N}}$ in terms of the arithmetic properties of the angle α (see [68] for definitions and details or Appendix B). If the type of the vector η is such that $1 \leq \eta < 1 + \frac{1}{r}$, $C_i(S), i = 1, 2$ can be chosen as the set of functions with bounded variation in the sense of Hardy and Krause.

If α is of type $\eta \geq 1 + \frac{1}{r}$, the classes $C_i(S), i = 1, \ldots, 3$ has to be restricted. A function f defined on $\mathbb{R}^r$, 1-periodic in each variable, which can be expanded into a multiple Fourier series $f(t) = \sum_{h \in \mathbb{Z}^r} c_h e^{2\pi i \langle h, t \rangle}$ is of class $\mathcal{E}^s, s > 1$, if there exists a constant M such that $|c_h| \leq M r^{-s}(h)$ for every $h \neq 0$, where

$$r(h) = \prod_{i=1}^{r} \max(1, |h_i|).$$

Then, if α is of type η,

$$\sup_{x \in \mathbb{T}^r} \left| \sum_{l=1}^{n} (f(T_\alpha^l x) - \int_{\mathbb{T}^r} f(t)dt) \right| = \mathcal{O}(1)$$

for all $f \in \mathcal{E}^{\eta + \lambda}$, where $\lambda > 0$ is arbitrary (see [143]). Then, if α is of type greater than $1 + \frac{1}{r}$, the classes $C_i(S), i = 1, \ldots, 3$ can be chosen as $\mathcal{E}^{\eta + \lambda}, \lambda > 0$ arbitrary.

4.2 DYNAMIC RANDOM WALKS GENERATED BY INDEPENDENT RANDOM VARIABLES

For every $j \in \{1, \ldots, d\}$, let $(\zeta_i^{(j)})_{i \geq 1}$ be independent sequences of independent random variables with values in $[0, \frac{1}{d}]$ such that $\mathbb{E}(\zeta_i^{(j)}) = \frac{1}{2d}$ and for every $j \in \{1, \ldots, d\}$,

$$\sum_{i \geq 1} \frac{(\log i)^2}{i} \text{Var}(\zeta_i^{(j)}) < \infty.$$

We are going to use Kolmogorov's strong law of large numbers which can be formulated as follows (see [170]):

LEMMA 1 *Let Y_1, Y_2,... be independent random variables with finite second moments. Assume there exist positive numbers b_n such that $b_n \to +\infty$ and*

$$\sum_{n=1}^{\infty} \frac{Var(Y_n)}{b_n^2} < \infty.$$

Let $S_n = \sum_{i=1}^{n} Y_i$.
Then,

$$\frac{S_n - \mathbb{E}(S_n)}{b_n} \to 0 \ \mathbb{P} - almost\ surely\ (a.s.).$$

In particular, if

$$\sum_{n=1}^{\infty} \frac{Var(Y_n)}{n^2} < \infty$$

then

$$\frac{S_n - \mathbb{E}(S_n)}{n} \to 0 \ \mathbb{P} - a.s.$$

By Lemma 1, for every $j \in \{1, \ldots, d\}$,

$$\frac{\log n}{\sqrt{n}} \sum_{i=1}^{n} (\zeta_i^{(j)} - \frac{1}{2d}) \to 0 \quad a.s.$$

Since the random variables $\zeta_i^{(j)}$ are mutually independent and bounded, again by Kolmogorov's criteria (cf. Lemma 1), for every $j, l \in \{1, \ldots, d\}$,

$$\frac{\log n}{\sqrt{n}} \sum_{i=1}^{n} (\zeta_i^{(j)} \zeta_i^{(l)} - \frac{1}{4d^2}) \to 0 \quad a.s.$$

The dynamic random walk S_n with the probabilities of transition

$$\mathbb{P}(X_i = z) = \begin{cases} \zeta_i^{(j)} & \text{if } z = e_j \\ \frac{1}{d} - \zeta_i^{(j)} & \text{if } z = -e_j \\ 0 & \text{otherwise} \end{cases}$$

verifies the local limit theorem given in Chapter 2 (Theorem 2.3) with the matrix $A = \frac{1}{d}I$. For example, we can take the random variables $\zeta_i^{(j)}$ with uniform distribution on the set $\{x \in \mathbb{R}; |x - \frac{1}{2d}| \leq (\log i)^{-\frac{3}{2}-\varepsilon}\}, \varepsilon > 0$. This means that we can perturb the probabilities of transition of the walk for a long time without changing the asymptotic behavior of the probability of return to the origin.

Chapter 2

LIMIT THEOREMS FOR DYNAMIC RANDOM WALKS

1. A STRONG LAW OF LARGE NUMBERS

Consider the dynamic $\mathbb{Z}^d$-random walk $(S_n)_{n\geq 0}$ defined in Chapter 1. Assume that $T\mu = \mu$ and that $(f_j)_{1\leq j\leq d}$ are measurable functions with values in $[0, \frac{1}{d}]$. We denote by $(X_i^{(1)}, \ldots, X_i^{(d)})$ the d-dimensional vector X_i.

Let $Y_i^{(j)} = X_i^{(j)} - (2f_j(T^i x) - \frac{1}{d})$.

Since
$$\sum_{n=1}^{+\infty} \frac{\mathbb{E}(Y_n^{(j)^2})}{n^2} < \infty,$$

then Kolmogorov's criteria (cf. Lemma 1) tells us that as $n \to \infty$,

$$\frac{1}{n}\sum_{i=1}^{n} Y_i^{(j)} \to 0, \quad \mathbb{P} - \text{a.s.}$$

Birkhoff's theorem (see Appendix A) tells us that as $n \to +\infty$,

$$\frac{1}{n}\sum_{i=1}^{n}(2f_j(T^i x) - \frac{1}{d}) \to 2\mathbb{E}(f_j|\mathcal{I}) - \frac{1}{d}, \quad \mu - \text{a.s.}$$

where $\mathcal{I}$ is the σ-algebra of the invariants (See Appendix A).

THEOREM 2.1 *For μ-almost every $x \in E$, as $n \to +\infty$,*

$$\frac{S_n}{n} \longrightarrow \left(2\mathbb{E}(f_j|\mathcal{I}) - \frac{1}{d}\right)_{1\leq j\leq d} \quad \mathbb{P} - \text{a.s.}$$

2. A CENTRAL LIMIT THEOREM

THEOREM 2.2 *Let us assume that for every $j, l \in \{1, \ldots, d\}$, $f_j \in \mathcal{C}_2(S)$, $f_j f_l \in \mathcal{C}_2(S)$ and $\int_E f_j d\mu = \frac{1}{2d}$. Then, for every $x \in E$, the sequence of processes $(\frac{1}{\sqrt{n}} S_{[nt]})_{t \geq 0}$ weakly converges in the Skorohod space $\mathcal{D} = \mathcal{D}([0, \infty[)$ (see [15]) to the d-dimensional Brownian motion*

$$B_t = (B_t^{(1)}, \ldots, B_t^{(d)})$$

with zero mean and covariance matrix At.

Proof:
Let us introduce the characteristic function of $\frac{S_{[nt]}}{\sqrt{n}}$

$$\phi_{\frac{S_{[nt]}}{\sqrt{n}}}(u) = \mathbb{E}(\exp(i < u, \frac{S_{[nt]}}{\sqrt{n}} >)).$$

By independence of the random vectors $X_i, i \geq 1$,

$$\phi_{\frac{S_{[nt]}}{\sqrt{n}}}(u) = \prod_{i=1}^{[nt]} \mathbb{E}(\exp(i \sum_{j=1}^{d} u_j \frac{X_i^{(j)}}{\sqrt{n}}))$$

$$= \prod_{i=1}^{[nt]} Q_n^{(i)}(u_1, \ldots, u_d)$$

where

$$Q_n^{(i)}(u_1, \ldots, u_d) = \frac{1}{d} \sum_{j=1}^{d} \cos(\frac{u_j}{\sqrt{n}}) + \frac{i}{d} \sum_{j=1}^{d} (2df_j(T^i x) - 1) \sin(\frac{u_j}{\sqrt{n}})$$

A direct calculation gives

$$\begin{aligned}
|Q_n^{(i)}(u)|^2 &= \frac{1}{d^2}(\sum_{j=1}^{d}(1 - \frac{u_j^2}{n}) + \mathcal{O}(n^{-2}))^2 + \frac{1}{d^2}\sum_{j=1}^{d}(2df_j(T^i x) - 1)^2 \frac{u_j^2}{n} \\
&\quad + \frac{1}{d^2}\sum_{j \neq l}(2df_j(T^i x) - 1)(2df_l(T^i x) - 1)\frac{u_j u_l}{n} + \mathcal{O}(n^{-3/2}) \\
&= 1 - \frac{1}{dn}\sum_{j=1}^{d} u_j^2 + \frac{1}{d^2 n}\sum_{j=1}^{d} u_j^2(2df_j(T^i x) - 1)^2 \\
&\quad + \frac{1}{d^2 n}\sum_{j \neq l}(2df_j(T^i x) - 1)(2df_l(T^i x) - 1)u_j u_l + \mathcal{O}(n^{-3/2})
\end{aligned}$$

and then

$$|\phi_{S_{[nt]}/\sqrt{n}}(u)| = \prod_{i=1}^{[nt]} |Q_n^{(i)}(u)|$$
$$= \exp(-\frac{t}{2} <u, Au> + o(1))$$

The imaginary part of the characteristic function can be rewritten as

$$\prod_{i=1}^{[nt]} \exp\left(i \arctan\left(\frac{\sum_{j=1}^{d}(2df_j(T^ix)-1)\sin(\frac{u_j}{\sqrt{n}})}{\sum_{j=1}^{d}\cos(\frac{u_j}{\sqrt{n}})}\right)\right)$$

$$= \exp\left(\frac{i}{d}\sum_{i=1}^{[nt]}\sum_{j=1}^{d}(2df_j(T^ix)-1)\frac{u_j}{\sqrt{n}} + o(1)\right) = 1 + o(1)$$

using the hypothesis (H_2) and the fact that for every j, the integral of f_j is equal to $1/2d$. The convergence of the finite dimensional distributions of $(S_{[nt]}/\sqrt{n})_t$ to the one of $(B_t)_t$ is obtained in a classical way. It remains to prove the tightness in $\mathcal{D}$ of the sequence $(S_{[nt]}/\sqrt{n})_t$. The classical criterion: Theorem 15.6 in Billingsley ([15]) in a d-dimensional form is used. Let $0 \le t_1 < t < t_2$, by independence of the random vectors X_i,

$$\frac{1}{n^2}\mathbb{E}(\|S_{[nt]} - S_{[nt_1]}\|^2 \cdot \|S_{[nt_2]} - S_{[nt]}\|^2)$$
$$= \frac{1}{n^2}\mathbb{E}(\|S_{[nt]} - S_{[nt_1]}\|^2) \times \mathbb{E}(\|S_{[nt_2]} - S_{[nt]}\|^2)$$

Now,

$$\frac{1}{n}\mathbb{E}(\|S_{[nt]} - S_{[nt_1]}\|^2)$$

$$= \frac{1}{n}\sum_{j=1}^{d}\mathbb{E}\left(\left(\sum_{i=[nt_1]+1}^{[nt]} X_i^{(j)}\right)^2\right)$$

$$= \frac{1}{n}\sum_{j=1}^{d}\left[\sum_{i=[nt_1]+1}^{[nt]} \mathrm{Var}(X_i^{(j)}) + \left(\sum_{i=[nt_1]+1}^{[nt]} \mathbb{E}(X_i^{(j)})\right)^2\right]$$

$$= \frac{(d-1)}{d}\frac{([nt] - [nt_1])}{n}$$

$$+ \frac{4}{dn}\sum_{j=1}^{d}\sum_{i=[nt_1]+1}^{[nt]} f_j(T^ix)(1 - df_j)(T^ix)$$

$$+ \frac{1}{n}\sum_{j=1}^{d}\left(\sum_{i=[nt_1]+1}^{[nt]}(2f_j(T^ix) - \frac{1}{d})\right)^2$$

Using (H_2), there exists a constant $C > 0$ such that for every n,

$$\left|\sum_{i=1}^{[nt]-[nt_1]} (f_j(T^{[nt_1]+i}x) - \frac{1}{2d})\right| \leq C\sqrt{[nt_2]-[nt_1]}$$

So, there exists a constant $C' > 0$ such that

$$\frac{1}{n}\mathbb{E}(\|S_{[nt]} - S_{[nt_1]}\|^2) \leq C'\frac{[nt_2]-[nt_1]}{n}$$

If $t_2 - t_1 \geq \frac{1}{n}$, $[nt_2] - [nt_1] \leq 2n(t_2 - t_1)$, otherwise $S_{[nt_2]} = S_{[nt]}$ or $S_{[nt_1]} = S_{[nt]}$ and the tightness of the sequence follows.

3. A LOCAL LIMIT THEOREM

We assume that the matrix A is positive definite.

THEOREM 2.3 *Let $(S_n)_{n \in \mathbb{N}}$ be a dynamic random walk generated by a dynamical system S and let $f_j \in C_1(S), j = 1,\ldots,d$ be functions with values in $[0, \frac{1}{d}]$ such that for every $j, l \in \{1,\ldots,d\}, f_j f_l \in C_1(S)$ and $\int_E f_j(x)d\mu(x) = \frac{1}{2d}$. Then, for almost every $x \in E$, S_n satisfies a local limit theorem, namely*

$$\mathbb{P}(S_{2n} = 0) \sim \frac{2}{\sqrt{\det A}(4\pi n)^{\frac{d}{2}}} \quad \text{as } n \to \infty.$$

Proof:
Let us introduce the characteristic function $\phi_n(\theta), \theta \in [-\pi,\pi)^d$, for the random vector S_{2n}:

$$\phi_n(\theta) = \mathbb{E}(\exp(i<\theta, S_{2n}>)),$$

reading in the present case

$$\phi_n(\theta) = \prod_{i=1}^{2n}\left(\frac{1}{d}\sum_{j=1}^{d}\cos\theta_j + i\frac{1}{d}\sum_{j=1}^{d}(2df_j(T^ix) - 1)\sin\theta_j\right). \quad (2.1)$$

The probability of return to the origin is expressed by the standard inversion formula

$$\mathbb{P}(S_{2n} = 0) = \frac{1}{(2\pi)^d}\int_{[-\pi,\pi[^d} \phi_n(\theta)d\theta. \quad (2.2)$$

Since the integrand in (2.2) is $\Pi = \pi(1,\ldots,1)$-periodic so

$$\frac{\sqrt{\det A}}{2}(4\pi n)^{\frac{d}{2}}\mathbb{P}(S_{2n} = 0) = \frac{\sqrt{\det A}}{2}(\frac{n}{\pi})^{\frac{d}{2}}\int_{[-\frac{\pi}{2},\frac{3\pi}{2}[^d} \phi_n(\theta)d\theta.$$

Let us define $B(x,r) = \{x \in \mathbb{Z}^d; |x| \le r\}$ where $|.|$ is the euclidean distance. We decompose the last integral on three integration domains: $B(0, \varepsilon_n)$, $B(\Pi, \varepsilon_n)$ and $D_n = [-\frac{\pi}{2}, \frac{3\pi}{2}[^d \setminus \{B(0, \varepsilon_n) \cup B(\Pi, \varepsilon_n)\}$ where $\varepsilon_n = \frac{\log n}{\sqrt{n}}$. Since the characteristic function is Π-periodic, its integral on the sets $B(0, \varepsilon_n)$ and $B(\Pi, \varepsilon_n)$ is the same. Finally, by the change of variables $\theta = \frac{u}{\sqrt{n}}$, we get

$$\frac{\sqrt{\det A}}{2}(4\pi n)^{\frac{d}{2}}\mathbb{P}(S_{2n} = 0) = J_1(n) + J_2(n)$$

where

$$J_1(n) = \sqrt{\frac{\det A}{\pi^d}} \int_{B(0, \log n)} \phi_n(\frac{u}{\sqrt{n}}) du$$

and

$$J_2(n) = \frac{1}{2}\sqrt{\frac{\det A}{\pi^d}} \int_{E_n} \phi_n(\frac{u}{\sqrt{n}}) du$$

where $E_n = [-\frac{\pi}{2}\sqrt{n}, \frac{3\pi}{2}\sqrt{n}[^d \setminus \{B(0, \log n) \cup B(\Pi\sqrt{n}, \log n)\}$.

LEMMA 2 *Under the conditions of Theorem 2.3, we have*

$$J_1(n) \underset{n \to \infty}{\to} 1.$$

Proof:
We denote by Q the quadratic form associated to the matrix A.

$$\forall \theta \in \mathbb{R}^d, Q(\theta) = \theta^t A \theta$$

The integral $J_1(n)$ can be decomposed into a dominant part equal to 1 and two correction terms,

$$J_1(n) = \sqrt{\frac{\det A}{\pi^d}} \int_{\mathbb{R}^d} e^{-Q(u)} du + I_1(n) + I_2(n)$$

where

$$I_1(n) = \sqrt{\frac{\det A}{\pi^d}} \int_{|u| \le \log n} (\phi_n(\frac{u}{\sqrt{n}}) - e^{-Q(u)}) du$$

$$I_2(n) = -\sqrt{\frac{\det A}{\pi^d}} \int_{|u| > \log n} e^{-Q(u)} du.$$

First, by hypothesis, since the quadratic form Q is positive definite, Q has d eigenvalues $\lambda_1, \ldots, \lambda_d$ such that $0 < \lambda_1 \le \lambda_2 \le \ldots \le \lambda_d$ and, by a change of variables,

$$\sqrt{\frac{\det A}{\pi^d}} \int_{\mathbb{R}^d} e^{-Q(u)} du = \sqrt{\frac{\det A}{(2\pi)^d}} \int_{\mathbb{R}^d} e^{-\frac{1}{2}Q(u)} du$$

$$= \sqrt{\frac{\det A}{(2\pi)^d}} \int_{\mathbb{R}^d} e^{-\frac{1}{2}\sum_{j=1}^{d} \lambda_j u_j^2} du$$

$$= 1$$

We now have to prove that $I_j(n), j=1,2$ go to 0 as $n \to \infty$. From the above result, the convergence of $I_2(n)$ to 0 is clear. We have to worry only about I_1. The characteristic function $\phi_n(\theta)$ is a complex number which can be rewritten as

$$\phi_n(\theta) = \exp(P_n(\theta) + iQ_n(\theta))$$

where

$$P_n(\theta) = \frac{1}{2} \sum_{i=1}^{2n} \log\left(\frac{1}{d^2}\left(\sum_{j=1}^{d} \cos\theta_j\right)^2 + \frac{1}{d^2}\left(\sum_{j=1}^{d}(2df_j(T^i x) - 1)\sin\theta_j\right)^2\right)$$

$$Q_n(\theta) = \sum_{i=1}^{2n} \arctan\left(\frac{\sum_{j=1}^{d}(2df_j(T^i x) - 1)\sin\theta_j}{\sum_{j=1}^{d} \cos\theta_j}\right)$$

Let $T_n(u) = Q(u) + P_n(\frac{u}{\sqrt{n}}) + iQ_n(\frac{u}{\sqrt{n}})$. Using the following inequality,

$$|e^z - 1| \leq |z| \, e^{|z|}, \forall z \in \mathbb{C}$$

and the fact that $\sqrt{\frac{\det(A)}{\pi^d}} \int_{\mathbb{R}^d} e^{-Q(u)} du = 1$, we obtain for $n \geq 1$,

$$|I_1(n)| \leq \sup_{u \in B(0, \log n)} |e^{T_n(u)} - 1|$$

$$\leq \sup_{u \in B(0, \log n)} |T_n(u)| \exp |T_n(u)|$$

For every $u \in B(0, \log n)$,

$$T_n(u) = \sum_{j,l=1}^{d} a_{jl} u_j u_l$$

$$+ \frac{1}{2} \sum_{i=1}^{2n} \log\left(\frac{1}{d^2}\left(\sum_{j=1}^{d} \cos(\frac{u_j}{\sqrt{n}})\right)^2 + \frac{1}{d^2}\left(\sum_{j=1}^{d}(2df_j(T^i x) - 1)\sin(\frac{u_j}{\sqrt{n}})\right)^2\right)$$

$$+ i \sum_{i=1}^{2n} \arctan\left(\frac{\sum_{j=1}^{d}(2df_j(T^i x) - 1)\sin(\frac{u_j}{\sqrt{n}})}{\sum_{j=1}^{d} \cos(\frac{u_j}{\sqrt{n}})}\right)$$

$$= \sum_{j,l=1}^{d} a_{jl}u_j u_l + \frac{1}{2}\sum_{i=1}^{2n} \log\left(\frac{1}{d^2}\left(\sum_{j=1}^{d}(1-\frac{u_j^2}{2n}+\mathcal{O}(\frac{(\log n)^4}{n^2}))\right)^2\right.$$

$$+ \frac{1}{d^2}\sum_{j=1}^{d}(2df_j(T^i x)-1)^2(\frac{u_j}{\sqrt{n}}+\mathcal{O}(\frac{(\log n)^3}{n^{\frac{3}{2}}}))^2$$

$$+ \frac{1}{d^2}\sum_{j\neq l}(2df_j(T^i x)-1)(2df_l(T^i x)-1)(\frac{u_j}{\sqrt{n}}+\mathcal{O}(\frac{(\log n)^3}{n^{\frac{3}{2}}}))$$

$$\times \left(\frac{u_l}{\sqrt{n}}+\mathcal{O}(\frac{(\log n)^3}{n^{\frac{3}{2}}})\right)\right)$$

$$+ i\sum_{i=1}^{2n} \arctan\left(\frac{\sum_{j=1}^{d}(2df_j(T^i x)-1)(\frac{u_j}{\sqrt{n}}+\mathcal{O}(\frac{(\log n)^3}{n^{\frac{3}{2}}}))}{\sum_{j=1}^{d}(1-\frac{u_j^2}{2n}+\mathcal{O}(\frac{(\log n)^4}{n^2}))}\right)$$

$$= \sum_{j,l=1}^{d} a_{jl}u_j u_l + \frac{1}{2}\sum_{i=1}^{2n}\log\left(1-\frac{1}{d^2}\sum_{j=1}^{d}(d-1+4df_j(T^i x))\right.$$

$$\times (1-df_j(T^i x))(\frac{u_j^2}{n}+\mathcal{O}(\frac{(\log n)^4}{n^2})))$$

$$- \frac{1}{d^2}\sum_{j\neq l}(2df_j(T^i x)+2df_l(T^i x)-1-4d^2 f_j(T^i x)f_l(T^i x))(\frac{u_j u_l}{n})$$

$$+ \mathcal{O}(\frac{(\log n)^4}{n^2}))\bigg)$$

$$+ \frac{i}{d}\sum_{i=1}^{2n}\sum_{j=1}^{d}(2df_j(T^i x)-1)\frac{u_j}{\sqrt{n}}+\mathcal{O}(\frac{(\log n)^3}{n^{\frac{3}{2}}})$$

Finally, using the definitions of a_{jl},

$$\sup_{u\in B(0,\log n)} |T_n(u)|$$

$$\leq \frac{(\log n)^2}{2n}\frac{1}{d^2}\sum_{j=1}^{d}\bigg|\sum_{i=1}^{2n} 4df_j(T^i x)(1-df_j(T^i x))$$

$$- 2n\int_E 4df_j(x)(1-df_j(x))d\mu(x)\bigg|$$

$$+ \frac{(\log n)^2}{2n}\frac{1}{d^2}\sum_{j\neq l}\bigg|\sum_{i=1}^{2n}(2df_j(T^i x)+2df_l(T^i x)-1-4d^2 f_j(T^i x)f_l(T^i x))$$

$$- 2n\int_E (1-4d^2 f_j(x)f_l(x))d\mu(x)\bigg| + \frac{\log n}{\sqrt{n}}\frac{1}{d}\sum_{j=1}^{d}\bigg|\sum_{i=1}^{2n}(2df_j(T^i x)-1)\bigg|$$

$$+ \mathcal{O}(\frac{(\log n)^3}{n^{\frac{3}{2}}})$$

Since the functions f_j and $f_j f_l$, $j,l \in \{1,\ldots,d\}$ satisfy the condition (H_1) and $\int_E f_j(x) d\mu(x) = \frac{1}{2d}$, so we conclude that

$$\sup_{u \in B(0,\log n)} | T_n(u) | \underset{n \to \infty}{\to} 0.$$

LEMMA 3 *Under the hypothesis of Theorem 2.3,*

$$J_2(n) \underset{n \to \infty}{\to} 0.$$

Proof:
To prove the lemma, we write

$$| J_2(n) | \leq \frac{1}{2} \sqrt{\frac{\det A}{\pi^d}} \int_{E_n} | \phi_n(\frac{u}{\sqrt{n}}) | du$$

where $E_n = [-\frac{\pi}{2}\sqrt{n}, \frac{3\pi}{2}\sqrt{n}]^d \setminus \{B(0,\log n) \cup B(\Pi\sqrt{n}, \log n)\}$. Now

$$| \phi_n(\frac{u}{\sqrt{n}}) | = \exp\left(\frac{1}{2} \sum_{i=1}^{2n} \log(G_i(\frac{u}{\sqrt{n}}))\right)$$

where

$$G_i(u) = \frac{1}{d^2}\Big(\sum_{j=1}^{d} \cos(u_j)\Big)^2 + \frac{1}{d^2}\Big(\sum_{j=1}^{d} (2df_j(T^i x) - 1) \sin(u_j)\Big)^2$$

The function G_i is positive and can be rewritten as

$$G_i(u) = \frac{1}{d^2} \sum_{j=1}^{d} \Big(1 - 4df_j(T^i x)(1 - df_j(T^i x)) \sin^2(u_j)\Big)$$

$$+ \frac{1}{d^2} \sum_{j \neq l} \Big(\cos(u_j) \cos(u_l) + (2df_j(T^i x) - 1)(2df_l(T^i x) - 1) \sin(u_j) \sin(u_l)\Big)$$

For every $u \in \mathbb{R}^d$, since the terms in the above sums are all less than one,

$$G_i(u) \leq \frac{d}{d^2} + \frac{d(d-1)}{d^2} = 1.$$

Applying the inequality $\log(1 - x) \leq -x$ for $x \in [0, 1[$ to the function $1 - G_i$,

$$| J_2(n) | \leq \frac{1}{2} \sqrt{\frac{\det A}{\pi^d}} \int_{E_n} \exp\Big(-\frac{1}{2} \sum_{i=1}^{2n} (1 - G_i(\frac{u}{\sqrt{n}}))\Big) du$$

Now

$$\sum_{i=1}^{2n} \left(1 - G_i(\frac{u}{\sqrt{n}})\right)$$

$$= \frac{1}{d^2} \sum_{i=1}^{2n} \sum_{j=1}^{d} 4df_j(T^i x)(1 - df_j(T^i x)) \sin^2(\frac{u_j}{\sqrt{n}})$$

$$+ \frac{1}{d^2} \sum_{i=1}^{2n} \sum_{j \neq l} \left(1 - \cos(\frac{u_j}{\sqrt{n}})\cos(\frac{u_l}{\sqrt{n}}) \right.$$

$$\left. - (2df_j(T^i x) - 1)(2df_l(T^i x) - 1) \sin(\frac{u_j}{\sqrt{n}}) \sin(\frac{u_l}{\sqrt{n}}) \right)$$

For every $x, y \in \mathbb{R}$,

$$1 - \cos(x)\cos(y) \geq \frac{1}{2}\Big(\sin^2(x) + \sin^2(y)\Big),$$

so

$$\sum_{i=1}^{2n} \left(1 - G_i(\frac{u}{\sqrt{n}})\right)$$

$$\geq \frac{1}{d^2} \sum_{i=1}^{2n} \sum_{j=1}^{d} \Big(d - 1 + 4df_j(T^i x)(1 - df_j(T^i x))\Big) \sin^2(\frac{u_j}{\sqrt{n}})$$

$$- \frac{1}{d^2} \sum_{i=1}^{2n} \sum_{j \neq l} (2df_j(T^i x) - 1)(2df_l(T^i x) - 1) \sin(\frac{u_j}{\sqrt{n}}) \sin(\frac{u_l}{\sqrt{n}})$$

$$= \sum_{i=1}^{2n} \sum_{j,l=1}^{d} \operatorname{Cov}(X_i^{(j)}, X_i^{(l)}) \sin(\frac{u_j}{\sqrt{n}}) \sin(\frac{u_l}{\sqrt{n}})$$

Then,

$$|J_2(n)| \leq \frac{1}{2}\sqrt{\frac{\det A}{\pi^d}} \int_{E_n} \exp\Big[-n\Big(Q(\sin(\frac{u}{\sqrt{n}}))$$

$$+ \sum_{j,l=1}^{d} (\frac{1}{2n} \sum_{i=1}^{2n} \operatorname{Cov}(X_i^{(j)}, X_i^{(l)}) - a_{jl}) \sin(\frac{u_j}{\sqrt{n}}) \sin(\frac{u_l}{\sqrt{n}}) \Big) \Big] du$$

Since the functions $f_j, f_j f_l$ satisfy the condition (H_1), it is easy to see that

$$\lim_{n \to \infty} \frac{1}{2n} \sum_{i=1}^{2n} \operatorname{Cov}(X_i^{(j)}, X_i^{(l)}) = a_{jl}.$$

Let ε be strictly positive, then there exists an n_0 such that for every $n \geq n_0$,

$$|J_2(n)| \leq \frac{1}{2}\sqrt{\frac{\det A}{\pi^d}} \int_{E_n} \exp\left[-n\left(Q(\sin(\frac{u}{\sqrt{n}})) - d\varepsilon|\sin(\frac{u}{\sqrt{n}})|^2\right)\right] du$$

where we have used the Cauchy-Schwarz inequality

$$\sum_{j=1}^{d} \sin(\frac{u_j}{\sqrt{n}}) \leq \sqrt{d}\left(\sum_{j=1}^{d} \sin(\frac{u_j}{\sqrt{n}})^2\right)^{\frac{1}{2}}$$

Since Q is positive definite, Q has d eigenvalues $\lambda_1, \ldots, \lambda_d$ such that $0 < \lambda_1 \leq \lambda_2 \leq \ldots \leq \lambda_d$ and we have the well-known inequality

$$\lambda_1 \cdot |v|^2 \leq Q(v) \leq \lambda_d \cdot |v|^2, \quad \forall v \in \mathbb{R}^d.$$

Then, since $\sin^2(x) \geq \frac{4}{\pi^2} x^2$ for $|x| \leq \frac{\pi}{2}$, we get

$$\begin{aligned} |J_2(n)| &\leq \frac{1}{2}\sqrt{\frac{\det A}{\pi^d}} \int_{E_n} \exp(-n(\lambda_1 - d\varepsilon)|\sin(\frac{u}{\sqrt{n}})|^2) du \\ &\leq C \cdot n^{\frac{d}{2}} \exp(-\frac{4}{\pi^2}(\log n)^2 (\lambda_1 - d\varepsilon)) = o(1) \end{aligned}$$

by choosing ε such that $\varepsilon < \frac{\lambda_1}{d}$.

4. A STRASSEN'S FUNCTIONAL LAW OF THE ITERATED LOGARITHM

Consider a dynamic $\mathbb{Z}$-random walk $(S_n)_{n \in \mathbb{N}}$ defined as in Chapter 1 ($d = 1$). We assume that $T\mu = \mu$ and the dynamical system is uniquely ergodic. The function f will be a Riemann integrable function such that $a = \int_E 4f(t)(1 - f(t))dt > 0$. Let us define for every $i \geq 1$, the random variables $Y_i = X_i - (2f(T^i x) - 1)$ and the sum $\tilde{S}_n = \sum_{i=1}^{n} Y_i$. Then we investigate the behavior of the functions $\psi_n(t)$ obtained by linear interpolation of the values

$$\psi_n\left(\frac{i}{n}\right) = T_n(i) \text{ where } T_n(i) = \frac{\tilde{S}_i}{\sqrt{2na \log \log n}}, \quad i = 1, \ldots, n; \quad \psi_n(0) = 0.$$

The set Σ will denote the set of absolutely continuous functions F defined on $[0, 1]$ with $F(0) = 0$ whose derivative satisfies

$$\int_0^1 F'(t)^2 dt \leq 1.$$

THEOREM 2.4 *For every $x \in E$, the set of limit functions of the sequence $\psi_3, \psi_4, \psi_5, \ldots$ under uniform convergence is almost certainly the set Σ.*

The proof will be deduced from the following lemmas.

LEMMA 4 *For any v with $0 < v < C_2 \cdot \sqrt{\frac{\log\log \operatorname{Var}(\tilde{S}_n)}{\operatorname{Var}(\tilde{S}_n)}}$, one has*

$$\mathbb{E}(e^{v\tilde{S}_n}) < e^{\frac{v^2}{2}\operatorname{Var}(\tilde{S}_n)(1+\delta_n)}, n \geq 1$$

where $\delta_n \to 0$ as $n \to \infty$.

Proof:
One has for every $i \geq 1$,

$$\mathbb{E}(e^{vY_i}) = \sum_{l=0}^{\infty} \frac{v^l}{l!}\mathbb{E}(Y_i^l) = 1 + \frac{v^2}{2}\operatorname{Var}(Y_i) + \sum_{l=3}^{\infty} \frac{v^l}{l!}\mathbb{E}(Y_i^l).$$

Therefore,

$$\mathbb{E}(e^{v\tilde{S}_n}) = \prod_{i=1}^{n} \mathbb{E}(e^{vY_i}) < \exp\left(\frac{v^2}{2}\operatorname{Var}(\tilde{S}_n) + \sum_{l=3}^{\infty} \frac{v^l}{l!} \sum_{i=1}^{n} \mathbb{E}(|Y_i|^l)\right).$$

Since T is uniquely ergodic, we have (see Appendix A),

$$\frac{1}{n}\operatorname{Var}(\tilde{S}_n) = \frac{1}{n}\sum_{i=1}^{n} 4f(T^i x)(1 - f(T^i x)) \to_{n \to +\infty} a, \text{ uniformly in } x \in E.$$

This result and the fact that the random variables Y_i are bounded by 2 imply that the sequence

$$\sqrt{\frac{\log\log \operatorname{Var}(\tilde{S}_n)}{\operatorname{Var}(\tilde{S}_n)}} \cdot \left(\frac{\sum_{i=1}^{n} \mathbb{E}(|Y_i|^l)}{\log\log \operatorname{Var}(\tilde{S}_n)}\right)^{1/l}$$

converges uniformly with respect to l with $l \geq 3$ to zero when n tends to $+\infty$. Thus the lemma follows.

LEMMA 5 (BERNSTEIN-KOLMOGOROV INEQUALITY) *For any $t < C_3 \cdot \log\log \operatorname{Var}(\tilde{S}_n)$, one has*

$$\mathbb{P}\left(\tilde{S}_n > \sqrt{2t\operatorname{Var}(\tilde{S}_n)}\right) < e^{-t(1+o(1))}.$$

Proof:
We have the classical inequality, for any $v > 0$,

$$\mathbb{P}\left(\tilde{S}_n > \frac{1}{v}(t + \log \mathbb{E}(e^{v\tilde{S}_n}))\right) < e^{-t}.$$

Lemma 4 with

$$v = \sqrt{\frac{2t}{\operatorname{Var}(\tilde{S}_n)}}$$

gives us

$$\log \mathbb{E}(e^{v\tilde{S}_n}) < \frac{v^2}{2}\operatorname{Var}(\tilde{S}_n)(1 + o(1)),$$

which proves the inequality.

The following inequality can be found in Kolmogorov [101].

LEMMA 6 *For any t satisfying $C_4 < t < C_3 \cdot \log \log \operatorname{Var}(\tilde{S}_n)$, one has*

$$\mathbb{P}(\tilde{S}_n > \sqrt{2t\operatorname{Var}(\tilde{S}_n)}) > e^{-t(1+o(1))}.$$

By combining Lemmas 5 and 6, we deduce

LEMMA 7 *For any $\beta > 1$ and $\epsilon > 0$ there exists an $n_o = n_o(\beta, \epsilon)$ such that, for all y satisfying $C_4 < |y| < \sqrt{C_5 \cdot \log \log \operatorname{Var}(\tilde{S}_n)}$, we have*

$$e^{-\frac{y^2}{2}(1+\epsilon)} < \mathbb{P}\left(\tilde{S}_n \in \left(y\sqrt{\operatorname{Var}(\tilde{S}_n)}, \beta y\sqrt{\operatorname{Var}(\tilde{S}_n)}\right)\right), n \geq n_o.$$

Proof of Theorem 2.4:
Let k be a sufficiently large integer. We restrict our attention to indices n divisible by k. Now we can split the sum $\tilde{S}_n$ into k sums of $\frac{n}{k}$ terms each,

$$\xi_1 = Y_1 + \ldots + Y_{\frac{n}{k}}, \quad \xi_2 = Y_{\frac{n}{k}+1} + \ldots + Y_{\frac{2n}{k}}, \ldots$$

Let $t_l \in [0,1]$ and $\beta > 1$ be given. If we apply Lemma 7 with $\frac{n}{k}, \xi_l, \sqrt{\frac{na}{k}} + o(1)$ instead of $n, \tilde{S}_n, \sqrt{\operatorname{Var}(\tilde{S}_n)}$ and with $y = t_l\sqrt{2k \log \log n}$, then

$$\mathbb{P}\left(\xi_l \in \left(y\sqrt{\frac{na}{k}}, \beta y\sqrt{\frac{na}{k}}\right)\right) > e^{-k \log \log n (1+\epsilon) t_l^2}.$$

Using that the random variables ξ_i are independent, we get

$$\mathbb{P}\left(\frac{\xi_l}{\sqrt{2na \log \log n}} \in (t_l, \beta t_l) \text{ for } l = 1, \ldots, k\right) > (\log n)^{-(1+\epsilon)k \sum_{l=1}^{k} t_l^2}.$$

Now we can choose
$$t_l = F\left(\frac{l+1}{k}\right) - F\left(\frac{l}{k}\right), \quad l = 0, 1, \ldots, k-1.$$

Let $\epsilon' > 0$ be given and set $I = \int_0^1 F'(t)^2 dt$. Then for k large enough, we get
$$\left|k \sum_{l=1}^{k} t_l^2 - I\right| < \epsilon'.$$

If $I < 1$, the series
$$\sum_{m=1}^{\infty} (\log q^m)^{-(1+\epsilon)I}, \quad q > 1,$$

diverges for sufficiently small ϵ and the proof is finished by applying the Borel-Cantelli Lemma and using the argument of [101]. If $I = 1$, we apply all this to a function $F^* \in \Sigma$ whose corresponding integral $I^* < 1$ and which is close to F under the metric of uniform convergence.

5. A FUNCTIONAL LARGE DEVIATION PRINCIPLE

In Section 1, a strong law of large numbers for the dynamic random walks was obtained for μ-almost every $x \in E$ from Kolmogorov's theorem assuming that the functions $f_1, \ldots, f_d$ are measurable. The limit vector is then given by $(2\mathbb{E}(f_j|\mathcal{I}) - 1)_{1 \le j \le d}$ where $\mathcal{I}$ is the invariant σ-algebra associated to the transformation T. So, $(S_n/n)_{n \ge 1}$ is a good candidate for a large deviation principle. Let Γ be a polish space endowed with the Borel σ-algebra $\mathcal{B}(\Gamma)$. A good rate function is a lower semi-continuous function $\Lambda^* : \Gamma \to [0, \infty]$ with compact level sets $\{x; \Lambda^*(x) \le \alpha\}, \alpha \in [0, \infty[$. Let $v = (v_n)_n \uparrow \infty$ be an increasing sequence of positive reals. A sequence of random variables $(Y_n)_n$ with values in Γ defined on a probability space $(\Omega, \mathcal{F}, \mathbb{P})$ is said to satisfy a large deviation principle (LDP) with speed $v = (v_n)_n$ and good rate function Λ^* if for every Borel set $B \in \mathcal{B}(\Gamma)$,

$$-\inf_{x \in B^\circ} \Lambda^*(x) \le \liminf_n \frac{1}{v_n} \log \mathbb{P}(Y_n \in B)$$
$$\le \limsup_n \frac{1}{v_n} \log \mathbb{P}(Y_n \in B) \le -\inf_{x \in \bar{B}} \Lambda^*(x).$$

THEOREM 2.5 *Let $f_1, \ldots, f_d$ be measurable functions defined on E with values in $[0, \frac{1}{d}]$. Then, for almost every point $x \in E$, the sequence $(\frac{S_n}{n})_{n \ge 1}$ satisfies a large deviation principle with speed n and good rate*

function
$$\Lambda^*(y) = \sup_{\lambda \in \mathbb{R}^d} \{<\lambda, y> - \Lambda(\lambda)\}$$
where
$$\Lambda(\lambda) = \mathbb{E}\Big(\log\Big(\sum_{j=1}^{d} e^{\lambda_j} f_j + (\frac{1}{d} - f_j)e^{-\lambda_j}\Big) \mid \mathcal{I}\Big),$$

$\mathcal{I}$ *being the σ-algebra generated by the fixed points of the transformation T.*

Let us assume E to be a compact metric space, $\mathcal{A}$ the associated Borel field and T a continuous transformation of E. If there exists an unique invariant measure μ i.e. $(E, \mathcal{A}, \mu, T)$ is uniquely ergodic, Theorem 2.5 holds when functions $f_1, \ldots, f_d$ are continuous and for every $x \in E$. In that case,
$$\Lambda(\lambda) = \int_E \log\Big(\sum_{j=1}^{d} e^{\lambda_j} f_j(t) + (\frac{1}{d} - f_j(t))e^{-\lambda_j}\Big) d\mu(t).$$

Under these stronger hypotheses on the dynamical system, we can extend Theorem 2.5 as follows.

Let us define for every $n \geq 1$,
$$S_n^*(t) = \frac{S_{[nt]}}{n}, \quad t \in [0,1].$$

The linear interpolation of $S_n^*(t)$, $t \in [0,1]$, is then defined by
$$\bar{S}_n(t) = S_n^*(t) + \Big(t - \frac{[nt]}{n}\Big) X_{[nt]+1}.$$

We will denote by ν_n and $\bar{\nu}_n$ the distributions of $S_n^*(.)$ and $\bar{S}_n(.)$ in $\mathcal{C}([0,1])$ the space of continuous functions defined on $[0,1]$. We denote by $\mathcal{AC}$ the space of absolutely continuous functions, i.e.,
$$\mathcal{AC} = \{\phi \in \mathcal{C}([0,1]); \sum_{l=1}^{k} |t_l - s_l| \to 0, s_l < t_l \leq s_{l+1} < t_{l+1}$$
$$\implies \sum_{l=1}^{k} |\phi(t_l) - \phi(s_l)| \to 0\}.$$

THEOREM 2.6 *Let $(E, \mathcal{A}, \mu, T)$ be an uniquely ergodic dynamical system. Let $f_1, \ldots, f_d$ be continuous functions defined on E with values in $[0, \frac{1}{d}]$, then for every $x \in E$, the laws $(\bar{\nu}_n)_{n \geq 1}$ satisfy in $\mathcal{C}([0,1])$ a large deviation principle with the good rate function*
$$I(x(.)) = \begin{cases} \int_0^1 \Lambda^*(\dot{x}(t)) \, dt, & \text{if } x(.) \in \mathcal{AC}, x(0) = 0 \\ +\infty, & \text{otherwise.} \end{cases}$$

5.1 PROOF OF THEOREM 2.5.

Using the independence of the random vectors $(X_i)_{i\geq 1}$ and the definition of their distribution, for every $\lambda \in \mathbb{R}^d$,

$$\frac{1}{n}\log \mathbb{E}(e^{<\lambda,S_n>}) = \frac{1}{n}\sum_{i=1}^{n}\log \mathbb{E}(e^{<\lambda,X_i>})$$

$$= \frac{1}{n}\sum_{i=1}^{n}\log\left(\sum_{j=1}^{d}(e^{\lambda_j}f_j(T^i x) + (\frac{1}{d} - f_j(T^i x))e^{-\lambda_j})\right)$$

Since $f_1, \ldots, f_d$ are measurable functions with values in $[0, \frac{1}{d}]$, for every $\lambda \in \mathbb{R}^d$, the function

$$t \in E \longrightarrow \log\left(\sum_{j=1}^{d}(e^{\lambda_j}f_j(t) + (\frac{1}{d} - f_j(t))e^{-\lambda_j})\right)$$

is well-defined and bounded by $\log\left(\frac{2}{d}(\sum_{j=1}^{d}\cosh \lambda_j)\right)$. Consequently, by Birkhoff's Theorem, for μ-almost every point $x \in E$, for every $\lambda \in \mathbb{R}^d$,

$$\frac{1}{n}\log \mathbb{E}(e^{<\lambda,S_n>}) \longrightarrow \mathbb{E}\left(\log\left(\sum_{j=1}^{d}e^{\lambda_j}f_j + (\frac{1}{d} - f_j)e^{-\lambda_j}\right)\bigg|\mathcal{I}\right) = \Lambda(\lambda)$$

as $n \to \infty$. The function Λ being finite and differentiable on $\mathbb{R}^d$, by Gärtner-Ellis Theorem (see [37]), the theorem follows.

5.2 PROOF OF THEOREM 2.6.

Throughout, $||x||$ denotes the Euclidean norm on $\mathbb{R}^d$ and $||.||_\infty$ the supremum norm on $L_\infty([0,1])$.

PROPOSITION 4 *The laws ν_n and $\bar{\nu}_n$ of $S_n^*(.)$ and $\bar{S}_n(.)$ are exponentially equivalent in $L_\infty([0,1])$.*

Proof:
Let us remark that

$$||\bar{S}_n - S_n^*||_\infty \leq \sup_{t\in[0,1]}\frac{||X_{[nt]+1}||}{n}$$

then for every $\eta > 0$, the set

$$\{\omega; ||\bar{S}_n - S_n^*||_\infty > \eta\}$$

is a subset of
$$\{\omega;\ \sup_{t\in[0,1]} ||X_{[nt]+1}|| > n\eta\}$$
and
$$\mathbb{P}(||\bar{S}_n - S_n^*||_\infty > \eta) \leq \sum_{k=1}^{n+1} \mathbb{P}(||X_k|| > n\eta)$$
$$\leq e^{-\lambda n\eta} \sum_{k=1}^{n+1} \mathbb{E}(e^{\lambda ||X_k||})$$

with $\lambda > 0$, by Markov's inequality. Now, $\mathbb{E}(e^{\lambda ||X_k||}) = e^\lambda$, so we deduce that
$$\frac{1}{n} \log \mathbb{P}(||\bar{S}_n - S_n^*||_\infty > \eta) \leq -\lambda\eta + \frac{\log(n+1)}{n} + \frac{\lambda}{n}$$
and consequently, for every $\eta > 0$,
$$\limsup_n \frac{1}{n} \log \mathbb{P}(||\bar{S}_n - S_n^*||_\infty > \eta) = -\infty.$$

PROPOSITION 5 *Let $\mathcal{P}$ be the set of all ordered finite subsets of the interval $]0,1]$ that is the set of k-tuples $t^k = \{t_0 = 0 < t_1 < t_2 < \ldots < t_k \leq 1\}$ with $k \geq 1$. Let $f : [0,1] \to \mathbb{R}^d$; for every $k \geq 1$, for every k-tuple $\{t_0 = 0 < t_1 < t_2 < \ldots < t_k \leq 1\}$, we define*
$$p_{t^k}(f) = (f(t_1), \ldots, f(t_k)) \in (\mathbb{R}^d)^k.$$
Then, the laws $\bar{\nu}_n \circ p_{t^k}^{-1}$ satisfy a large deviation principle in $(\mathbb{R}^d)^k$ with the good rate function
$$\Lambda_k^*(y) = \sum_{l=1}^k (t_l - t_{l-1})\Lambda^*\left(\frac{y_l - y_{l-1}}{t_l - t_{l-1}}\right).$$

Proof:
Let $t^k \in \mathcal{P}$ for some $k \geq 1$. Let us define the random vector
$$Y_n^k = (S_n^*(t_1), S_n^*(t_2) - S_n^*(t_1), \ldots, S_n^*(t_k) - S_n^*(t_{k-1})).$$
For every $\lambda \in (\mathbb{R}^d)^k$, from the independence of the increments of the dynamic random walk, we derive
$$\Lambda_k(\lambda) = \lim_{n\to\infty} \frac{1}{n} \log \mathbb{E}(e^{n<\lambda, Y_n^k>})$$
$$= \sum_{l=1}^k \lim_{n\to\infty} \frac{1}{n} \log \mathbb{E}(e^{<\lambda_l, S_{[nt_l]} - S_{[nt_{l-1}]}>})$$
$$= \sum_{l=1}^k (t_l - t_{l-1})\Lambda(\lambda_l)$$

where Λ is defined in Theorem 2.5. The last equality is obtained by remarking that each increment of the dynamic random walk $S_{[nt_l]} - S_{[nt_{l-1}]}$ is a new dynamic random walk associated to the dynamical system $(E, \mathcal{A}, \mu, T)$, to the functions $f_1, \ldots, f_d$ and to the point $T^{[nt_{l-1}]}x$. The hypothesis of unique ergodicity on the dynamical system and Theorem 2.5 permits us to conclude. The function Λ_k being finite and differentiable, from Gärtner-Ellis Theorem, we deduce that the random vectors Y_n^k satisfy in $(\mathbb{R}^d)^k$ a large deviation principle with good rate function

$$\begin{aligned}\Lambda_k^*(y) &= \sup_{\lambda \in (\mathbb{R}^d)^k} \{<\lambda, y> - \Lambda_k(\lambda)\} \\ &= \sup_{\lambda \in (\mathbb{R}^d)^k} \left\{ \sum_{l=1}^k (t_l - t_{l-1}) \frac{<\lambda_l, y_l>}{t_l - t_{l-1}} - \sum_{l=1}^k (t_l - t_{l-1}) \Lambda(\lambda_l) \right\} \\ &= \sum_{l=1}^k (t_l - t_{l-1}) \sup_{\lambda_l \in \mathbb{R}^d} \left\{ \frac{<\lambda_l, y_l>}{t_l - t_{l-1}} - \Lambda(\lambda_l) \right\} \\ &= \sum_{l=1}^k (t_l - t_{l-1}) \Lambda^* \left(\frac{y_l}{t_l - t_{l-1}} \right)\end{aligned}$$

By the contraction principle (see Dembo [37]), we derive a large deviation principle in $(\mathbb{R}^d)^k$ for the random vectors $(S_n^*(t_1), S_n^*(t_2), \ldots, S_n^*(t_k))$ with good rate function

$$\Lambda_k^*(y) = \sum_{l=1}^k (t_l - t_{l-1}) \Lambda^* \left(\frac{y_l - y_{l-1}}{t_l - t_{l-1}} \right).$$

The probability measures $(\nu_n \circ p_{t^k}^{-1})_n$ and $(\bar{\nu}_n \circ p_{t^k}^{-1})_n$ being exponentially equivalent in $(\mathbb{R}^d)^k$ by Proposition 4, we deduce that $(\bar{\nu}_n \circ p_{t^k}^{-1})_n$ satisfies a large deviation principle in $(\mathbb{R}^d)^k$ with the good rate function Λ_k^*.

PROPOSITION 6 *The laws $\bar{\nu}_n$ are exponentially tight in the space $(\mathcal{C}([0,1]), \|.\|_\infty)$ of continuous functions with values in $\mathbb{R}^d$ such that $f(0) = 0$.*

Proof:
For every $\alpha > 0$, we define the d sets

$$K_\alpha^j = \{\phi \in \mathcal{AC}; \phi(0) = 0, \int_0^1 \Lambda^{j,*}(\dot{\phi}_j(t)) dt \leq \alpha\}, \ 1 \leq j \leq d$$

where $\phi_j(.)$ is the jth component of $\phi : [0,1] \to \mathbb{R}^d$ and $\Lambda^{j,*}$ is the Fenchel-Legendre transform of $\Lambda^j(\lambda) = \log \mathbb{E}(\exp <\lambda, \tilde{X}_j>)$ where $\tilde{X}_j$

is a random vector with distribution

$$\mathbb{P}(\tilde{X}_j = z) = \begin{cases} \int_E f_j d\mu & \text{if } z = e_j \\ 1 - \int_E f_j d\mu & \text{if } z = -e_j \\ 0 & \text{otherwise.} \end{cases}$$

Let $K_\alpha = \bigcap_{j=1}^{d} K_\alpha^j$. Then, we get by considering for almost every $t \in [0, 1[$,

$$\frac{d\bar{S}_n(t)}{dt} = X_{[nt]+1}$$

that

$$\bar{\nu}_n(K_\alpha^c) \leq \sum_{j=1}^{d} \mathbb{P}\left(\frac{1}{n}\sum_{i=1}^{n} \Lambda^{j,*}(X_i^j) > \alpha\right).$$

By Markov's inequality, we obtain for every $\delta > 0$,

$$\bar{\nu}_n(K_\alpha^c) \leq e^{-\delta \alpha n} \sum_{j=1}^{d} \mathbb{E}\left(e^{\delta \sum_{i=1}^{n} \Lambda^{j,*}(X_i^j)}\right)$$

$$\leq d e^{-\delta \alpha n} \max_{j=1}^{d} \prod_{i=1}^{n} \mathbb{E}\left(e^{\delta \Lambda^{j,*}(X_i^j)}\right)$$

and then,

$$\frac{1}{n}\log \bar{\nu}_n(K_\alpha^c) \leq -\delta\alpha + \max_{j=1}^{d} \frac{1}{n}\sum_{i=1}^{n} \log \mathbb{E}\left(e^{\delta \Lambda^{j,*}(X_i^j)}\right) + \frac{1}{n}\log(d).$$

Then, by Birkhoff's Theorem and Jensen's inequality,

$$\limsup_n \frac{1}{n}\log \bar{\nu}_n(K_\alpha^c)$$

$$\leq -\delta\alpha + \max_{j=1}^{d} \int_E \log\left(e^{\delta \Lambda^{j,*}(e_j)} f_j(t) + e^{\delta \Lambda^{j,*}(-e_j)}(\frac{1}{d} - f_j(t))\right) d\mu(t)$$

$$\leq -\delta\alpha + \max_{j=1}^{d} \left[\log \int_E \left(e^{\delta \Lambda^{j,*}(e_j)} f_j(t) + e^{\delta \Lambda^{j,*}(-e_j)}(1 - f_j(t))\right) d\mu(t)\right]$$

$$\leq -\delta\alpha + \max_{j=1}^{d} \log \mathbb{E}\left(e^{\delta \Lambda^{j,*}(\tilde{X}_j)}\right)$$

Now, for every $j = 1, \ldots, d$,

$$\mathbb{E}\left(e^{\delta \Lambda^{j,*}(\tilde{X}_j)}\right) \leq \frac{2}{1-\delta}$$

for some $0 < \delta < 1$ (see Lemma 5.1.14 in [37]). So, we obtain the inequality
$$\limsup_n \frac{1}{n} \log \bar{\nu}_n(K_\alpha^c) \leq -\delta\alpha + \log\left(\frac{2}{1-\delta}\right)$$
Consequently,
$$\lim_{\alpha \to \infty} \limsup_n \frac{1}{n} \log \bar{\nu}_n(K_\alpha^c) = -\infty.$$

By Arzelà-Ascoli's Theorem, the proposition is proved since the set K_α is a bounded set of equicontinuous functions.

From Proposition 5, by Dawson-Gärtner's Theorem, the probability measures $\bar{\nu}_n$ defined on the space of functions f defined on $[0,1]$ with values in $\mathbb{R}^d$ such that $f(0) = 0$ equipped with the topology of pointwise convergence satisfy a large deviation principle with the good rate function
$$I'(f) = \sup_{t^k} \Lambda_k^*(f(t^k))$$
where $f(t^k) = (f(0) = 0, f(t_1), \ldots, f(t_k))$.

Using the properties of the function Λ^* (nonnegativity, convexity and lower semicontinuity), we can prove that $I = I'$ on the subset $\mathcal{AC}$ of $\mathcal{C}([0,1])$ and $I = +\infty$ otherwise (see p. 180-181 in [37] for technical details). From this remark and Proposition 6, Corollary 4.2.6 in [37] gives us Theorem 2.6.

5.3 A COMPARISON BETWEEN DYNAMIC RANDOM WALKS AND SIMPLE RANDOM WALKS.

The large deviation principle for the sequence $(\frac{S'_n}{n})_{n\geq 1}$ when $(S'_n)_{n\geq 0}$ is the simple $\mathbb{Z}^d$-random walk is well-known and corresponds to Theorem 2.5 in the case when the functions f_j are all equal to $\frac{1}{2d}$. Let us denote the logarithmic moment generating function of the sequence $(\frac{S'_n}{n})_{n\geq 1}$ by

$$\Lambda_{\frac{1}{2d}}(\lambda) = \log\left(\frac{1}{2d} \sum_{j=1}^d (e^{\lambda_j} + e^{-\lambda_j})\right)$$

and $\Lambda^*_{\frac{1}{2d}}$ the associated Fenchel-Legendre transform. Our aim is to compare the large deviation principle obtained in Theorem 2.5 for the dynamic random walks with the simple $\mathbb{Z}^d$-random walk one. In the sequel, we will assume that for every $j = 1, \ldots, d$, the function $4df_j(1 - df_j)$ is μ−almost everywhere non zero and that

$$\mathbb{E}(f_j \mid \mathcal{I}) = \frac{1}{2d} \quad \mu - \text{almost everywhere}.$$

PROPOSITION 7 *i)* $\Lambda^* \geq \Lambda^*_{\frac{1}{2d}}$ μ-almost everywhere.

ii) The equality $\Lambda = \Lambda_{\frac{1}{2d}}$ holds $\mu-$ almost everywhere if and only if the functions $f_1, \ldots, f_d$ are all equal to $\frac{1}{2d}$ μ-almost everywhere.

Remark:

From this result, we deduce that if the functions $f_1, \ldots, f_d$ are not all equal to $\frac{1}{2d}$, the large deviations of the associated dynamic $\mathbb{Z}^d$-random walk is quite different from the ones of the classical simple $\mathbb{Z}^d$-random walk even in the case where the integrals of the functions $f_1, \ldots, f_d$ are all equal to $\frac{1}{2d}$. More precisely, for simplicity, assume $(E, \mathcal{A}, \mu, T)$ to be uniquely ergodic, if $f_1, \ldots, f_d$ are continuous and $\int_E f_j d\mu = \frac{1}{2d}$ for every $j = 1, \ldots, d$, then $(S_n/n)_{n\geq 1}$ and $(S'_n/n)_{n\geq 1}$ converges $\mathbb{P}$−almost surely as $n \to \infty$ to the vector null but the probability for the dynamic $\mathbb{Z}^d$−random walk S_n to do large deviations from 0 is smaller than the one obtained for the simple one. In some sense, in that case, the dynamic $\mathbb{Z}^d$−random walk is more concentrated around the vector null than the simple one.

Proof:

Let us prove the first assertion. By Jensen's inequality, for every $\lambda \in \mathbb{R}^d$,

$$\Lambda(\lambda) = \mathbb{E}\Big(\log \Big(\sum_{j=1}^d e^{\lambda_j} f_j + (\frac{1}{d} - f_j) e^{-\lambda_j} \Big) \mid \mathcal{I} \Big)$$

$$\leq \log \Big(\mathbb{E}\Big(\sum_{j=1}^d e^{\lambda_j} f_j + (\frac{1}{d} - f_j) e^{-\lambda_j} \Big) \mid \mathcal{I} \Big)$$

$$= \log \Big(\frac{1}{2d} \sum_{j=1}^d (e^{\lambda_j} + e^{-\lambda_j}) \Big) = \Lambda_{\frac{1}{2d}}(\lambda).$$

Thus, for every $y \in \mathbb{R}^d$, $\Lambda^*(y) \geq \Lambda^*_{\frac{1}{2d}}(y)$ μ-almost everywhere.

Let us come to the second assertion. The equality $\Lambda = \Lambda_{\frac{1}{2d}}$ means that for every $\lambda \in \mathbb{R}^d$,

$$\mathbb{E}\left(\log \left(\frac{\sum_{j=1}^d e^{\lambda_j} f_j + e^{-\lambda_j}(\frac{1}{d} - f_j)}{\frac{1}{2d} \sum_{j=1}^d (e^{\lambda_j} + e^{-\lambda_j})} \right) \mid \mathcal{I} \right) = 0 \quad \mu - \text{a.e..} \quad (2.3)$$

Now fix an integer $k \in \{1, \ldots, d\}$ and consider the vectors λ such that $\lambda_i = 0$ if $i \neq k$ and λ_k is not fixed. Then, equation (2.3) can be rewritten as

$$\mathbb{E}\left(\log \left(\frac{2d e^{\lambda_k} f_k + e^{-\lambda_k} 2(1 - df_k) + 2(d-1)}{e^{\lambda_k} + e^{-\lambda_k} + 2(d-1)} \right) \mid \mathcal{I} \right) = 0 \quad \mu-\text{a.e..} \quad (2.4)$$

Firstly, let λ_k go to $+\infty$ in the equality (2.4), then
$$\mathbb{E}(\log(2df_k) \mid \mathcal{I}) = 0 \quad \mu-\text{a.e..}$$
Secondly, let λ_k go to $-\infty$ in the equality (2.4) gives us
$$\mathbb{E}(\log(2(1-df_k)) \mid \mathcal{I}) = 0 \quad \mu-\text{a.e..}$$
By summing these two equalities, we get
$$\mathbb{E}\Big(\log(4df_k(1-df_k)) \mid \mathcal{I}\Big) = 0 \quad \mu-\text{a.e..}$$
Since for every $x \in [0,1]$, $0 \le 4x(1-x) \le 1$ and $df_k \in [0,1]$, the integrand in the above conditional expectation is less or equal to 0. So, this conditional expectation is null if and only if $4df_k(1-df_k) = 1 \; \mu-\text{a.e.}$ or equivalently, $f_k = \frac{1}{2d} \; \mu-\text{a.e.}$. By doing the same reasoning for each $k \in \{1, \ldots, d\}$, we prove ii).

5.4 EXAMPLE: THE ROTATION ON THE TORUS.

Let $(\mathbb{T}^r, \mathcal{B}(\mathbb{T}^r), \lambda, T_\alpha)$ be the dynamical system where T_α is the rotation on the r-dimensional torus $\mathbb{T}^r$ associated to the r-dimensional vector $\alpha = (\alpha_1, \ldots, \alpha_r)$ defined by $x \to x + \alpha \mod 1$, and λ is the Lebesgue measure on $\mathbb{T}^r$. We consider the dynamic $\mathbb{Z}^d$-random walk associated to this particular dynamical system. Twofold motivations are related to this example:
1. Explicit calculations are possible in this case.
2. When the angle of rotation is a rational vector, we get a periodic dynamical system which models many practical resource sharing problems (including distributed algorithms in computer science, option pricing in financial markets, dam management problems ...).

Two cases are considered:

- When the angle is an *irrational* vector, the dynamical system is uniquely ergodic and Theorems 2.5 and 2.6 hold for continuous functions $f_1, \ldots, f_d$ defined on $\mathbb{T}^r$ with values in $[0, \frac{1}{d}]$ and for every $x \in \mathbb{T}^r$. In that particular setting,
$$\Lambda^*(x) = \sup_{\lambda \in \mathbb{R}^d} \{<\lambda, x> - \Lambda(\lambda)\}$$
where
$$\Lambda(\lambda) = \int_{\mathbb{T}^r} \log \Big(\sum_{j=1}^d e^{\lambda_j} f_j(t) + (\frac{1}{d} - f_j(t))e^{-\lambda_j} \Big) dt.$$

- Let $\alpha = (\alpha_1, \ldots, \alpha_r)$ be a *rational* vector with

$$\alpha_i = \frac{p_i}{q_i}, q = s.c.m.(q_1, \ldots, q_r)$$

and $x \in \mathbb{T}^r$ is chosen such that for every $j = 1, \ldots, d$,

$$\sum_{i=1}^{q} f_j(T_\alpha^i x) = \frac{q}{2d}.$$

Then, for every n, there exist two integers m_n and r_n such that $n = m_n q + r_n, 0 \le r_n \le q - 1$ and

$$\sum_{i=1}^{n} \left(f_j(T_\alpha^i x) - \frac{1}{2d} \right) = \sum_{j=0}^{m_n - 1} \sum_{i=1}^{q} \left(f_j(T_\alpha^{i+jq} x) - \frac{1}{2d} \right)$$
$$+ \sum_{i=1}^{r_n} \left(f_j(T_\alpha^{i+m_n q} x) - \frac{1}{2d} \right) = \mathcal{O}(1)$$

uniformly in $x \in \mathbb{T}^r$ so conclusions of Theorem 2.6 still hold in the rational angle case.

More general sequences than $(T_\alpha^i x)_{i \ge 1}$ can be chosen to drive the dynamic of the random walk like Van der Corput's or Halton's sequences (see [68] for further details).

Chapter 3

RECURRENCE AND TRANSIENCE

1. INTRODUCTION

DEFINITION 3.1 *A point z of $\mathbb{Z}^d$ is called recurrent for the dynamic random walk $(S_n)_{n\geq 0}$ if and only if*

$$\mathbb{P}_z(S_n = z, \text{ infinitely often}) = 1$$

with the notation $\mathbb{P}_z(\, . \,) = \mathbb{P}_z(\, . \mid S_0 = z \,)$.
Otherwise, it is called transient.

DEFINITION 3.2 *The dynamic random walk $(S_n)_{n\geq 0}$ is called recurrent (respectively transient) if and only if every point z of $\mathbb{Z}^d$ is recurrent (respectively transient).*

The main difficulty in the study of the properties of recurrence-transience of the dynamic random walks is that we can not use arguments from the renewal theory due to the temporal inhomogeneity of the walk. We prove below that under some assumptions, the dynamic random walk S_n is recurrent in dimension $d = 1, 2$ and transient if $d \geq 3$. The proofs rests on the central limit theorem 2.2 and on the local limit theorem 2.3. In this chapter, the dynamical system $S = (E, \mathcal{A}, \mu, T)$ is assumed ergodic. We begin by studying the one-dimensional case.

2. THE ONE-DIMENSIONAL CASE

THEOREM 3.1 *1- If $\int_E f(t)dt \neq \frac{1}{2}$, then for μ-almost every $x \in E$, $(S_n)_{n \in \mathbb{N}}$ is transient.*
2- Assume that $f, f^2 \in \mathcal{C}_2(S)$, $\int_E f(t)dt = \frac{1}{2}$ and $a = \int_E 4f(1-f)\,d\mu > 0$, then for every $x \in E$, $(S_n)_{n \in \mathbb{N}}$ is recurrent.

Proof:
The first assertion is quite easy to prove using the strong law of large numbers for the dynamic random walk proved in Chapter 2. For μ-almost every $x \in E$, as $n \to +\infty$, $S_n \sim n(2\int_E f d\mu - 1)$ $\mathbb{P}$-almost surely. Since $2\int_E f d\mu - 1 \neq 0$, the dynamic random walk tends to $\pm\infty$ as $n \to \infty$ which implies the transience.

In order to prove the recurrence of the Markov chain $(S_n)_{n \in \mathbb{N}}$, we only have to prove the recurrence at point 0, since $(S_n)_{n \in \mathbb{N}}$ is homogeneous in space. Let us prove that the event

$$A = \{\limsup_n \frac{S_n}{\sqrt{n}} = \infty, \liminf_n \frac{S_n}{\sqrt{n}} = -\infty\}$$

has probability one. Let

$$\begin{aligned} A_c &= \{\limsup_n \frac{S_n}{\sqrt{n}} > c\} \cap \{\liminf_n \frac{S_n}{\sqrt{n}} < -c\} \\ &= A'_c \cap A''_c, c > 0. \end{aligned}$$

Then, $A_c \searrow A, c \to \infty$ and A, A_c, A'_c, A''_c are in the tail algebra $\mathcal{G} = \cap \mathcal{G}_n$, where $\mathcal{G}_n = \sigma(X_n, X_{n+1}, \ldots)$. Let us show that $\mathbb{P}(A'_c) = \mathbb{P}(A''_c) = 1$ for each $c > 0$. Because of the Kolmogorov's zero-one law, since $A'_c \in \mathcal{G}$ and $A''_c \in \mathcal{G}$, it is sufficient to show only that $\mathbb{P}(A'_c) > 0$ and $\mathbb{P}(A''_c) > 0$. The events $\limsup\{\frac{S_n}{\sqrt{n}} > c\}$ and $\limsup\{\frac{S_n}{\sqrt{n}} < -c\}$ being respectively included into $\{\limsup \frac{S_n}{\sqrt{n}} > c\}$ and $\{\liminf \frac{S_n}{\sqrt{n}} < -c\}$, we obtain that

$$\mathbb{P}(A'_c) \geq \limsup_n \mathbb{P}(\frac{S_n}{\sqrt{n}} > c)$$

and

$$\mathbb{P}(A''_c) \geq \limsup_n \mathbb{P}(\frac{S_n}{\sqrt{n}} < -c).$$

Now, we use the central limit theorem for the dynamic random walk which is proved in Chapter 2:

$$\frac{S_n}{\sqrt{n}} \to \mathcal{N}(0, a), \text{ in distribution.}$$

Therefore,

$$\limsup_n \mathbb{P}(\frac{S_n}{\sqrt{n}} > c) > 0$$

and

$$\limsup_n \mathbb{P}(\frac{S_n}{\sqrt{n}} < -c) > 0.$$

Thus, $\mathbb{P}(A_c) = 1$, for all $c > 0$ and
$$\mathbb{P}(A) = \lim_{c \to \infty} \mathbb{P}(A_c) = 1.$$
From the definition of the event A, we obtain that
$$\mathbb{P}(\limsup_n \{S_n = 0\}) = 1,$$
so, the recurrence of $(S_n)_{n \in \mathbb{N}}$ is proved.

3. THE HIGHER-DIMENSIONAL CASE

In higher dimension, if all the integrals $\int_E f_j d\mu, j = 1, \ldots, d$ are not equal to $\frac{1}{2d}$, the transience follows using the strong of large numbers for the dynamic random walks. The only interesting case is the one where all these integrals are equal to $\frac{1}{2d}$. In this section, the matrix A is assumed to be positive definite. The following result holds for dimension 2.

THEOREM 3.2 *Let $f_j \in \mathcal{C}_3(S), j = 1, 2$ be functions with values in $[0, \frac{1}{2}]$ such that for every $j, l \in \{1, 2\}, f_j f_l \in \mathcal{C}_3(S)$ and $\int_E f_j d\mu = \frac{1}{4}$. Then, for every $x \in E$, $(S_n)_{n \geq 0}$ is recurrent.*

Proof:
We will use the next lemma (see [177] p. 317) generalizing the classical Borel-Cantelli lemma to the case where the considered events are not necessarily independent.

LEMMA 8 *Let $(A_n)_{n \geq 0}$ be any sequence of events in a probability space. If $\sum_{n=0}^{\infty} \mathbb{P}(A_n) = +\infty$, and if for some $c > 0$,*
$$\liminf_{n \to \infty} \frac{\sum_{k,l=0}^{n} \mathbb{P}(A_k \cap A_l)}{\left\{\sum_{k=0}^{n} \mathbb{P}(A_k)\right\}^2} \leq c,$$
then
$$\mathbb{P}(\limsup_n A_n) \geq \frac{1}{c}.$$

We evidently take $A_n = \{S_n = 0\}$ and estimate
$$\frac{\sum_{k,l=0}^{n} \mathbb{P}(\{S_k = 0\} \cap \{S_l = 0\})}{\left\{\sum_{k=0}^{n} \mathbb{P}(S_k = 0)\right\}^2} - 1$$

$$= \frac{1}{\sum_{k=0}^{n} \mathbb{P}(S_k = 0)} - \frac{\sum_{k=0}^{n} \mathbb{P}(S_k = 0)^2}{\left\{\sum_{k=0}^{n} \mathbb{P}(S_k = 0)\right\}^2} \quad (3.1)$$

$$+ 2\frac{\sum_{k=0}^{[(n-1)/2]} \mathbb{P}(S_k = 0) \sum_{l=1}^{[(n-k)/2]} \left[\mathbb{P}(\tilde{S}_{2l}^{(k)} = 0) - \mathbb{P}(S_{2l} = 0)\right]}{\left\{\sum_{k=0}^{n} \mathbb{P}(S_k = 0)\right\}^2}$$

where $(\tilde{S}_{2l}^{(k)})_{l\in\mathbb{N}}$ is defined in the same way as the Markov chain $(S_n)_{n\in\mathbb{N}}$, but with transition probabilities given by

$$\mathbb{P}(\tilde{S}_l^{(k)} = z | \tilde{S}_{l-1}^{(k)} = 0) = \begin{cases} f_j(T^{l+2k}x) & \text{if } z = e_j \\ \frac{1}{2} - f_j(T^{l+2k}x) & \text{if } z = -e_j \\ 0 & \text{otherwise.} \end{cases}$$

The local limit theorem (Section 3. of Chapter 2),

$$\mathbb{P}(S_{2n} = 0) \sim \frac{1}{\sqrt{\det A(2\pi n)}} \quad \text{as } n \to \infty,$$

implies that there exists a strictly positive constant C such that

$$\sum_{k=0}^{n} \mathbb{P}(S_k = 0) \sim C \log n \text{ as } n \to \infty.$$

Consequently, the first two terms in the right-hand side formula (3.1) tends to 0 as n goes to infinity. The third term can be majored by

$$2\frac{\sum_{l=0}^{n} \sup_{k\geq 0} \left|\mathbb{P}(\tilde{S}_{2l}^{(k)} = 0) - \mathbb{P}(S_{2l} = 0)\right|}{\sum_{k=0}^{n} \mathbb{P}(S_k = 0)}. \quad (3.2)$$

Let $\phi_l^{(k)}$ be the characteristic function of $\tilde{S}_{2l}^{(k)}$ and we still denote by Q the quadratic form associated to the matrix A. For every $k \geq 0$,

$$\mathbb{P}(\tilde{S}_{2l}^{(k)} = 0) = \frac{2}{\sqrt{\det A}(4\pi l)^{\frac{d}{2}}} + R_1(k,l) + R_2(k,l) + R_3(k,l),$$

where $R_i(k,l), i=1,2,3$, are given by

$$R_1(k,l) = \frac{1}{(2\pi^2 l)} \int_{|u|\leq \log l} (\phi_l^{(k)}(\frac{u}{\sqrt{l}}) - e^{-Q(u)})du$$

$$R_2(k,l) = -\frac{1}{(2\pi^2 l)} \int_{|u|>\log l} e^{-Q(u)}du$$

$$R_3(k,l) = \frac{1}{(4\pi^2 l)} \int_{E_l} \phi_l^{(k)}(\frac{u}{\sqrt{l}})du.$$

where $E_l = [-\frac{\pi}{2}\sqrt{l}, \frac{3\pi}{2}\sqrt{l}]^2 \setminus \{B(0,\log l) \cup B(\Pi\sqrt{l}, \log l)\}$.
Then, since $R_2(k,l)$ does not depend on k, we have

$$\mathbb{P}(\tilde{S}_{2l}^{(k)} = 0) - \mathbb{P}(S_{2l} = 0) = \frac{1}{(2\pi^2 l)} \int_{|u|\leq \log l} (\phi_l^{(k)}(\frac{u}{\sqrt{l}}) - e^{-Q(u)})du$$
$$- \frac{1}{(2\pi^2 l)} \int_{|u|\leq \log l} (\phi_l(\frac{u}{\sqrt{l}}) - e^{-Q(u)})du$$
$$+ \frac{1}{(4\pi^2 l)} \int_{E_l} \phi_l^{(k)}(\frac{u}{\sqrt{l}})du - \frac{1}{(4\pi^2 l)} \int_{E_l} \phi_l(\frac{u}{\sqrt{l}})du.$$

There exists a constant $C > 0$ such that for every $k \geq 0$,

$$|\frac{1}{(2\pi^2 l)} \int_{|u|\leq \log l} (\phi_l^{(k)}(\frac{u}{\sqrt{l}}) - e^{-Q(u)})du| \leq \frac{C}{l} \sup_{u \in B(0,\log l)} |e^{T_l(u)} - 1|$$
$$\leq \frac{C}{l} \sup_{u \in B(0,\log l)} |T_l(u)| \exp|T_l(u)|$$

where $T_n(u)$ is defined as in Section 3, Chapter 2. Following calculations performed in Section 3 of Chapter 2, we get

$$\sup_{u\in B(0,\log l)} |T_l(u)|$$
$$\leq \frac{(\log l)^2}{8l} \sum_{j=1}^{2} \Big| \sum_{i=1}^{2l} 8f_j(T^{i+2k}x)(1-2f_j(T^{i+2k}x))$$
$$- 16l \int_E f_j(x)(1-2f_j(x))d\mu(x)\Big|$$
$$+ \frac{(\log l)^2}{8l} \sum_{j\neq l} \Big| \sum_{i=1}^{2l} (4f_j(T^{i+2k}x) + 4f_l(T^{i+2k}x) - 1 - 16f_j(T^{i+2k}x)f_l(T^{i+2k}x))$$
$$- 2l \int_E (1-16f_j(x)f_l(x))d\mu(x)\Big| + \frac{\log l}{\sqrt{l}} \frac{1}{2} \sum_{j=1}^{2} \Big| \sum_{i=1}^{2l} (4f_j(T^{i+2k}x) - 1)\Big|$$
$$+ \mathcal{O}(\frac{(\log l)^3}{l^{\frac{3}{2}}})$$

Then, since f_i and $f_i f_j$, $i,j = 1,2$ belong to the class $\mathcal{C}_3(S)$,

$$\sup_{k \geq 0} \sum_{l=0}^{n} \frac{1}{l} \int_{|u| \leq \log l} \left| \phi_l^{(k)}(\frac{u}{\sqrt{l}}) - e^{-Q(u)} \right| du = o(\log(n)).$$

Moreover, following the ideas developed in the proof of Lemma 3 from Section 3 of Chapter 2, we deduce that there exists a constant $C' > 0$ such that

$$\frac{1}{l} \sup_{k \geq 0} \left| \int_{E_l} \phi_l^{(k)}(\frac{u}{\sqrt{l}}) du \right| \leq C' \exp(-\frac{4}{\pi^2}(\log l)^2 (\lambda_1 - 2\epsilon))$$

where λ_1 is the smallest eigenvalue of Q and ϵ is chosen less than $\lambda_1/2$. Then,

$$\sum_{l=0}^{n} \frac{1}{(4\pi^2 l)} \sup_{k \geq 0} \left| \int_{E_l} \phi_l^{(k)}(\frac{u}{\sqrt{l}}) \right| du < +\infty.$$

Since the denominator in (3.2) is equivalent (up to a multiplicative constant) to $\log(n)$, the third term in the right-hand side of (3.1) tends to 0 as n goes to infinity.

The local limit theorem combined with Lemma 8 (with $c = 1$) implies that

$$\mathbb{P}(\limsup_{n} \{S_n = 0\}) = 1$$

and then the recurrence of the two-dimensional dynamic random walk.

The transience in dimension $d \geq 3$ is much more easier to establish.

THEOREM 3.3 *Let $f_j \in \mathcal{C}_1(S), j = 1, \ldots, d$ be functions with values in $[0, \frac{1}{d}]$ such that for every $j, l \in \{1, \ldots, d\}$, $f_j f_l \in \mathcal{C}_1(S)$ and $\int_E f_j d\mu = \frac{1}{2d}$. Then, for μ-almost every $x \in E$, $(S_n)_{n \geq 0}$ is transient.*

Proof:
From the local limit theorem (Section 3 of Chapter 2), we have

$$\mathbb{P}_0(S_{2n} = 0) \sim \frac{2}{\sqrt{\det A}(4\pi n)^{\frac{d}{2}}} \quad \text{as } n \to \infty.$$

Then, when $d \geq 3$,

$$\sum_{n=0}^{\infty} \mathbb{P}_0(S_n = 0) < +\infty$$

and from Borel-Cantelli lemma,

$$\mathbb{P}_0(\limsup_{n} \{S_n = 0\}) = 0$$

so $\mathbb{P}$-almost surely, we only visit 0 a finite number of times.

Chapter 4

DYNAMIC RANDOM WALKS IN A RANDOM SCENERY

Let $\xi_x, x \in \mathbb{Z}^d$, be a family of i.i.d. $\mathbb{R}$-valued random variables, with zero mean and finite positive variance σ^2. These random variables play the rôle of random scenery. Let $(X_i)_{i\geq 1}$ be a sequence of random variables with values in $\mathbb{Z}^d$, independent of the ξ's, and $S_n = X_1 + \ldots + X_n$ for $n \geq 1$, $S_0 = 0$ the associated random walk. We are interested in the asymptotic behavior of the "cumulative sum"

$$Z_n = \sum_{k=0}^{n} \xi_{S_k}.$$

When $(S_n)_{n\in\mathbb{N}}$ is a standard $\mathbb{Z}^d$-random walk (a sum of i.i.d. random variables) then $(\xi_{S_n})_{n\in\mathbb{N}}$ is a stationary sequence and $(\xi_{S_k})_{k\geq 0}$ satisfies a strong law of large numbers by Birkhoff's theorem. In the case $d = 1$, Kesten and Spitzer [94] proved that when X and ξ belong to the domains of attraction of different stable laws of indices $1 < \alpha \leq 2$ and $0 < \beta \leq 2$, respectively, then there exists a $\delta > \frac{1}{2}$ such that $n^{-\delta} Z_{[nt]}$ converges weakly as $n \to \infty$ to a self-similar process with stationary increments, δ being related to α and β by $\delta = 1 - \alpha^{-1} + (\alpha\beta)^{-1}$. The case $0 < \alpha < 1$ and β arbitrary is easier; they showed then that $n^{-\frac{1}{\beta}} Z_{[nt]}$ converges weakly, as $n \to \infty$, to a stable process with index β. Bolthausen [17] showed that when $(S_n)_{n\in\mathbb{N}}$ is a recurrent $\mathbb{Z}^2$-random walk, $(n \log n)^{-\frac{1}{2}} Z_{[nt]}$ satisfies a functional central limit theorem. Lewis [113] established a self-normalized law of the iterated logarithm (l.i.l.) for random walk in random scenery. Then, in [114], under some assumptions on the random walk, namely a strong law for the range and the number of self-intersection of the random walk, he obtained a l.i.l. with deterministic normalizers. This result covers the transient random walks and the recurrent random walk in two-dimensions, but not the simple, symmetric random walk on $\mathbb{Z}$. Khoshnevisan and Lewis solved this problem in [96] by proving a l.i.l. for stable processes in random scenery, then by constructing the simple random walk from the Brownian motion in a classical way, they get a l.i.l. for a simple, symmetric random walk in random scenery with a stronger normalization than the natural one. Further results about this subject can also be found in paper [31]. Recent developments in quantun information and communication theories show a strong connection between the model of $\mathbb{Z}$-random walk evolving in a random scenery and a random walk moving on randomly directed versions of the two-dimensional lattice (see [23], [24], [77] for details or Chapter 10). Another motivation for studying the random walks in random sceneries is that this model appears in the study of polymers represented by the random walk $(S_k)_{k\geq 0}$ evolving in a disordered media which creates random electrical charges on the polymer. These random electrical charges correspond to the random variables $\xi_i, i \in \mathbb{Z}^d$ (the random landscape). The total electrical energy describing the interactions of the polymer with itself can be written as a double sum $\sum_{i,j} h(\xi_{S_i}, \xi_{S_j})$ for a convenient function h. This physical model is quite realistic as the considered proteins are very long charged molecules and their stereo-chemical shape is the one minimising the total electrical energy. The results obtained in papers [22] (dimension $d \geq 2$) and [76] (dimension $d = 1$) constitutes a first step for the study of this problem (see [21], [133] where two different models are defined). Our aim in this chapter is to show that the previous theoretical results about classical random walks in random sceneries can be extended to the case of dynamic random walks.

When Z_n is driven by a dynamic random walk, we are able to describe the evolution of the process on the product space $\mathbb{R}^{\mathbb{Z}^d} \times (\mathbb{Z}^d)^{\mathbb{N}} \times E$. For every $x \in E$, we denote by $\mathbb{P}_x$ the product measure on the cartesian product of the set of sceneries and the set of dynamics paths $(\mathbb{R}^{\mathbb{Z}^d} \times (\mathbb{Z}^d)^{\mathbb{N}}, \mathcal{F})$, where $\mathcal{F}$ is the σ-field generated by the cylinder sets. Now, for $A \in \mathcal{A}$ and $F \in \mathcal{F}$, let

$$\mathbb{P}(F \times A) = \int_A \mathbb{P}_x(F) d\mu(x).$$

$\mathbb{P}$ is a probability measure on the space $(\mathbb{R}^{\mathbb{Z}^d} \times (\mathbb{Z}^d)^{\mathbb{N}} \times E, \mathcal{F} \otimes \mathcal{A})$. By a straightforward calculation we can prove that $(\xi_{S_k})_{k\geq 0}$ is a stationary sequence. Hence by the ergodic theorem, with probability one,

$$\frac{Z_n}{n} \to 0.$$

When $x \in E$ is fixed, this reasoning does not hold anymore. In the paper [68], $(S_n)_{n\in\mathbb{N}}$ was chosen as a dynamic $\mathbb{Z}$-random walk generated by an irrational rotation on the torus. For every point x fixed in the torus, it was proved under some assumptions on f and the irrational angle α that $(\xi_{S_k})_{k\geq 0}$ satisfies a weak law of large numbers. This result can also be obtained by techniques coming from ergodic theory (see [160] for ideas). In the next two sections, we prove strong laws of large numbers and functional limit theorems for general dynamic random walk in a random scenery. We distinguish the case where the dynamic random walk is either recurrent or transient. In a last section, we study in details a particular dynamical system namely the rotation on the multi-dimensional torus.

1. THE RECURRENT CASE

In probability theory one of the most famous theorems is the strong law of large numbers: take an infinite sequence of integrable i.i.d. random variables then the average of these random variables converge almost surely to the mean of their distribution. For standard $\mathbb{Z}^d$−random walk in random scenery the strong law of large numbers follows directly from Birkhoff's Theorem due to ergodicity and stationarity. In case of dynamic random walks, the sequence $(\xi_{S_k})_{k\geq 0}$ is no more stationary and a new approach has to be achieved. More precisely, we prove a maximal inequality for the sequence $(\xi_{S_k})_{k\geq 0}$. The main tool in the proof of this inequality is that conditioned on the random walk, the sequence of the random variables $(\xi_{S_k})_{k\geq 0}$ is associated. We recall in the next section the definition and the interesting properties of associated random variables.

1.1 SOME RESULTS ABOUT ASSOCIATED RANDOM VARIABLES

DEFINITION 4.1 *A finite collection of random variables $(Y_i)_{1\leq i\leq m}$ is associated if for any two coordinatewise nondecreasing functions f_1, f_2 on $\mathbb{R}^m$ such that $\tilde{f}_i = f_i(Y_1, \ldots, Y_m), i = 1, 2$ has finite variance, $\text{Cov}(\tilde{f}_1, \tilde{f}_2) \geq 0$. An infinite collection of random variables is associated if every finite subcollection is associated.*

Any sequence of independent random variables is associated. Newman and Wright [140] proved the following maximal inequality

THEOREM 4.1 *If $Y_i, 1 \leq i \leq m$ are associated, mean zero, finite variance random variables and $M_m = \max(T_1, \ldots, T_m)$ where $T_n = Y_1 + \ldots + Y_n$, then*

$$\mathbb{E}(M_m^2) \leq \text{Var}(T_m) \qquad (4.1)$$

and for any $\varepsilon > 0$, for any $m \geq 1$,

$$\mathbb{P}(\max_{1 \leq n \leq m} |M_n| \geq \varepsilon) \leq C(\varepsilon) \text{Var}(T_m) \qquad (4.2)$$

where $C(\varepsilon)$ only depends on ε.

Proof:
Let us prove (4.1): Let us define $K_m = \min(Y_2 + \ldots + Y_m, Y_3 + \ldots + Y_m, \ldots, Y_m, 0)$, $L_m = \max(Y_2, Y_2+Y_3, \ldots, Y_2+\ldots+Y_m)$, $J_m = \max(0, L_m)$. Remark that $K_m = Y_2 + \ldots + Y_m - J_m$ is a nondecreasing function of the Y_i's so that $\text{Cov}(Y_1, K_m) \geq 0$. Since $M_m = Y_1 + J_m$, we have

$$\begin{aligned}
\mathbb{E}(M_m^2) = \mathbb{E}((Y_1 + J_m)^2) &= \text{Var}(Y_1) + 2\text{Cov}(Y_1, J_m) + \mathbb{E}(J_m^2) \\
&= \text{Var}(Y_1) + 2\text{Cov}(Y_1, Y_2 + \ldots + Y_m) \\
&\quad - 2\text{Cov}(Y_1, K_m) + \mathbb{E}(J_m^2) \\
&\leq \text{Var}(Y_1) + 2\text{Cov}(Y_1, Y_2 + \ldots + Y_m) \\
&\quad + \mathbb{E}(L_m^2) \qquad (4.3)
\end{aligned}$$

by remarking that $J_m^2 \leq L_m^2$. We conclude by induction on m: $\mathbb{E}(L_m^2) \leq \text{Var}(Y_2 + \ldots + Y_m)$ implies using (4.3) that

$$\mathbb{E}(M_m^2) \leq \text{Var}(Y_1 + \ldots + Y_m) = \text{Var}(T_m).$$

Proof of (4.2): Let us define for any $m \geq 1$, $M_m^* = \max_{1 \leq n \leq m} M_n$. For every $0 < \varepsilon' < \varepsilon$,

$$\begin{aligned}
\mathbb{P}(M_m^* \geq \varepsilon) &\leq \mathbb{P}(M_m \geq \varepsilon') \\
&\quad + \mathbb{P}(M_{m-1}^* \geq \varepsilon; M_{m-1}^* - M_m > \varepsilon - \varepsilon')
\end{aligned}$$

The random variables $Y_i, 1 \leq i \leq m$ being associated, so are the two random variables M_{m-1}^* and $M_m - M_{m-1}^*$ since they are nondecreasing functions of the Y_i's. Remark also that if two random variables X and Y are associated, then for any $x, y \in \mathbb{R}$,

$$\mathbb{P}(X > x; Y > y) - \mathbb{P}(X > x)\mathbb{P}(Y > y) \geq 0.$$

So,

$$\begin{aligned}
\mathbb{P}(M_m^* \geq \varepsilon) &\leq \mathbb{P}(M_m \geq \varepsilon') \\
&\quad + \mathbb{P}(M_{m-1}^* \geq \varepsilon)\mathbb{P}(M_{m-1}^* - M_m \geq \varepsilon - \varepsilon') \\
&\leq \frac{\mathbb{E}(M_m^2)}{\varepsilon'^2} + \frac{\mathbb{E}((M_{m-1}^* - M_m)^2)}{(\varepsilon - \varepsilon')^2} \\
&\leq C(\varepsilon)\mathbb{E}(M_m^2)
\end{aligned}$$

by choosing for example $\varepsilon' = \varepsilon/2$. The same inequality holds for the random variables $-Y_i, 1 \leq i \leq m$. Thus, from inequality (4.1),

$$\mathbb{P}(\max_{1 \leq n \leq m} |M_n| \geq \varepsilon) \leq C(\varepsilon)\mathbb{E}(M_m^2) \leq C(\varepsilon)\text{Var}(T_m).$$

1.2 THE STRONG LAW OF LARGE NUMBERS

THEOREM 4.2 *Let $(S_n)_{n \in \mathbb{N}}$ be a dynamic random walk generated by a dynamical system S and let $f_j \in \mathcal{C}_3(S), j = 1, \ldots, d$ be functions with values in $[0, \frac{1}{d}]$ such that for every $j, l \in \{1, \ldots, d\}, f_j f_l \in \mathcal{C}_3(S)$ and $\int_E f_j d\mu = \frac{1}{2d}$. The matrix A associated to the f_i's is assumed to be positive definite. Then, if $d = 1$ (resp. $d \geq 2$), for any $\gamma > \frac{3}{4}$ (resp. $\gamma > \frac{1}{2}$), for every $x \in E$ fixed,*

$$\frac{Z_n}{n^\gamma} \underset{n \to \infty}{\to} 0 \quad a.s.$$

Proof:
We define

$$V_n = \sum_{i,j=0}^{n} \mathbf{1}_{\{S_i = S_j\}}.$$

LEMMA 9
$$\text{Var}(Z_n) = \sigma^2 \mathbb{E}(V_n).$$

Proof:
Obviously, $\mathbb{E}(Z_n) = 0$. Then

$$\text{Var}(Z_n) = \sum_{k,l=0}^{n} \mathbb{E}(\xi(S_k)\xi(S_l))$$
$$= \sum_{k \neq l} \mathbb{E}(\xi(S_k)\xi(S_l)) + \sigma^2(n+1).$$

Let $\mathcal{F}$ be the σ-field generated by the r.v. $X_1, X_2, \ldots$. We have

$$\mathbb{E}(\xi(S_k)\xi(S_l)|\mathcal{F}) = \sum_{i,j \in \mathbb{Z}^d, i \neq j} \mathbf{1}_{\{S_k = i\}} \mathbf{1}_{\{S_l = j\}} \mathbb{E}(\xi(i)\xi(j)) + \sigma^2 \mathbf{1}_{\{S_k = S_l\}}$$
$$= \sigma^2 \mathbf{1}_{\{S_k = S_l\}}.$$

Consequently,

$$\text{Var}(Z_n) = \sigma^2 [\sum_{k \neq l} \mathbb{P}(S_k = S_l) + (n+1)]$$
$$= \sigma^2 \mathbb{E}(V_n).$$

Lemma 10

$$\mathbb{E}(V_n) \sim \begin{cases} \frac{4(2^{\frac{3}{2}}-1)}{3\sqrt{\pi \det A}} n^{\frac{3}{2}} & \text{if } d=1 \\ \frac{1}{\pi\sqrt{\det A}} n \log n & \text{if } d=2 \\ C n & \text{if } d \geq 3 \end{cases}$$

Proof:

$$\begin{aligned} \mathbb{E}(V_n) &= \sum_{k \neq l} \mathbb{P}(S_k = S_l) + (n+1) \\ &= 2 \sum_{l=1}^{n} \sum_{k=0}^{l-1} \mathbb{P}(S_k = S_l) + (n+1) \end{aligned}$$

Since $(S_n)_{n \in \mathbb{N}}$ is homogeneous in space, for every $l \geq 0$,

$$\begin{aligned} \mathbb{P}(S_{k+2l} = S_k) &= \mathbb{P}(S_{k+2l} = 0 | S_k = 0) \text{ if } k \text{ even} \\ \mathbb{P}(S_{k+2l} = S_k) &= \mathbb{P}(S_{k+2l} = e_1 | S_k = e_1) \text{ if } k \text{ odd.} \end{aligned}$$

Therefore,

$$\begin{aligned} \mathbb{E}(V_n) &= (n+1) + 2 \sum_{k=0}^{[\frac{(n-1)}{2}]} \sum_{l=1}^{[\frac{(n-k)}{2}]} \mathbb{P}(S_{2(k+l)} = 0 | S_{2k} = 0) \\ &+ 2 \sum_{k=0}^{[\frac{(n-2)}{2}]} \sum_{l=1}^{[\frac{(n-k-1)}{2}]} \mathbb{P}(S_{2k+1+2l} = e_1 | S_{2k+1} = e_1) \end{aligned}$$

We are only interested in the asymptotic behavior of the first double sum in the above expression; the second one can be treated in the same way. Let $\phi_l^{(k)}$ be the characteristic function of $\tilde{S}_{2l}^{(k)}$, where $(\tilde{S}_l^{(k)})_{l \in \mathbb{N}}$ is defined in the same way as the Markov chain $(S_n)_{n \in \mathbb{N}}$, but with transition probabilities given by

$$\mathbb{P}(\tilde{S}_l^{(k)} = z | \tilde{S}_{l-1}^{(k)} = 0) = \begin{cases} f_j(T^{l+2k}x) & \text{if } z = e_j \\ \frac{1}{d} - f_j(T^{l+2k}x) & \text{if } z = -e_j \\ 0 & \text{otherwise} \end{cases}$$

For every $k \geq 0$,

$$\mathbb{P}(S_{2(k+l)} = 0 | S_{2k} = 0) = \frac{2}{\sqrt{\det A}(4\pi l)^{\frac{d}{2}}} + R_1(k,l) + R_2(k,l) + R_3(k,l),$$

where $R_i(k,l), i = 1,2,3$, are given by

$$R_1(k,l) = \frac{2}{(4\pi^2 l)^{\frac{d}{2}}} \int_{|u| \leq \log l} (\phi_l^{(k)}(\frac{u}{\sqrt{l}}) - e^{-Q(u)}) du$$

$$R_2(k,l) = -\frac{2}{(4\pi^2 l)^{\frac{d}{2}}} \int_{|u| > \log l} e^{-Q(u)} du$$

$$R_3(k,l) = \frac{1}{(4\pi^2 l)^{\frac{d}{2}}} \int_{E_l} \phi_l^{(k)}(\frac{u}{\sqrt{l}}) du.$$

where $E_l = [-\frac{\pi}{2}\sqrt{l}, \frac{3\pi}{2}\sqrt{l}]^d \setminus \{B(0, \log l) \cup B(\Pi\sqrt{l}, \log l)\}$.
When $d = 1$, we have the following equivalence

$$\sum_{k=0}^{[\frac{(n-1)}{2}]} \sum_{l=1}^{[\frac{(n-k)}{2}]} \frac{1}{\sqrt{l}} \sim \frac{(2^{\frac{3}{2}} - 1)}{3} n^{\frac{3}{2}}$$

and we proved in [68] that for $i = 1,2,3$,

$$\sum_{k=0}^{[\frac{(n-1)}{2}]} \sum_{l=1}^{[\frac{(n-k)}{2}]} R_i(k,l) = o(n^{\frac{3}{2}}).$$

We treat in details the case $d = 2$. The case $d \geq 3$ can be solved by the same techniques. We have

$$\sum_{k=0}^{[\frac{(n-1)}{2}]} \sum_{l=1}^{[\frac{(n-k)}{2}]} \frac{1}{l} \sim \frac{1}{2} n \log n.$$

So we only have to prove that for $i = 1,2,3$,

$$\sum_{k=0}^{[\frac{(n-1)}{2}]} \sum_{l=1}^{[\frac{(n-k)}{2}]} R_i(k,l) = o(n \log n).$$

We denote any constant by the same letter C.
Applying the method developed in the proof of Lemma 2 of Section 3 of Chapter 2, we have

$$|R_1(k,l)| \leq \frac{C}{l} \sup_{u \in B(0, \log l)} |T_l^{(k)}(u)| \exp |T_l^{(k)}(u)|$$

where

$$T_l^{(k)}(u)$$
$$= Q(u) + \frac{1}{2}\sum_{i=1}^{2l}\log\left(\frac{1}{d^2}\Big(\sum_{j=1}^{d}\cos(\frac{u_j}{\sqrt{l}})\Big)^2\right.$$
$$+ \left.\frac{1}{d^2}\Big(\sum_{j=1}^{d}(2df_j(T^{2k+i}x)-1)\sin(\frac{u_j}{\sqrt{l}})\Big)^2\right)$$
$$+ \sum_{i=1}^{2l}\arctan\left(\frac{\sum_{j=1}^{d}(2df_j(T^{2k+i}x)-1)\sin(\frac{u_j}{\sqrt{l}})}{\sum_{j=1}^{d}\cos(\frac{u_j}{\sqrt{l}})}\right)$$

Since the functions f_j and f_jf_l satisfy the condition (H_3), there exists $\varepsilon > 0$, a positive constant C, and an integer $l_0 \geq 1$ such that

$$\left|\frac{1}{n\log n}\sum_{k=0}^{[\frac{(n-1)}{2}]}\sum_{l=1}^{[\frac{(n-k)}{2}]}R_1(k,l)\right| \leq \frac{C}{\log n}\sum_{l\geq l_0}\frac{1}{l(\log l)^{1+\varepsilon}} = o(1).$$

And,

$$\left|\frac{1}{n\log n}\sum_{k=0}^{[\frac{(n-1)}{2}]}\sum_{l=1}^{[\frac{(n-k)}{2}]}R_2(k,l)\right| \leq \frac{C}{\log n}\sum_{l\geq 1}\frac{1}{l}\int_{|u|>\log l}e^{-Q(u)}du.$$

If $0 < \lambda_1 \leq \ldots \leq \lambda_d$ are the eigenvalues of the quadratic form Q, we have the inequality, $\lambda_1.|u|^2 \leq Q(u) \leq \lambda_d.|u|^2, \forall u \in \mathbb{R}^d$. Then,

$$\left|\frac{1}{n\log n}\sum_{k=0}^{[\frac{(n-1)}{2}]}\sum_{l=1}^{[\frac{(n-k)}{2}]}R_2(k,l)\right| \leq \frac{C}{\log n}\sum_{l\geq 1}\frac{1}{l}\int_{|u|>\log l}e^{-\lambda_1.|u|^2}du$$
$$\leq \frac{C}{\log n}\sum_{l\geq 1}\frac{1}{l}\int_{\rho>\log l}e^{-\lambda_1.\rho^2}\rho d\rho$$
$$\leq \frac{C}{\log n}\sum_{l\geq 1}\frac{1}{l}e^{-\lambda_1(\log l)^2} = o(1)$$

since λ_1 is strictly positive.

Finally, following the proof of Lemma 3 of Section 3, Chapter 2, since the functions f_j and f_jf_l satisfy the hypothesis (H_3), we can choose ε, $0 < \varepsilon < \frac{\lambda_1}{d}$ uniformly in the variable k, and an integer $l_0 \geq 1$ such that

$$\left|\frac{1}{n\log n}\sum_{k=0}^{[\frac{(n-1)}{2}]}\sum_{l=1}^{[\frac{(n-k)}{2}]}R_3(k,l)\right| \leq \frac{C}{\log n}\sum_{l\geq l_0}\exp(-\frac{4(\log l)^2}{\pi^2}(\lambda_1 - d\varepsilon))$$
$$= o(1).$$

We complete now the proof of Theorem 4.2.

It is not difficult to prove that any sequence of independent random variables is associated. Moreover, if the sequence $(Y_i)_{i\geq 1}$ is associated, so is the sequence $(\tilde{Y}_i)_{i\geq 1}$ where each $\tilde{Y}_i$ is some Y_i. Since the random variables $\xi_x, x \in \mathbb{Z}^d$ are independent, conditioned on $\mathcal{F} = \sigma(X_1, X_2, \ldots)$ the σ-field generated by the dynamic random walk, the sequence of the random variables $\xi_{S_k}, k \geq 0$ is thus associated. Let $\rho > 1$ and for every $k \geq 0, n_k = [\rho^k]$. Using inequality (4.2), we obtain

$$\mathbb{P}(\max_{0 \leq n \leq n_{k+1}} |Z_n| \geq \varepsilon n_k^\gamma | \mathcal{F}) \leq C(\varepsilon) \frac{V_{n_{k+1}}}{n_k^{2\gamma}}.$$

Let $d = 1$ (the case $d \geq 2$ can be treated in the same way). Using Lemma 10, there exists a constant $C > 0$ and k_0 such that for every $k \geq k_0$,

$$\begin{aligned}\mathbb{P}(\max_{n_k \leq n < n_{k+1}} |Z_n| \geq \varepsilon n^\gamma) &\leq \mathbb{P}(\max_{0 \leq n \leq n_{k+1}} |Z_n| \geq \varepsilon n_k^\gamma) \\ &\leq C \frac{\mathbb{E}(V_{n_{k+1}})}{n_k^{2\gamma}} \\ &\leq \frac{C n_{k+1}^{\frac{3}{2}}}{n_k^{2\gamma}}.\end{aligned}$$

Since $\sum_{k \geq 0} \frac{n_{k+1}^{\frac{3}{2}}}{n_k^{2\gamma}} < \infty$, by Borel-Cantelli lemma, we conclude that $\frac{Z_n}{n^\gamma}$ converges almost surely to 0.

1.3   A FUNCTIONAL LIMIT THEOREM ($D = 1$)

Using the methods developed by Kesten and Spitzer [94], we prove a functional limit theorem for the "cumulative random scenery"

$$\begin{aligned}Z_n &= \sum_{k=0}^{n} \xi_{S_k} \\ &= \sum_{x \in \mathbb{Z}} \xi(x) N_n(x) \quad (4.4)\end{aligned}$$

where $(S_n)_{n \geq 0}$ is a dynamic $\mathbb{Z}$-random walk generated by a dynamical system $S = (E, \mathcal{A}, \mu, T)$ and a function f such that $f, f^2 \in \mathcal{C}_3(S)$, $\int_E f\, d\mu = \frac{1}{2}$ and $a = 4 \int_E f(1-f)\, d\mu > 0$. Here, $N_n(x)$ is the number of visits of $(S_n)_{n \geq 0}$ at point x in the time interval $[0, n]$. By Lemmata 9 and 10,

$$\operatorname{Var}(Z_n) \sim \frac{4\sigma^2(2^{3/2} - 1)}{3\sqrt{a\pi}} n^{\frac{3}{2}}.$$

That suggests looking for a process $(\Delta_t)_{t\geq 0}$ which is the weak limit of the sequence of stochastic processes

$$D_t^n = \frac{Z_{nt}}{n^{\frac{3}{4}}}, \quad t \geq 0, \quad n = 1, 2, 3, \ldots$$

where Z_s is defined as the linear interpolation

$$Z_s = Z_n + (s-n)(Z_{n+1} - Z_n), \text{ when } n \leq s \leq n+1.$$

The limiting process $(\Delta_t)_{t\geq 0}$ may be described as the process obtained from the random walk in a random scenery when $\mathbb{Z}$ is changed into $\mathbb{R}$, the dynamic random walk $(S_n)_{n\geq 0}$ into a Brownian motion $(B_t^{(a)})_{t\geq 0}$ with zero mean and variance at (see Theorem 2.2) and the random scenery $(\xi_x)_{x\in\mathbb{Z}}$ into a white noise, time derivative in the distributional sense of a Brownian motion $(Z(x))_{x\in\mathbb{R}}$ with zero mean and variance $\sigma^2 t$. By analogy with (4.4), replacing $N_n(x)$ by $L_t(x)$ the local time at x of the Brownian motion $(B_t^{(a)})_{t\geq 0}$, the limiting process $(\Delta_t)_{t\geq 0}$ can be represented by the following stochastic integral

$$\Delta_t = \int_{-\infty}^{\infty} L_t(x) dZ(x).$$

Since random scenery is defined on the whole $\mathbb{Z}$ axis, the Brownian motion $(Z(x))_{x\in\mathbb{R}}$ is to be defined with real time. Therefore we introduce a pair of independent Brownian motions $Z_+(x), Z_-(x), x \geq 0$ so that the limiting process can be rewritten

$$\Delta_t = \int_0^{\infty} L_t(x) dZ_+(x) + \int_0^{\infty} L_t(-x) dZ_-(x). \tag{4.5}$$

We prove the

THEOREM 4.3 *Let us assume that* $f, f^2 \in C_2(S)$, *then for every* $x \in E$, $(D_t^n)_{t\geq 0}$ *converges weakly in* $C([0,\infty))$ *to the process* $(\Delta_t)_{t\geq 0}$ *defined in (4.5). This process has a continuous version which is self-similar with index* $\frac{3}{4}$ *and has stationary increments.*

1.3.1   PRELIMINARIES

For each $x \in \mathbb{Z}$, we define $\tau(x) = \inf\{n \geq 0; S_n = x\}$ and consider the linear interpolation of the occupation times $N_n(x)$ of the random walk

$$N_s(x) = N_n(x) + (s-n)(N_{n+1}(x) - N_n(x)), n \leq s \leq n+1.$$

The next lemma gives us the asymptotic behavior of the sequence of occupation times $N_s(x)$ as $s \to \infty$.

LEMMA 11
$$\lim_{s \to \infty} s^{-\frac{3}{4}} \sup_{x \in \mathbb{Z}} N_s(x) = 0 \text{ in probability}.$$

Proof:
If $\epsilon > 0$ is fixed, we can write

$$\mathbb{P}(\sup_{x \in \mathbb{Z}} s^{-\frac{3}{4}} N_s(x) > \epsilon) \leq \mathbb{P}(N_s(x) > 0 \text{ for some } x \text{ such that } |x| > A\sqrt{s})$$
$$+ \sum_{|x| \leq A\sqrt{s}} \mathbb{P}(N_s(x) > \epsilon s^{\frac{3}{4}}). \quad (4.6)$$

First, we prove that the first term in the right hand side of (4.6) tends to zero as A tends to infinity.

$$\mathbb{P}(N_s(x) > 0 \text{ for some } x \text{ such that } |x| > A\sqrt{s})$$
$$\leq \sup_{s \geq 1} \mathbb{P}(|S_n| > A\sqrt{s} \text{ for some } n \leq s+1)$$
$$= \sup_{s \geq 1} \mathbb{P}(\max_{n \leq s+1} |S_n| > A\sqrt{s}).$$

Now, by Theorem 2.2,
$$\frac{S_{[nt]}}{\sqrt{n}} \underset{n \to \infty}{\to} (B_t^{(a)})_{t \geq 0}, \text{ in distribution}.$$

Therefore, denoting $(B_t)_{t \geq 0}$ the standard Brownian motion, we have

$$\lim_{s \to \infty} \mathbb{P}(\max_{n \leq s+1} |S_n| > A\sqrt{s}) \leq \lim_{n \to \infty} \mathbb{P}(\max_{t \leq 1} \frac{|S_{[nt]}|}{\sqrt{n}} > A)$$
$$= \mathbb{P}(\sup_{t \leq 1} |B_t^{(a)}| > A)$$
$$= \mathbb{P}(\sup_{t \leq 1} |B_t| > \frac{A}{a})$$
$$= 2\mathbb{P}(B_1 > \frac{A}{a}) \underset{A \to \infty}{\to} 0.$$

Secondly, we have
$$N_s(x) \leq \sum_{j=0}^{[s+1]} \mathbf{1}_{\{\tau(x)=j\}} \sum_{m=j}^{[s+1]} \mathbf{1}_{\{S_m=x\}}.$$

Hence,
$$\mathbb{P}(N_s(x) > \epsilon s^{\frac{3}{4}}) \leq \sum_{j=0}^{[s+1]} \mathbb{P}(\tau(x) = j; \sum_{m=j}^{[s+1]} \mathbf{1}_{\{S_m=x\}} > \epsilon s^{\frac{3}{4}}).$$

The event $\{\tau(x) = j\}$ is measurable with respect to $\mathcal{F} = \sigma(X_1, X_2, \ldots, X_j)$. Hence, on $\{\tau(x) = j\}$,

$$\sum_{m=j}^{[s+1]} \mathbf{1}_{\{S_m = x\}} = \sum_{m=0}^{[s+1]-j} \mathbf{1}_{\{S_{m+j} = S_j\}}$$

$$\leq \sum_{m=0}^{[s+1]} \mathbf{1}_{\{S_{m+j} = S_j\}}.$$

Then,

$$\mathbb{P}(N_s(x) > \epsilon s^{\frac{3}{4}}) \leq \sum_{j=0}^{[s+1]} \mathbb{P}(\tau(x) = j) \mathbb{P}(\sum_{m=0}^{[s+1]} \mathbf{1}_{\{S_{m+j} = S_j\}} > \epsilon s^{\frac{3}{4}})$$

and

$$\sum_{|x| \leq A\sqrt{s}} \mathbb{P}(N_s(x) > \epsilon s^{\frac{3}{4}}) \leq (2A\sqrt{s} + 1) \sum_{j=0}^{[s+1]} \mathbb{P}(\sum_{m=0}^{[s+1]} \mathbf{1}_{\{S_{m+j} = S_j\}} > \epsilon s^{\frac{3}{4}}).$$

By Tchebychev's inequality, for every integer $\nu \geq 1$,

$$\mathbb{P}(\sum_{m=0}^{[s+1]} \mathbf{1}_{\{S_{m+j} = S_j\}} > \epsilon s^{\frac{3}{4}}) \leq \frac{1}{\epsilon^\nu s^{\frac{3\nu}{4}}} \mathbb{E}((\sum_{m=0}^{[s+1]} \mathbf{1}_{\{S_{m+j} = S_j\}})^\nu)$$

and

$$\mathbb{E}((\sum_{m=0}^{[s+1]} \mathbf{1}_{\{S_{m+j} = S_j\}})^\nu) = \mathbb{E}(\sum_{m_1=0}^{[s+1]} \cdots \sum_{m_\nu=0}^{[s+1]} \mathbf{1}_{\{S_{m_1+j} = S_j\}} \cdots \mathbf{1}_{\{S_{m_\nu+j} = S_j\}})$$

$$\leq \nu! \sum_{m_1=0}^{[s+1]} \cdots \sum_{m_\nu=0}^{[s+1]} \mathbb{P}(S_{m_1+j} = S_j) \quad (4.7)$$

$$\times \quad \mathbb{P}(S_{m_2+m_1+j} = S_{m_1+j})$$

$$\cdots \quad \mathbb{P}(S_{m_\nu+\ldots+m_1+j} = S_{m_{\nu-1}+\ldots+m_1+j})$$

$$+ \quad \text{diagonal terms.}$$

Let $\phi_m^{(j)}$ be the characteristic function of $\tilde{S}_{2m}^{(j)}$, where $(\tilde{S}_m^{(j)})_{m \in \mathbb{N}}$ is defined in the same way as the Markov chain $(S_n)_{n \in \mathbb{N}}$ but with transition probabilities given by

$$\mathbb{P}(\tilde{S}_m^{(j)} = l + 1 | \tilde{S}_{m-1}^{(j)} = l) = f(\tau_\alpha^{m+2j} x)$$

$$= 1 - \mathbb{P}(\tilde{S}_m^{(j)} = l - 1 | \tilde{S}_{m-1}^{(j)} = l).$$

We have
$$\mathbb{P}(S_{2(j+m)} = 0 | S_{2j} = 0) = (a\pi)^{-\frac{1}{2}} \frac{1}{\sqrt{m}} + \sum_{i=1}^{3} R_i(j,m)$$

where $R_i(j,m), i = 1,2,3$, are given by

$$R_1(j,m) = \frac{1}{\pi}\frac{1}{\sqrt{m}} \int_{|u| \leq \log m} (\phi_m^{(j)}(\frac{u}{\sqrt{m}}) - e^{-au^2}) du$$

$$R_2(j,m) = -\frac{1}{\pi}\frac{1}{\sqrt{m}} \int_{|u| > \log m} e^{-au^2} du$$

$$R_3(j,m) = \frac{1}{\pi}\frac{1}{\sqrt{m}} \int_{\log m < |u| \leq \frac{\pi}{2}\sqrt{m}} \phi_m^{(j)}(\frac{u}{\sqrt{m}}) du.$$

The correction terms $R_i(j,m), i = 1,2,3$, satisfy independently of j

$$\sum_{m=1}^{[s+1]} R_i(j,m) = o(s^{1/2}), \quad i = 1,2,3.$$

Then, the dominant term in (4.7) is

$$\sum_{m_1=1}^{[s+1]} \cdots \sum_{m_\nu=1}^{[s+1]} \frac{1}{\sqrt{m_1 \cdots m_\nu}} \sim [s+1]^{\frac{\nu}{2}}, \text{ as } s \to \infty.$$

It remains to properly choose $\nu > 6$ so that

$$\sum_{|x| \leq A\sqrt{s}} \mathbb{P}(N_s(x) > \epsilon s^{\frac{3}{4}}) = o(1), \text{ as } s \to \infty.$$

LEMMA 12 *There exists a constant C independent of x such that for large n,*
$$\mathbb{E}(N_n(x)^2) \leq Cn.$$

Proof:
We have

$$\mathbb{E}(N_n(x)^2) = \sum_{k=0}^{n} \mathbb{P}(S_k = x) + 2 \sum_{k<l} \mathbb{P}(S_k = S_l = x)$$

$$= \sum_{k=0}^{n} \mathbb{P}(S_k = x) + 2 \sum_{k=0}^{n} \sum_{l=1}^{n-k} \mathbb{P}(S_{k+l} = 0 | S_k = 0)\mathbb{P}(S_k = x)$$

We have the following decomposition

$$\mathbb{P}(S_k = x) = (a\pi k)^{-\frac{1}{2}} e^{-\frac{x^2}{4ka}} + \sum_{j=1}^{3} I_j(k)$$

where $I_j(k), j = 1, 2, 3$, are given by

$$I_1(k) = \frac{1}{\pi}\frac{1}{\sqrt{k}}\int_{|u|\leq \log k} e^{-i\frac{x}{\sqrt{k}}u}(\phi_k^{(0)}(\frac{u}{\sqrt{k}}) - e^{-au^2})du$$

$$I_2(k) = -\frac{1}{\pi}\frac{1}{\sqrt{k}}\int_{|u|> \log k} e^{-i\frac{x}{\sqrt{k}}u}e^{-au^2}du$$

$$I_3(k) = \frac{1}{\pi}\frac{1}{\sqrt{k}}\int_{\log k < |u| \leq \frac{\pi}{2}\sqrt{k}} e^{-i\frac{x}{\sqrt{k}}u}\phi_k^{(0)}(\frac{u}{\sqrt{k}})du.$$

In the same manner,

$$\mathbb{P}(S_{k+l} = 0 | S_k = 0) = (a\pi l)^{-\frac{1}{2}} + \sum_{j=1}^{3} R_j(l, k)$$

where $R_j(l, k), j = 1, 2, 3$, are given by

$$R_1(l, k) = \frac{1}{\pi}\frac{1}{\sqrt{l}}\int_{|u|\leq \log l} (\phi_l^{(k)}(\frac{u}{\sqrt{l}}) - e^{-au^2})du$$

$$R_2(l, k) = -\frac{1}{\pi}\frac{1}{\sqrt{l}}\int_{|u|> \log l} e^{-au^2}du$$

$$R_3(l, k) = \frac{1}{\pi}\frac{1}{\sqrt{l}}\int_{\log l < |u| \leq \frac{\pi}{2}\sqrt{l}} \phi_l^{(k)}(\frac{u}{\sqrt{l}})du.$$

The general idea of the proof of Theorem 4.3 comes from Kesten and Spitzer's paper but the estimations of the terms I_i and R_i need some very technical details, namely the estimations of the ergodic sums of the dynamical system. This study was already done to prove Lemma 10 and we use these results to prove that the dominant term in $\mathbb{E}(N_n(x)^2)$ is

$$\frac{1}{a\pi}\sum_{k=0}^{n}\sum_{l=1}^{n-k}\frac{e^{-\frac{x^2}{4ka}}}{\sqrt{kl}} = \mathcal{O}(n),$$

and that the other terms are in $o(n)$. For instance, there exists $l_0 \geq 1$ and a positive constant M such that

$$\frac{1}{n}\sum_{k=0}^{n}\sum_{l=1}^{n-k}(a\pi k)^{-\frac{1}{2}}e^{-\frac{x^2}{4ka}}R_1(l,k)$$

$$= \frac{1}{n}\sum_{k=0}^{n}\sum_{l=1}^{n-k}\frac{e^{-\frac{x^2}{4ka}}}{\pi\sqrt{a\pi kl}}\int_{|u|\leq \log l}(\phi_l^{(k)}(\frac{u}{\sqrt{l}}) - e^{-au^2})du$$

$$\leq \frac{1}{\pi\sqrt{a\pi n}}\sum_{k=0}^{n}\frac{1}{\sqrt{k}}\sum_{l=l_0+1}^{n}\frac{1}{\sqrt{l}}\sum_{p=0}^{\infty}\frac{1}{p!}(M\frac{\log^{3+\epsilon}2l}{\sqrt{l}})^{p+1} + o(1)$$

$$\underset{n \to \infty}{\to} 0$$

So the lemma is proved.

LEMMA 13 *For any distinct $t_1, t_2, \ldots, t_k \geq 0$ and $\theta_1, \theta_2, \ldots, \theta_k \in \mathbb{R}$, the joint distribution of*

$$n^{-\frac{3}{2}} \sum_{x \in \mathbb{Z}} (\sum_{j=1}^{k} \theta_j N_{nt_j}(x))^2$$

converges, as $n \to \infty$, to the joint distribution of

$$\int_{-\infty}^{\infty} (\sum_{j=1}^{k} \theta_j L_{t_j}(x))^2 dx.$$

Proof:
We define, for $-\infty < a < b < \infty$,

$$T_t^n(a,b) = \frac{1}{n} \sum_{a \leq n^{-\frac{1}{2}} x < b} N_{nt}(x).$$

$T_t^n(a,b)$ is just the fraction of time spent by the process $(\frac{S_{[nt]}}{\sqrt{n}})_{t \geq 0}$ in the interval $[a,b]$, during the time interval $[0, nt]$.

We know that the Brownian motion $(B_t^{(a)})_{t \geq 0}$ possesses a local time $L_t(x)$ which is jointly continuous in t and x (see [153] for example) and the analogue of $T_t^n(a,b)$ for this process is

$$\Lambda_t(a,b) = \int_a^b L_t(x) dx.$$

The weak convergence of $(\frac{S_{[nt]}}{\sqrt{n}})_{t \geq 0}$ to the process $(B_t^{(a)})_{t \geq 0}$ gives us the convergence in distribution of $T_t^n(a,b)$ to $\Lambda_t(a,b)$. More generally, the joint distribution of $T_{t_i}^n(a_i, b_i), 1 \leq i \leq k$, converges to the joint distribution of $\Lambda_{t_i}(a_i, b_i), 1 \leq i \leq k$. In order to prove the lemma, we approximate $n^{-\frac{3}{2}} \sum_{x \in \mathbb{Z}} (\sum_{j=1}^{k} \theta_j N_{nt_j}(x))^2$ by a combination of $T_{t_j}^n$ which converges in distribution to the corresponding combination of the Λ_{t_j}. Then, we prove that this combination approximates $\int_{-\infty}^{\infty} (\sum_{j=1}^{k} \theta_j L_{t_j}(x))^2 dx$. We decompose the set of all possible indices into small slices where sharp estimates can be made. We define,

$$a(l,n) = \tau l \sqrt{n}, l \in \mathbb{Z},$$

$$T(l,n) = \sum_{j=1}^{k} \theta_j T_{t_j}^n(l\tau, (l+1)\tau)$$

$$= \frac{1}{n}\sum_{j=1}^{k}\theta_j \sum_{a(l,n)\leq y<a(l+1,n)} N_{nt_j}(y),$$

$$U = U(\tau,M,n) = n^{-\frac{3}{2}} \sum_{\substack{x<-M\tau\sqrt{n} \\ \text{or } x\geq M\tau\sqrt{n}}} (\sum_{j=1}^{k}\theta_j N_{nt_j}(x))^2,$$

$$V = V(\tau,M,n) = \frac{1}{\tau}\sum_{-M\leq l<M}(T(l,n))^2.$$

We have

$$n^{-\frac{3}{2}}\sum_{x\in\mathbb{Z}}(\sum_{j=1}^{k}\theta_j N_{nt_j}(x))^2 - U(\tau,M,n) - V(\tau,M,n) =$$

$$n^{-\frac{3}{2}}\sum_{-M\leq l<M}\sum_{a(l,n)\leq x<a(l+1,n)}((\sum_{j=1}^{k}\theta_j N_{nt_j}(x))^2 - n^2[a(l+1,n)-a(l,n)]^{-2}$$
$$\times (T(l,n))^2).$$

We want to show that this sum tends in probability to zero as $n\to\infty$. The L^1-convergence gives us the result. Firstly, fix τ, M and n and consider

$$\mathbb{E}(|(\sum_{j=1}^{k}\theta_j N_{nt_j}(x))^2 - n^2[a(l+1,n)-a(l,n)]^{-2}(T(l,n))^2|)$$

$$= \mathbb{E}(|\sum_{j=1}^{k}\theta_j N_{nt_j}(x) - n[a(l+1,n)-a(l,n)]^{-1}T(l,n)|$$

$$\times |\sum_{j=1}^{k}\theta_j N_{nt_j}(x) + n[a(l+1,n)-a(l,n)]^{-1}T(l,n)|)$$

$$\leq \mathbb{E}(|\sum_{j=1}^{k}\theta_j N_{nt_j}(x) - n[a(l+1,n)-a(l,n)]^{-1}T(l,n)|^2)^{\frac{1}{2}}$$

$$\times \mathbb{E}(|\sum_{j=1}^{k}\theta_j N_{nt_j}(x) + n[a(l+1,n)-a(l,n)]^{-1}T(l,n)|^2)^{\frac{1}{2}}.$$

First,

$$\mathbb{E}(|\sum_{j=1}^{k}\theta_j N_{nt_j}(x) + n[a(l+1,n)-a(l,n)]^{-1}T(l,n)|^2)$$

$$\leq [a(l+1,n) - a(l,n)]^{-2}\mathbb{E}((\sum_{j=1}^{k}\sum_{a(l,n)\leq y<a(l+1,n)}|\theta_j|(N_{nt_j}(x)+N_{nt_j}(y)))^2)$$

$$\leq [a(l+1,n) - a(l,n)]^{-1}\sum_{i=1}^{k}\theta_i^2\sum_{j=1}^{k}\sum_{a(l,n)\leq y<a(l+1,n)}\mathbb{E}((N_{nt_j}(x)+N_{nt_j}(y))^2)$$

$$\leq \sum_{i=1}^{k}\theta_i^2\sum_{j=1}^{k}\max_{\substack{a(l,n)\leq y<a(l+1,n)\\y\neq x}}\mathbb{E}((N_{nt_j}(x)+N_{nt_j}(y))^2).$$

In the same manner as previously,

$$\mathbb{E}(|\sum_{j=1}^{k}\theta_j N_{nt_j}(x) - n[a(l+1,n)-a(l,n)]^{-1}T(l,n)|^2)$$

$$\leq \sum_{i=1}^{k}\theta_i^2\sum_{j=1}^{k}\max_{\substack{a(l,n)\leq y<a(l+1,n)\\y\neq x}}\mathbb{E}((N_{nt_j}(x)-N_{nt_j}(y))^2).$$

Using the previous lemma, we have for large n,

$$\mathbb{E}(|n^{-\frac{3}{2}}\sum_{x\in\mathbb{Z}}(\sum_{j=1}^{k}\theta_j N_{nt_j}(x))^2 - U(\tau,M,n) - V(\tau,M,n)|) \leq C(2M+1)\tau$$

Moreover, we have

$$\mathbb{P}(U(\tau,M,n)\neq 0)$$
$$\leq \mathbb{P}(N_{nt_j}(x) > 0 \text{ for some } x \text{ such that } |x| > M\tau\sqrt{n} \text{ and } 1\leq j\leq k)$$
$$\leq \mathbb{P}(N_{\max(nt_j)}(x) > 0 \text{ for some } x \text{ such that } |x| > \frac{M\tau}{\sqrt{\max(t_j)}}\sqrt{\max(nt_j)}).$$

Doing the same reasoning as in the proof of Lemma 11, we can choose $M\tau$ so large that $\mathbb{P}(U(\tau,M,n)\neq 0)$ is small. Then, we have proved that for each $\eta > 0$, we can choose τ, M and large n such that

$$\mathbb{P}(|n^{-\frac{3}{2}}\sum_{x\in\mathbb{Z}}(\sum_{j=1}^{k}\theta_j N_{nt_j}(x))^2 - V(\tau,M,n)| > \eta) \leq 2\eta.$$

By the convergence of the joint distribution of $T_{t_i}^n(a_i,b_i), 1\leq i\leq k$, to the joint distribution of $\Lambda_{t_i}(a_i,b_i), 1\leq i\leq k$, $V(\tau,M,n)$ converges in distribution, when $n\to\infty$, to

$$\frac{1}{\tau}\sum_{-M\leq l<M}(\sum_{j=1}^{k}\theta_j\Lambda_{t_j}(l\tau,(l+1)\tau))^2 = \frac{1}{\tau}\sum_{-M\leq l<M}(\sum_{j=1}^{k}\theta_j\int_{l\tau}^{(l+1)\tau}L_{t_j}(x)dx)^2.$$

The function $x \to L_t(x)$ being continuous and having a.s compact support,

$$\frac{1}{\tau}\sum_{-M \leq l < M}(\sum_{j=1}^{k}\theta_j \int_{l\tau}^{(l+1)\tau} L_{t_j}(x)dx)^2 \to \int_{-\infty}^{\infty}(\sum_{j=1}^{k}\theta_j L_{t_j}(x))^2 dx,$$

as $\tau \to 0, M\tau \to \infty$. Then, the lemma is proved.

LEMMA 14 *The joint distributions of Δ_t are given for distinct $t_1, t_2, \ldots, t_k \geq 0$ and $\theta_1, \theta_2, \ldots, \theta_k \in \mathbb{R}$, by*

$$\mathbb{E}(\exp(i\sum_{j=1}^{k}\theta_j \Delta_{t_j})) = \mathbb{E}(\exp(-\frac{\sigma^2}{2}\int_{-\infty}^{\infty}(\sum_{j=1}^{k}\theta_j L_{t_j}(x))^2 dx)).$$

Proof:
Let $(x_l^n)_{l \geq 0}$ be sequences which satisfy $0 = x_0^n < x_1^n < \ldots,$

$$\lim_{l \to \infty} x_l^n = \infty \text{ and } \lim_{n \to \infty}\max_l(x_{l+1}^n - x_l^n) = 0.$$

The function $(x,t) \to L_t(x)$ being continuous, almost surely with compact support for fixed t, we can choose $(x_l^n)_{l \geq 0}$ such that

$$\int L_t(x)dZ_+(x) = \lim_{n \to \infty}\sum_{l=0}^{\infty}L_t(x_l^n)[Z_+(x_{l+1}^n) - Z_+(x_l^n)],$$

with probability one. $Z_+(x_{l+1}^n) - Z_+(x_l^n)$ are independent with characteristic function

$$\exp(-(x_{l+1}^n - x_l^n)\frac{\theta^2 \sigma^2}{2}).$$

Then, by definition,

$$\mathbb{E}(\exp(i\sum_{j=1}^{k}\theta_j \int_0^{\infty}L_{t_j}(x)dZ_+(x)))$$

$$= \lim_{n \to \infty}\mathbb{E}(\exp(i\sum_{j=1}^{k}\theta_j\sum_{l=0}^{\infty}L_{t_j}(x_l^n)[Z_+(x_{l+1}^n) - Z_+(x_l^n)]))$$

$$= \lim_{n \to \infty}\mathbb{E}(\exp(-\sum_{l=0}^{\infty}(x_{l+1}^n - x_l^n)(\sum_{j=1}^{k}\theta_j L_{t_j}(x_l^n))^2 \frac{\sigma^2}{2}))$$

$$= \mathbb{E}(\exp(-\frac{\sigma^2}{2}\int_0^{\infty}(\sum_{j=1}^{k}\theta_j L_{t_j}(x))^2 dx)).$$

We treat Z_- in the same manner as Z_+ and using (4.5), we obtain the lemma.

1.3.2 PROOF OF THEOREM 4.3

We prove the convergence of the finite dimensional distributions of D_t^n to those of Δ_t using the results of the previous section.

PROPOSITION 8 *The finite dimensional distributions of D_t^n converge to those of Δ_t, as $n \to \infty$.*

Proof:
For any distinct $t_1, t_2, \ldots, t_k \geq 0$ and $\theta_1, \theta_2, \ldots, \theta_k \in \mathbb{R}$,

$$\sum_{j=1}^{k} \theta_j D_{t_j}^n = n^{-\frac{3}{4}} \sum_{x \in \mathbb{Z}} \sum_{j=1}^{k} \theta_j N_{nt_j}(x) \xi(x)$$

Denoting by $\lambda(\theta) = \mathbb{E}(e^{i\theta \xi(x)})$ the characteristic function of $\xi(x)$, we have

$$\mathbb{E}(\exp(i \sum_{j=1}^{k} \theta_j D_{t_j}^n)) = \mathbb{E}(\prod_{x \in \mathbb{Z}} \lambda(n^{-\frac{3}{4}} \sum_{j=1}^{k} \theta_j N_{nt_j}(x))).$$

Let $A_n = \{\omega; n^{-\frac{3}{4}} \sup_{x \in \mathbb{Z}} |\sum_{j=1}^{k} \theta_j N_{nt_j}(x)| > \epsilon\}$. For all positive ϵ, we can write

$$\left| \mathbb{E}(\exp(i \sum_{j=1}^{k} \theta_j D_{t_j}^n)) - \mathbb{E}(\exp(-\frac{\sigma^2}{2} \sum_{x \in \mathbb{Z}} n^{-\frac{3}{2}} (\sum_{j=1}^{k} \theta_j N_{nt_j}(x))^2)) \right|$$

$$\leq \int_{A_n} |\prod_{x \in \mathbb{Z}} \lambda(n^{-\frac{3}{4}} \sum_{j=1}^{k} \theta_j N_{nt_j}(x)) - \exp(-\frac{\sigma^2}{2} \sum_{x \in \mathbb{Z}} n^{-\frac{3}{2}} (\sum_{j=1}^{k} \theta_j N_{nt_j}(x))^2)| d\mathbb{P}$$

$$+ \int_{A_n^c} |\prod_{x \in \mathbb{Z}} \lambda(n^{-\frac{3}{4}} \sum_{j=1}^{k} \theta_j N_{nt_j}(x)) - \exp(-\frac{\sigma^2}{2} \sum_{x \in \mathbb{Z}} n^{-\frac{3}{2}} (\sum_{j=1}^{k} \theta_j N_{nt_j}(x))^2)| d\mathbb{P}.$$

$$\leq 2\mathbb{P}(A_n)$$

$$+ \int_{A_n^c} |\prod_{x \in \mathbb{Z}} \frac{\lambda(n^{-\frac{3}{4}} \sum_{j=1}^{k} \theta_j N_{nt_j}(x))}{\exp(-\frac{\sigma^2}{2} \sum_{x \in \mathbb{Z}} n^{-\frac{3}{2}} (\sum_{j=1}^{k} \theta_j N_{nt_j}(x))^2)} - 1| d\mathbb{P}.$$

The first term in the right hand side of the above inequality tends to zero by vertue of Lemma 11. Moreover, by definition, $(\xi(\alpha))_{\alpha \in \mathbb{Z}}$ is a family of i.i.d $\mathbb{R}$-valued random variables with zero mean and finite positive variance σ^2, then $(\xi(\alpha))_{\alpha \in \mathbb{Z}}$ satisfy a central limit theorem and hence, $\lambda(\theta) \sim 1 - \frac{\sigma^2 \theta^2}{2}$, as $\theta \to 0$, so the second term also vanishes in the limit. Then, we get

$$\lim_{n \to \infty} \mathbb{E}(\exp(i \sum_{j=1}^{k} \theta_j D_{t_j}^n))$$

$$= \lim_{n \to \infty} \mathbb{E}(\exp((-\frac{\sigma^2}{2} \sum_{x \in \mathbb{Z}} n^{-\frac{3}{2}} (\sum_{j=1}^{k} \theta_j N_{nt_j}(x))^2)))$$

$$= \mathbb{E}(\exp(-\frac{\sigma^2}{2} \int_{-\infty}^{\infty} (\sum_{j=1}^{k} \theta_j L_{t_j}(x))^2 dx)), \text{ by Lemma 12}$$

$$= \mathbb{E}(\exp(i \sum_{j=1}^{k} \theta_j \Delta_{t_j})), \text{ by Lemma 13.}$$

Then, the proposition is proved.

In order to prove weak convergence of $(D_t^n)_{t \geq 0}$ to $(\Delta_t)_{t \geq 0}$ in $\mathcal{C}([0,\infty))$, we prove the tightness of the family $(D_t^n)_{t \geq 0}, n = 1, 2, 3, \ldots$

PROPOSITION 9 *The family $(D_t^n)_{t \geq 0}, n = 1, 2, 3, \ldots$ is tight in $\mathcal{C}([0,\infty))$.*

Proof:
By Theorem 12.3 of Billingsley [15], it suffices to prove that there exists $K_0 > 0$ such that for all $t_1, t_2 \in [0,T], T < \infty$, for all $n \geq 1$ and all positive η,

$$\mathbb{P}(|D_{t_2}^n - D_{t_1}^n| \geq \eta) \leq \frac{1}{\eta^2} K_0 |t_2 - t_1|^{\frac{3}{2}}. \tag{4.8}$$

We suppose $t_2 > t_1$.

$$\mathbb{E}(|D_{t_2}^n - D_{t_1}^n|^2) = n^{-\frac{3}{2}} \sum_{x \in \mathbb{Z}} \mathbb{E}((N_{nt_2}(x) - N_{nt_1}(x))^2) \mathbb{E}(\xi^2(x))$$

$$= \sigma^2 n^{-\frac{3}{2}} \sum_{x \in \mathbb{Z}} \mathbb{E}((N_{nt_2}(x) - N_{nt_1}(x))^2).$$

Then,

$$\mathbb{E}(\sum_{x \in \mathbb{Z}} (N_{nt_2}(x) - N_{nt_1}(x))^2)$$

$$\leq \mathbb{E}(\sum_{x \in \mathbb{Z}} (N_{[nt_2]+1}(x) - N_{[nt_1]}(x))^2)$$

$$= \sum_{x \in \mathbb{Z}} \mathbb{E}((\sum_{k=[nt_1]+1}^{[nt_2]+1} \mathbf{1}_{\{S_k = x\}})^2)$$

$$= \sum_{x \in \mathbb{Z}} \sum_{k=[nt_1]+1}^{[nt_2]+1} \mathbb{P}(S_k = x) + 2 \sum_{x \in \mathbb{Z}} \mathbb{E}(\sum_{\substack{k,l=[nt_1]+1 \\ k<l}}^{[nt_2]+1} \mathbf{1}_{\{S_k=x\}} \mathbf{1}_{\{S_l=x\}})$$

$$= [nt_2] - [nt_1] + 1$$

$$+ 2 \sum_{k=0}^{[nt_2]-[nt_1]} \sum_{l=1}^{[nt_2]-[nt_1]-k} \mathbb{P}(S_{[nt_1]+1+k+l} = 0 | S_{[nt_1]+1+k} = 0).$$

Now, $[nt_2] - [nt_1] \leq [n(t_2 - t_1) + 2]$. From Lemma 10, there exists a constant $C > 0$ such that

$$\mathbb{E}(\sum_{x \in \mathbb{Z}} (N_{nt_2}(x) - N_{nt_1}(x))^2) \leq [nt_2] - [nt_1] + 1$$

$$+ 2 \sum_{k=0}^{[n(t_2-t_1)+2]} \sum_{l=1}^{[n(t_2-t_1)+2]-k} \mathbb{P}(S_{[nt_1]+1+k+l} = 0 | S_{[nt_1]+1+k} = 0)$$

$$\sim C([n(t_2 - t_1) + 2])^{\frac{3}{2}}, \text{ as } n \to \infty.$$

If $(t_2 - t_1)n \geq 1$, $[n(t_2 - t_1) + 2]^{\frac{3}{2}} \leq (3n)^{\frac{3}{2}}(t_2 - t_1)^{\frac{3}{2}}$. Then, there exists $K_1 > 0$ such that

$$\mathbb{E}(|D_{t_2}^n - D_{t_1}^n|^2) \leq K_1(t_2 - t_1)^{\frac{3}{2}}.$$

If $0 \leq (t_2 - t_1)n < 1$, $N_{k+1}(x) - N_k(x) = 1_{\{S_{k+1}=x\}}$. If $S_{k+1} = x$, then $N_{k+1}(x) - N_k(x) = 1$. So, for $k \leq s_1 < s_2 < k+1$,

$$\sum_{x \in \mathbb{Z}} (N_{s_2}(x) - N_{s_1}(x))^2 = (s_2 - s_1)^2$$

and then,

$$\mathbb{E}(\sum_{x \in \mathbb{Z}} (N_{nt_2}(x) - N_{nt_1}(x))^2) = n^2(t_2 - t_1)^2$$

$$\leq n^{\frac{3}{2}}(t_2 - t_1)^{\frac{3}{2}}.$$

Therefore,

$$\mathbb{E}(|D_{t_2}^n - D_{t_1}^n|^2) \leq \sigma^2(t_2 - t_1)^{\frac{3}{2}}.$$

Using Tchebychev's inequality and taking $K_0 = \max(K_1, \sigma^2)$, the inequality (4.8) is proved.

2. THE TRANSIENT CASE

We are now interested in the case when the dynamic random walk $(S_n)_{n \geq 0}$ evolves on $\mathbb{Z}$ and is transient. The notations in this section are the same as the ones used in the previous section.

2.1 STRONG LAW OF LARGE NUMBERS, CENTRAL LIMIT THEOREM

We begin by proving the following strong law of large numbers for transient one-dimensional dynamic random walk in random scenery.

THEOREM 4.4 *Let $(S_n)_{n \geq 0}$ be a dynamic random walk generated by a dynamical system S and $f \in \mathcal{C}_3(S)$ be a function with values in $[0, 1]$*

such that $f^2 \in \mathcal{C}_3(S)$, $\int_E f \, d\mu = p \neq \frac{1}{2}$ and $a = 4 \int_E f(1-f) \, d\mu > 0$.
Then, for any $\gamma > \frac{1}{2}$, for every $x \in E$ fixed,

$$\frac{Z_n}{n^\gamma} \underset{n \to +\infty}{\to} 0 \quad a.s..$$

Proof:

From Lemma 9, if $V_n = \sum_{i,j=0}^{n} \mathbf{1}_{\{S_i = S_j\}}$, then

$$\mathrm{Var}(Z_n) = \sigma^2 \mathbb{E}(V_n)$$

and

$$\mathbb{E}(V_n) = (n+1) + 2 \sum_{k=0}^{[\frac{(n-1)}{2}]} \sum_{l=1}^{[\frac{(n-k)}{2}]} \mathbb{P}(S_{2(k+l)} = 0 | S_{2k} = 0).$$

Let us determine the asymptotic behavior of the sequence $\mathbb{E}(V_n)$ for large n. We use the techniques developed in the proof of Lemma 10. Let us define

$$\mu_{k,n} = \sum_{l=1}^{2n} (2f(T^{2k+l}x) - 1)$$

and

$$\sigma_{k,n} = \sum_{l=1}^{2n} 4f(T^{2k+l}x)(1 - f(T^{2k+l}x)).$$

For every $k \geq 0$ fixed, the following decomposition holds

$$\sum_{l=0}^{n} \mathbb{P}(S_{2(k+l)} = 0 | S_{2k} = 0) = \sum_{i=1}^{4} R_i(k,n)$$

where

$$R_1(k,n) = \sqrt{\frac{2}{\pi}} \sum_{l=0}^{n} \frac{1}{\sigma_{k,l}^{1/2}} \exp\left(-\frac{\mu_{k,l}^2}{2\sigma_{k,l}}\right)$$

$$R_2(k,n) = \sqrt{\frac{2}{\pi}} \sum_{l=0}^{n} \frac{1}{(2\pi\sigma_{k,l})^{1/2}} \int_{|u| \leq \log l} \left[\mathbb{E}\left(\exp\left(iu \frac{\tilde{S}_{2l}^{(k)} - \mu_{k,l}}{\sqrt{\sigma_{k,l}}}\right)\right)\right.$$
$$\left. - \exp(-u^2/2)\right] \exp\left(\frac{iu\mu_{k,l}}{\sqrt{\sigma_{k,l}}}\right) du$$

$$R_3(k,n) = -\sqrt{\frac{2}{\pi}} \sum_{l=0}^{n} \frac{1}{(2\pi\sigma_{k,l})^{1/2}} \int_{|u| > \log l} \exp(-u^2/2) \exp\left(\frac{iu\mu_{k,l}}{\sqrt{\sigma_{k,l}}}\right) du$$

$$R_4(k,n) = \sqrt{\frac{2}{\pi}} \sum_{l=0}^{n} \frac{1}{(2\pi\sigma_{k,l})^{1/2}} \int_{\log l < |u| \leq \frac{\pi}{2}\sqrt{\sigma_{k,l}}} \mathbb{E}\left(\exp\left(iu\left(\frac{\tilde{S}_{2l}^{(k)} - \mu_{k,l}}{\sqrt{\sigma_{k,l}}}\right)\right)\right)$$
$$\times \exp\left(\frac{iu\mu_{k,l}}{\sqrt{\sigma_{k,l}}}\right) du.$$

Since f and f^2 satisfy the hypothesis (H_3), for every $i = 1,\ldots,4$, we obtain (see the proof of Lemma 10) that

$$|R_i(k,n)| < C_i < \infty,$$

C_i being a constant independent of k. As in the proof of Lemma 10, using that, conditioned on the dynamic random walk, $(\xi_{S_k})_{k\geq 0}$ is associated, we establish a maximal inequality for Z_n and conclude by Borel-Cantelli lemma.

As in the recurrent case, it is natural to study the statistical fluctuations of Z_n around 0. The following functional limit theorem holds.

THEOREM 4.5 *Assume that there exists $\epsilon > 0$ such that f takes its values in $]\epsilon, 1-\epsilon[$, $\int_E \log\left(\frac{4f(1-f)}{(\sqrt{5}-2)(1-\epsilon)^4}\right) d\mu$ and $\int_E \log\left(\frac{1-f}{f}\right) d\mu$ are strictly negative. Then, for almost every $x \in E$, the process $(Z_{nt}/\sqrt{n})_{t\geq 0}$ weakly converges in $\mathcal{C}([0,\infty))$ to a Brownian motion with zero mean and variance at.*

Proof:
The proof of the theorem rests on the following lemma

LEMMA 15
$$\lim_{s \to \infty} s^{-\frac{1}{2}} \sup_{x \in \mathbb{Z}} N_s(x) = 0 \text{ in probability.}$$

Proof:
Let $\epsilon > 0$, $A > 0$,

$$\mathbb{P}(\sup_{x \in \mathbb{Z}} s^{-\frac{1}{2}} N_s(x) > \epsilon)$$
$$\leq \mathbb{P}(N_s(x) > 0 \text{ for some } x \text{ such that } |x - s(2p-1)| > A\sqrt{s})$$
$$+ \sum_{|x-s(2p-1)| \leq A\sqrt{s}} \mathbb{P}(N_s(x) > \epsilon s^{\frac{1}{2}}).$$

As in the proof of Lemma 11, we have

$$\mathbb{P}(N_s(x) > 0 \text{ for some } x \text{ such that } |x - s(2p-1)| > A\sqrt{s})$$
$$\leq \sup_{s \geq 1} \mathbb{P}(|S_n - n(2p-1)| > A\sqrt{s} \text{ for some } n \leq s+1)$$
$$= \sup_{s \geq 1} \mathbb{P}(\max_{n \leq s+1} |S_n - n(2p-1)| > A\sqrt{s}).$$

Following the method developed in the proof of Theorem 2.2, we have

$$\frac{S_{[nt]} - [nt](2p-1)}{\sqrt{n}} \xrightarrow[n \to \infty]{\mathcal{D}} (B_t^{(a)})_{t \geq 0}.$$

Then, denoting $(B_t)_{t \geq 0}$ the standard Brownian motion,

$$\lim_{s \to \infty} \mathbb{P}(\max_{n \leq s+1} |S_n - n(2p-1)| > A\sqrt{s})$$

$$\leq \lim_{n \to \infty} \mathbb{P}\left(\max_{t \leq 1} \frac{|S_{[nt]} - [nt](2p-1)|}{\sqrt{n}} > A\right)$$

$$= \mathbb{P}(\sup_{t \leq 1} |B_t^{(a)}| > A)$$

$$= \mathbb{P}(\sup_{t \leq 1} |B_t| > \frac{A}{a})$$

$$\leq 2\mathbb{P}(|B_1| > \frac{A}{a}) \xrightarrow[A \to \infty]{} 0.$$

The following inequality proved in the recurrent case still holds

$$\sum_{|x-s(2p-1)| \leq A\sqrt{s}} \mathbb{P}(N_s(x) > \epsilon\sqrt{s}) \leq$$

$$(2A\sqrt{s}+1) \sum_{j=0}^{[s+1]} \mathbb{P}\left(\sum_{m=0}^{[s+1]} \mathbf{1}_{\{S_{m+j}=S_j\}} > \epsilon\sqrt{s}\right)$$

From Tchebychev's inequality, for every $\nu \geq 1$,

$$\mathbb{P}\left(\sum_{m=0}^{[s+1]} \mathbf{1}_{\{S_{m+j}=S_j\}} > \epsilon\sqrt{s}\right) \leq \frac{1}{\epsilon^\nu s^{\frac{\nu}{2}}} \mathbb{E}\left(\left(\sum_{m=0}^{[s+1]} \mathbf{1}_{\{S_{m+j}=S_j\}}\right)^\nu\right)$$

and

$$\mathbb{E}\left(\left(\sum_{m=0}^{[s+1]} \mathbf{1}_{\{S_{m+j}=S_j\}}\right)^\nu\right)$$

$$= \mathbb{E}\left(\sum_{m_1=0}^{[s+1]} \cdots \sum_{m_\nu=0}^{[s+1]} \mathbf{1}_{\{S_{m_1+j}=S_j\}} \cdots \mathbf{1}_{\{S_{m_\nu+j}=S_j\}}\right)$$

$$\leq \nu! \sum_{m_1=0}^{[s+1]} \cdots \sum_{m_\nu=0}^{[s+1]} \mathbb{P}(S_{m_1+j}=S_j)\mathbb{P}(S_{m_2+m_1+j}=S_{m_1+j})$$

$$\cdots \mathbb{P}(S_{m_\nu+\ldots+m_1+j}=S_{m_{\nu-1}+\ldots+m_1+j}) + \text{diagonal terms}$$

$$\leq \text{Constant}$$

Consequently,
$$\sum_{|x-s(2p-1)|\leq A\sqrt{s}} \mathbb{P}(N_s(x) > \epsilon\sqrt{s}) \leq Cs^{\frac{3-\nu}{2}}.$$

By choosing $\nu > 3$ we get the lemma.

Let us come back to the proof of the theorem. The characteristic function of the random variable ξ being still denoted by λ, we have

$$\mathbb{E}\Big(\exp(i\theta\frac{Z_{nt}}{\sqrt{n}})\Big) = \mathbb{E}\Big(\prod_{x\in\mathbb{Z}}\lambda(\theta\frac{N_{nt}(x)}{\sqrt{n}})\Big).$$

Since $\lambda(\theta) \sim 1 - \frac{\sigma^2\theta^2}{2}$, as $\theta \to 0$, using Lemma 15, we get (see the proof of Proposition 8) that

$$\lim_{n\to\infty}\mathbb{E}\Big(\exp(i\theta\frac{Z_{nt}}{\sqrt{n}})\Big) = \lim_{n\to\infty}\mathbb{E}\Big(\exp(-\frac{\sigma^2|\theta|^2}{n}\sum_{x\in\mathbb{Z}}N_{nt}^2(x))\Big).$$

Remark that V_n is equal to $\sum_{x\in\mathbb{Z}}N_n^2(x)$, so it is enough to show that $(V_n/n)_n$ converges in probability to a constant. Let us begin to show that $\mathrm{Var}(V_n) = o(n^2)$. The variance of V_n is equal to

$$4\sum_{0\leq i_1<j_1\leq n}\sum_{0\leq i_2<j_2\leq n}[\mathbb{P}(S_{i_1}=S_{j_1};S_{i_2}=S_{j_2})-\mathbb{P}(S_{i_1}=S_{j_1})\mathbb{P}(S_{i_2}=S_{j_2})].$$

The unique non-vanishing terms in this sum correspond to 4-tuples (i_1, j_1, i_2, j_2) of indices such that either $0 \leq i_1 \leq i_2 < j_1 < j_2 \leq n$ or $0 \leq i_1 < i_2 < j_2 \leq j_1 \leq n$. Let us denote by I_1 and I_2 these sets. We have

$$\mathrm{Var}(V_n) \leq 8[\sum_{I_1}\mathbb{P}(S_{i_1}=S_{j_1};S_{i_2}=S_{j_2}) + \sum_{I_2}\mathbb{P}(S_{i_1}=S_{j_1};S_{i_2}=S_{j_2})]$$
$$+ 4\sum_{0\leq i<j\leq n}\mathbb{P}(S_i=S_j).$$

Let us denote

$$a_1(n) = \sum_{I_1}\mathbb{P}(S_{i_1}=S_{j_1};S_{i_2}=S_{j_2})$$
$$\leq \sum_{z\in\mathbb{Z}}\sum_{m_1=0}^{n}\sum_{m_2=0}^{n}\mathbb{P}(\tilde{S}_{m_2}^{(m_1)}=z)\times\sum_{m_3=0}^{n}\mathbb{P}(\tilde{S}_{m_3}^{(m_1+m_2)}=-z)$$
$$\times\sum_{m_4=1}^{n}\mathbb{P}(\tilde{S}_{m_4}^{(m_1+m_2+m_3)}=z)$$

64   DYNAMIC RANDOM WALKS

Now, for every $z \in \mathbb{Z}$, for every $l \geq 0$,

$$\sum_{m=|z|}^{n} \mathbb{P}(\tilde{S}_m^{(l)} = z)$$

$$= \sum_{m=|z|}^{n} \sum_{\substack{A \subseteq \{1,\ldots,m\} \\ |A| = \frac{m+z}{2}}} \prod_{j \in A} f(T^{l+j}x) \prod_{j \in A^c} (1 - f(T^{l+j}x))$$

$$= \sum_{m=|z|}^{n} \sum_{\substack{A \subseteq \{1,\ldots,m\} \\ |A| = \frac{m+z}{2}}} \exp\Big(\sum_{j \in A} \log(f(T^{l+j}x)) - |A| \int_E \log f(t)\, d\mu(t)\Big)$$

$$\times \exp\Big(\sum_{j \in A^c} \log(1 - f(T^{l+j}x)) - |A^c| \int_E \log(1 - f(t))\, d\mu(t)\Big)$$

$$\times \exp\Big(|A| \int_E \log f(t)\, d\mu(t)\Big) \times \exp\Big(|A^c| \int_E \log(1 - f(t))\, d\mu(t)\Big).$$

By hypothesis, $\epsilon < f < 1 - \epsilon$, then

$$\Big|\sum_{j \in A} \log(f(T^{l+j}x)) - |A| \int_E \log f(t)\, d\mu(t)\Big| \leq 2|A||\log(1 - \epsilon)|$$

and

$$\Big|\sum_{j \in A^c} \log(1 - f(T^{l+j}x)) - |A^c| \int_E \log(1 - f(t))\, d\mu(t)\Big| \leq 2|A^c||\log(1 - \epsilon)|.$$

So,

$$\sum_{m=|z|}^{n} \mathbb{P}(\tilde{S}_m^{(l)} = z) \leq \Big(\frac{\tilde{p}}{\tilde{q}}\Big)^{\frac{z}{2}} \sum_{m=|z|}^{n} C_m^{\frac{m+z}{2}} (\tilde{p}\tilde{q})^{\frac{m}{2}}, \quad (4.9)$$

where

$$\tilde{p} = \exp\Big(2|\log(1 - \epsilon)| + \int_E \log f(t)\, d\mu(t)\Big)$$

$$\tilde{q} = \exp\Big(2|\log(1 - \epsilon)| + \int_E \log(1 - f(t))\, d\mu(t)\Big).$$

Since $0 < 4\tilde{p}\tilde{q} < \sqrt{5} - 2$, the previous series converges and

$$\Big(\frac{\tilde{p}}{\tilde{q}}\Big)^{\frac{z}{2}} \sum_{m=|z|}^{\infty} C_m^{\frac{m+z}{2}} (\tilde{p}\tilde{q})^{\frac{m}{2}} = \begin{cases} \frac{1}{\sqrt{1-4\tilde{p}\tilde{q}}} \Big(\frac{1-\sqrt{1-4\tilde{p}\tilde{q}}}{2\tilde{q}}\Big)^z, & \text{if } z \geq 0 \\ \frac{1}{\sqrt{1-4\tilde{p}\tilde{q}}} \Big(\frac{1+\sqrt{1-4\tilde{p}\tilde{q}}}{2\tilde{q}}\Big)^z, & \text{if } z \leq 0. \end{cases}$$

Consequently,

$$a_1(n) \leq \frac{(n+1)}{(1-4\tilde{p}\tilde{q})^{\frac{3}{2}}} \left[\sum_{z \in \mathbb{Z}^+} \left(\frac{(1-\sqrt{1-4\tilde{p}\tilde{q}})^2}{1+\sqrt{1-4\tilde{p}\tilde{q}}} \times \frac{1}{2\tilde{q}} \right)^z \right.$$
$$\left. + \sum_{z \in (\mathbb{Z}^+)^*} \left(\frac{1-\sqrt{1-4\tilde{p}\tilde{q}}}{(1+\sqrt{1-4\tilde{p}\tilde{q}})^2} \times 2\tilde{q} \right)^z \right] \quad (4.10)$$

Using the fact that $0 < 4\tilde{p}\tilde{q} < \sqrt{5} - 2$ and that the inequality $2x < (1+\sqrt{1-x})^3$ is satisfied for every $x \in [0, \sqrt{5}-2[$, we deduce that both series in inequality (4.10) converge. Therefore,

$$a_1(n) = \mathcal{O}(n).$$

Equivalently, if we define

$$a_2(n) = \sum_{I_2} \mathbb{P}(S_{i_1} = S_{j_1}; S_{i_2} = S_{j_2}),$$

a similar proof leads to

$$a_2(n) = \mathcal{O}(n).$$

Moreover, from inequality (4.9), we claim that

$$\sum_{0 \leq i < j \leq n} \mathbb{P}(S_i = S_j) = o(n^2).$$

So, $\text{Var}(V_n) = o(n^2)$, hence $\left(\frac{V_n - \mathbb{E}(V_n)}{n} \right)_n$ tends to 0 in probability when n tends to infinity. The problem is now to prove that the sequence $\left(\frac{\mathbb{E}(V_n)}{n} \right)_n$ converges to a strictly positive constant when $n \to \infty$. A direct computation gives

$$\frac{\mathbb{E}(V_n)}{n} = \frac{n+1}{n} + \frac{2}{n} \sum_{m=0}^{n} \sum_{j=1}^{m} \mathbb{P}(\tilde{S}_j^{(n-m)} = 0)$$

$$= \frac{n+1}{n} + \frac{2}{n} \sum_{m=0}^{n} \sum_{j=1}^{m} \sum_{\substack{A \subseteq \{1,\ldots,j\} \\ |A| = \frac{j}{2}}} \exp\left(\sum_{l \in A} \log(f(T^{n-m+l}x)) \right)$$

$$\times \exp\left(\sum_{l \in A^c} \log(1 - f(T^{n-m+l}x)) \right).$$

From inequality (4.9), the second term in the previous sum is a Césaro sum of a bounded sequence. It is not enough to conclude. The convergence seems difficult to get because of the lacunary sequences which appear in this sequence. Nevertheless, we are able to integrate this Césaro mean on the set E. We then have a Césaro mean of a convergent series and we conclude that the sequence $\left(\frac{\mathbb{E}(V_n)}{n}\right)_n$ converges to a strictly positive constant for almost every $x \in \mathbb{T}^d$.

The tightness of the sequence $(Z_{nt}/\sqrt{n})_{t \geq 0}$ can easily be deduced from inequality (4.9).

Remarks:

1. Hypotheses on the function f can be relaxed. The theorem is still verified if both geometrical series in inequality (4.10) converge.

2. The proof of a functional limit theorem for a random walk in a random scenery strongly rests on a local limit theorem for the random walk (see the proof of Theorem 4.3). In the transient case, the probability of return to the origin is more difficult to handle than in the recurrent one, therefore stronger hypotheses on function f have to be done and the functional limit theorem holds only for almost every $x \in E$.

3. Let $x \in E$ fixed, it would be interesting to construct a function f for which $(\mathbb{E}(V_n)/n)_n$ is a bounded sequence with many adherence values, it would imply that $(Z_{nt}/\sqrt{n})_{t \geq 0}$ has not a Gaussian limit.

3. A PARTICULAR DYNAMICAL SYSTEM: THE ROTATION ON THE TORUS

Let S be the dynamical system $(\mathbb{T}^r, \mathcal{B}(\mathbb{T}^r), \lambda, T_\alpha)$ where T_α is the rotation on the r-dimensional torus $\mathbb{T}^r$ associated with the r-dimensional irrational vector $\alpha = (\alpha_1, \ldots, \alpha_r)$ defined by $x \to x + \alpha \bmod 1$, and λ is the Lebesgue measure on $\mathbb{T}^r$. When α is an *irrational* vector, from Propositions 2 and 3 from Section 4.1 of Chapter 1, the next theorem is easily deduced from theorems of the previous section.

THEOREM 4.6 *Let* $f : \mathbb{T}^r \to [0,1]$ *and* $\alpha = (\alpha_1, \ldots, \alpha_r)$ *an irrational vector.*

- *When* $r = 1$, f *is assumed to be with bounded variation and the partial quotients of* α *to verify* $a_m < m^{1+\epsilon}, \epsilon > 0$, *for* m *large enough.*

- *When* $r > 1$, f *is assumed to be with bounded variation in the sense of Hardy and Krause and* α *to be of type* η *with* $1 \leq \eta < 1 + \frac{1}{r}$.

Assume that $\int_{\mathbb{T}^r} f \, dx = p \neq \frac{1}{2}$ *and* $a = 4 \int_{\mathbb{T}^r} f(1-f) \, dx > 0$. *Then, for any* $\gamma > \frac{1}{2}$, *for every* $x \in \mathbb{T}^r$,

$$\frac{Z_n}{n^\gamma} \underset{n \to +\infty}{\to} 0 \text{ a.s..}$$

If there exists $\epsilon > 0$ such that f takes its values in $]\epsilon, 1-\epsilon[$, the integrals $\int_{\mathbb{T}^r} \log\left(\frac{4f(1-f)}{(\sqrt{5}-2)(1-\epsilon)^4}\right) dx$ and $\int_{\mathbb{T}^r} \log\left(\frac{1-f}{f}\right) dx$ are strictly negative, then, for almost every $x \in \mathbb{T}^r$, the process $(\frac{Z_{nt}}{\sqrt{n}})_{t \geq 0}$ weakly converges in $\mathcal{C}([0, \infty))$ to a Brownian motion with zero mean and variance at.

When α is a *rational* vector, we prove the following theorem

THEOREM 4.7 *Let $f : \mathbb{T}^r \to [0,1]$ and $\alpha = (\alpha_1, \ldots, \alpha_r)$ where $\alpha_i = \frac{p_i}{q_i}$. We denote by $\tilde{q}$ the s.c.m. of $q_1, \ldots, q_r$. For any $x \in \mathbb{T}^r$ such that*
$$\sum_{k=1}^{\tilde{q}} f(T_\alpha^k x) = p\tilde{q} \text{ and } a = 4 \sum_{k=1}^{\tilde{q}} f(T_\alpha^k x)(1 - f(T_\alpha^k x)) > 0,$$

1. *When $p = \frac{1}{2}$, $(S_n)_{n \in \mathbb{N}}$ is recurrent, otherwise it is transient.*

2. *Moreover,*
$$\frac{Z_n}{n} \underset{n \to +\infty}{\to} 0 \text{ a.s..}$$

Proof:
Because of the periodicity, the vectors $(X_{n\tilde{q}+1}, \ldots, X_{(n+1)\tilde{q}})$, $n \geq 0$, have the same distribution. So, if we denote $Y_j = \sum_{l=1}^{\tilde{q}} X_{j\tilde{q}+l}, j = 0, \ldots, n-1$, we can write
$$S_0 = 0, \quad S_{n\tilde{q}} = \sum_{j=0}^{n-1} Y_j, n \geq 1$$

where $(Y_j)_{j \geq 0}$ is a family of random variables independent, identically distributed with mean $\sum_{l=1}^{\tilde{q}}(2f(T_\alpha^l x) - 1) = (2p-1)\tilde{q}$ and finite variance $\sum_{l=1}^{\tilde{q}} 4f(T_\alpha^l x)(1 - f(T_\alpha^l x))$. Therefore, when $p = \frac{1}{2}$ (respectively $p \neq \frac{1}{2}$), $(S_{n\tilde{q}})_{n \geq 0}$ is a recurrent (respectively transient) $\mathbb{Z}$-random walk embedded in the random walk $(S_n)_{n \geq 0}$, which implies the recurrence (respectively the transience) of the Markov chain $(S_n)_{n \geq 0}$.

For every $i = 0, \ldots, \tilde{q} - 1$, $S_{n\tilde{q}+i} - S_i$ is a sum of independent, identically distributed random variables with mean $(2p-1)\tilde{q}$ and finite variance. Therefore, for every $i = 0, \ldots, \tilde{q} - 1$, $(\xi_{S_{n\tilde{q}+i}})_{n \geq 0}$ is a stationary sequence and by the Birkhoff's theorem, as $n \to \infty$,

$$\frac{1}{n} \sum_{k=0}^{n} \xi_{S_{k\tilde{q}+i}} \to 0 \text{ a.s..} \tag{4.11}$$

Now, there exists two sequences of integers $(m_n)_{n\geq 0}, (r_n)_{n\geq 0}$ such that $n = m_n\tilde{q} + r_n, 0 \leq r_n \leq \tilde{q} - 1$. And,

$$Z_n = \sum_{i=1}^{r_n}\sum_{k=0}^{m_n} \xi_{S_{k\tilde{q}+i}} + \sum_{i=r_n+1}^{\tilde{q}-1}\sum_{k=0}^{m_n-1} \xi_{S_{k\tilde{q}+i}} + \sum_{i=0}^{m_n} \xi_{S_{i\tilde{q}}}.$$

Then,

$$\left|\frac{Z_n}{n}\right| \leq \sum_{i=1}^{\tilde{q}-1}\frac{1}{n}\left|\sum_{k=0}^{m_n} \xi_{S_{k\tilde{q}+i}}\right| + \sum_{i=1}^{\tilde{q}-1}\frac{1}{n}\left|\sum_{k=0}^{m_n-1} \xi_{S_{k\tilde{q}+i}}\right| + \frac{1}{n}\left|\sum_{i=0}^{m_n} \xi_{S_{i\tilde{q}}}\right|.$$

Using (4.11), the second assertion of the theorem is proved.

THEOREM 4.8 *Let $f : \mathbb{T}^r \to [0,1]$, $\alpha = (\alpha_1, \ldots, \alpha_r)$ with $\alpha_i = \frac{p_i}{q_i}$. We denote by $\tilde{q}$ the s.c.m. of $q_1, \ldots, q_r$. For any $x \in \mathbb{T}^r$ such that $\sum_{k=1}^{\tilde{q}} f(T_\alpha^k x) = p\tilde{q}$ with $p \neq \frac{1}{2}$ and $a = 4\sum_{k=1}^{\tilde{q}} f(T_\alpha^k x)(1 - f(T_\alpha^k x)) > 0$, the process $\left(\frac{Z_{nt}}{\sqrt{n}}\right)_{t\geq 0}$ weakly converges in $\mathcal{C}([0,\infty))$ to a Brownian motion with zero mean and variance at.*

Proof:
The proof strictly follows the one of Theorem 4.5. The only difference is that, in the rational case, the almost sure convergence of the sequence $(V_n/n)_n$ to a strictly positive constant follows directly from the periodicity of the dynamical system. We have $V_n = \sum_{k=0}^{n} N_n(S_k)$. Since the random walk $(S_n)_{n\geq 0}$ is transient, we can replace $N_n(S_k)$ by $N_\infty(S_k)$. For every $r = 0, \ldots, \tilde{q}-1$, $(S_{k\tilde{q}+r} - S_r)_{k\geq 0}$ is a transient random walk then $(N_\infty(S_{k\tilde{q}+r}))_{k\geq 0}$ is a stationary and ergodic sequence. From Birkhoff's theorem, as $n \to +\infty$,

$$\frac{1}{n}\sum_{k=0}^{n} N_\infty(S_{k\tilde{q}+r}) \to \mathbb{E}(N_\infty(S_r)) \quad \text{a.s..}$$

If we write $n = m_n\tilde{q} + r_n, 0 \leq r_n \leq \tilde{q} - 1$, we have

$$\frac{1}{n}\sum_{k=0}^{n} N_\infty(S_k) = \frac{1}{n}\sum_{r=0}^{\tilde{q}-1}\sum_{k=0}^{m_n-1} N_\infty(S_{k\tilde{q}+r}) + \frac{1}{n}\sum_{r=0}^{r_n} N_\infty(S_{m_n\tilde{q}+r}).$$

The random walk being transient,

$$N_\infty(S_{m_n\tilde{q}+r}) < +\infty, \quad \text{a.s.}$$

So, as $n \to +\infty$,

$$\frac{1}{n}\sum_{k=0}^{n} N_\infty(S_k) \to \sum_{r=0}^{\tilde{q}-1} \mathbb{E}(N_\infty(S_r)) \quad \text{a.s..}$$

The convergence of the finite-dimensional distributions of the process $(\frac{Z_{nt}}{\sqrt{n}})_{t\geq 0}$ to those of the Brownian motion follows.

The proof of the tightness of the sequence $(X_t^n)_{t\geq 0} = (\frac{Z_{nt}}{\sqrt{n}})_{t\geq 0}, n = 1, 2, 3, \ldots$ in $\mathcal{C}([0, \infty))$ follows from a result of [15] (Problem 7 page 102). Let $s, t \in \mathbb{R}$ such that $0 \leq s < t \leq T < \infty$.

$$\mathbb{E}(|X_t^n - X_s^n|^2) = \frac{\sigma^2}{n} \sum_{x \in \mathbb{Z}} \mathbb{E}((N_{nt}(x) - N_{ns}(x))^2)$$

$$\leq \frac{\sigma^2}{n} \sum_{x \in \mathbb{Z}} \mathbb{E}((N_{[nt]+1}(x) - N_{[ns]}(x))^2)$$

$$= \frac{\sigma^2}{n} \sum_{x \in \mathbb{Z}} \mathbb{E}((\sum_{k=[ns]+1}^{[nt]+1} \mathbf{1}_{S_k=x})^2)$$

$$= \frac{\sigma^2}{n}([nt] - [ns] + 1)$$

$$+ 2\frac{\sigma^2}{n} \sum_{k=0}^{[nt]-[ns]} \sum_{l=1}^{[nt]-[ns]-k} \mathbb{P}(S_{[ns]+1+k+l} = 0 | S_{[ns]+1+k} = 0)$$

Let $k \geq 0$. There exist l, r such that $k = l\tilde{q} + r, 0 \leq r \leq \tilde{q} - 1$. Then,

$$\sum_{m=0}^{[s+1]} \mathbb{P}(S_{k+m} = 0 | S_k = 0) = \sum_{m=0}^{[s+1]} \mathbb{P}(S_{r+m} = 0 | S_r = 0)$$

$$\leq C(r) < \infty,$$

So, for every $k \geq 0$,

$$\sum_{m=0}^{[s+1]} \mathbb{P}(S_{k+m} = 0 | S_k = 0) \leq C = \max C(r) < \infty. \qquad (4.12)$$

From (4.12), there exists a constant $C_1 > 0$ such that

$$\mathbb{E}(|X_t^n - X_s^n|^2) \leq C_1 \frac{([nt] - [ns] + 1)}{n}.$$

If $(t-s)n \geq 1$, then

$$[nt] - [ns] + 1 \leq [n(t-s) + 2] + 1$$
$$\leq 4n(t-s).$$

Consequently,

$$\mathbb{E}(|X_t^n - X_s^n|^2) \leq 4C_1(t-s).$$

If $0 \leq (t-s)n < 1$, $N_{k+1}(x) - N_k(x) = 1_{\{S_{k+1}=x\}}$. If $S_{k+1} = x$, then $N_{k+1}(x) - N_k(x) = 1$. So, for $k \leq s_1 < s_2 < k+1$,

$$\sum_{x \in \mathbb{Z}} (N_{s_2}(x) - N_{s_1}(x))^2 = (s_2 - s_1)^2$$

and then,

$$\mathbb{E}(\sum_{x \in \mathbb{Z}} (N_{nt}(x) - N_{ns}(x))^2) = n^2(t-s)^2$$

$$\leq n(t-s).$$

Therefore,

$$\mathbb{E}(|X_t^n - X_s^n|^2) \leq \sigma^2(t-s).$$

So there exists a constant $C_2 = \max(4C_1, \sigma^2) > 0$ such that for every $s, t \in [0, T]$, $s < t$ and for every $n \geq 1$,

$$\mathbb{E}(|X_t^n - X_s^n|^2) \leq C_2(t-s). \tag{4.13}$$

The increments of the process $(X_t^n)_{t \geq 0}$ being independent in the limit, there exists $C_3 > 0$ such that for any $n \geq 1$, for any $t, t_1, t_2 \in \mathbb{R}$ such that $0 \leq t_1 < t < t_2 \leq T < \infty$,

$$\mathbb{E}(|X_t^n - X_{t_1}^n|^2 |X_{t_2}^n - X_t^n|^2) \leq C_3 \mathbb{E}(|X_t^n - X_{t_1}^n|^2) \mathbb{E}(|X_{t_2}^n - X_t^n|^2).$$

From inequality (4.13), there exists $C_4 > 0$ such that

$$\mathbb{E}(|X_t^n - X_{t_1}^n|^2 |X_{t_2}^n - X_t^n|^2) \leq C_4(t_2 - t_1)^2,$$

so the tightness is proved.

References

This chapter is based on [67, 68, 69, 70].

Chapter 5

ERGODIC THEOREMS

1. INTRODUCTION

Generalizations and improvements of Von Neumann and Birkoff's theorems (see Appendix A) have appeared continually since 1932. One of them was to determine the sequences of integers $(a_k)_{k\geq 0}$ for which Von Neumann and Birkoff's theorems still hold, that is, for every function $f \in L^2(\mu)$, (resp. $f \in L^1(\mu)$)

$$\lim_{n\to\infty} \frac{1}{n} \sum_{k=0}^{n-1} f(T^{a_k}x) = E(f|\mathcal{I}) \quad \text{in } L^2-\text{sense} \quad (\text{resp. } L^1, \ \mu-\text{a.e.})$$

where $\mathcal{I}$ is the invariant σ-algebra.
Bourgain ([18, 19, 20]) proved that the sequence $(P(k))_{k\geq 0}$ where P is a polynomial with integer coefficients and degree greater than 2 is a good sequence for the pointwise ergodic theorem for the functions $L^p, p > 1$. Beside its mathematical interest, this question is fundamental in mathematical physics when the physically interesting variables can not be measured along the full sequence of integers but only along a deterministic or random subsequence $(a_k)_{k\geq 0}$. The case when $(a_k)_{k\geq 0}$ are random times where the measurements on the system are realized was studied by Lacey, Petersen, Rudolph and Wierdl ([109]). They assumed that the time between two measurements are independent of each other and with the same distribution. In this chapter we study the more realistic case where these intermediary are time-dependent and independent. They are naturally modelled by dynamic random walks.

2. PRINCIPAL RESULTS

Let us consider a complete probability space $(\Omega, \mathcal{F}, \mathbb{P})$ and a sequence $(S_k)_{k \geq 1}$ of random vectors defined on this space with values in $\mathbb{Z}^d$, $d \geq 1$. Let $(Y, \mathcal{A}, \mu, T)$ be a measurable dynamical system, where $(Y, \mathcal{A}, \mu)$ is a probability space and T is an action of $\mathbb{Z}^d$ defined on Y such that $T\mu = \mu$. Let us introduce the notion of universally representative sequences. A sequence of random vectors $(S_k)_{k \geq 1}$ with values in $\mathbb{Z}^d$, $d \geq 1$ defined on a probability space $(\Omega, \mathcal{F}, \mathbb{P})$ is universally representative for L^p, $p > 1$ if there exists $\Omega_0 \subseteq \Omega$ of probability one such that for every $\omega \in \Omega_0$:
For every dynamical system $(Y, \mathcal{A}, \mu, T)$ and $g \in L^p(\mu)$ we have

$$\mu\left\{y: \lim_{n \to +\infty} \frac{1}{n} \sum_{k=1}^{n} g(T^{S_k(\omega)} y) \text{ exists}\right\} = 1.$$

For example, when $d = 1$, the sequence $\{p_k + \theta_k,\ k \geq 1\}$ where p_k is the k-th prime number and $\{\theta_k,\ k \geq 1\}$ a sequence of independent, identically distributed random variables with a moment of strictly positive order, is universally representative for L^p, $p > 1$ (see [161] for more explanations).

M. Lacey, K. Petersen, D. Rudolph and M. Wierdl (see Theorem 5 in [109]) also proved:

THEOREM 5.1 *Let $(X_k)_{k \geq 1}$ be a sequence of independent, identically distributed random variables such that $\mathbb{E}X_1 \neq 0$ and $\mathbb{E}(X_1^2) < \infty$. Then the sequence*

$$S = \left\{\sum_{j=1}^{k} X_j,\ k \geq 1\right\}$$

is universally representative for L^p, $p > 1$.

In the case $d \geq 2$, they obtained the following result (see Theorem 7 in [109]):
When $g \in L^2(\mu)$, then

$$\lim_{\substack{n \to \infty \\ n \in \mathcal{N}}} \frac{1}{n} \sum_{k=1}^{n} g \circ T^{S_k} \text{ exists } \mu - \text{a.e.}$$

where $\mathcal{N} = \{[2^{t \log t}],\ t \in \mathbb{N}^*\}$.

There also exists a recent paper of D. Schneider (see [164]) that explains the behavior of the previous averages for almost-everywhere convergence in terms of conditions on the spectral measure of the operator T.

Let us recall another definition.

DEFINITION 5.1 *A sequence of random vectors $(S_k)_{k\geq 1}$ with values in $\mathbb{Z}^d$, defined on a probability space $(\Omega, \mathcal{F}, \mathbb{P})$ is universally 2-representative in mean, if there exists $\Omega_0 \subseteq \Omega$ of probability one, such that for every $\omega \in \Omega_0$:*
For every dynamical system $(Y, \mathcal{A}, \mu, T)$ and $g \in L^2(\mu)$ we have

$$\lim_{n\to+\infty} \frac{1}{n} \sum_{k=1}^{n} g \circ T^{S_k(\omega)} \text{ exists in } L^2(\mu).$$

In this situation we have the following result (see [71]):

THEOREM 5.2 *Let $(X_k)_{k\geq 1}$ be a sequence of independent, identically distributed random vectors defined on $(\Omega, \mathcal{F}, \mathbb{P})$, with values in $\mathbb{Z}^d$. Let us assume that there exists $\delta > 0$ such that $\mathbb{E}\|X_1\|_{\mathbb{R}^d}^\delta < \infty$. Then the sequence of random vectors $S = \{\sum_{j=1}^{k} X_j, \ k \geq 1\}$ is universally 2-representative in mean.*

Let us now introduce a new dynamic random walk on $\mathbb{Z}^d$. Let $X_i = (X_i^{(1)}, \ldots, X_i^{(d)}), i \geq 1$, be a sequence of independent random vectors defined on a probability space $(\Omega, \mathcal{F}, \mathbb{P})$ with values in $\{-1, +1\}^d$ such that for every $i \geq 1$, the random variables $X_i^{(j)}, 1 \leq j \leq d$ are independent. Let $f_1, \ldots, f_d$ be functions defined on $\mathbb{T}^r$ with values in $[0, 1]$ and τ_α the rotation on the r-dimensional torus $\mathbb{T}^r$, associated to the r-dimensional irrational vector $\alpha = (\alpha_1, \ldots, \alpha_r)$, defined by $x = (x_1, \ldots, x_r) \mapsto (x_1 + \alpha_1 \mod 1, \ldots, x_r + \alpha_r \mod 1)$. For every $i \geq 1$ and every $1 \leq j \leq d$, the law of the random variable $X_i^{(j)}$ is given by

$$\mathbb{P}(X_i^{(j)} = +1) = f_j(\tau_\alpha^i x) \text{ where } x \in \mathbb{T}^r \text{ is fixed}$$
$$= 1 - \mathbb{P}(X_i^{(j)} = -1).$$

We write

$$S_0 = 0, \quad S_n = \sum_{i=1}^{n} X_i \text{ for } n \geq 1$$

for the $\mathbb{Z}^d$-random walk generated by the family $(X_i)_{i\in\mathbb{N}}$. In dimension one, this new dynamic random walk corresponds to the one defined in Chapter 1 when T is a rotation on the torus $\mathbb{T}^r$. Our main results are summarized in the following theorems.

THEOREM 5.3 *Let $f_1, \ldots, f_d$ be Riemann integrable functions defined on $\mathbb{T}^r$ with values in $[0,1]$ such that for every $j \in \{1, \ldots, d\}$,*

$$\int_{\mathbb{T}^r} f_j(t)(1 - f_j(t))dt > 0.$$

Then, for every $x \in \mathbb{T}^r$ and for every irrational vector α, the dynamic random walk $(S_n)_{n \in \mathbb{N}}$ is universally 2-representative in mean.

Remarks:
1. This result holds when T is a $\mathbb{Z}^d$-action of contractions.
2. The fact that the dynamic random walk $(S_n)_{n \in \mathbb{N}}$ is universally 2-representative in mean means that the set Ω_0 in Definition 5.1 is the same for every dynamical system $(Y, \mathcal{A}, \mu, T)$. In fact, we have even more here: the set Ω_0 is the same for any set of functions $f_j, 1 \leq j \leq d$ satisfying the conditions of the above theorem.

In the case $d = 1$, with stronger hypotheses on the function f and using the approximation of the irrational vector α by rationals, we show that the dynamic random walk is not universally representative for $L^p, p \geq 1$.

THEOREM 5.4 *Let $f : \mathbb{T}^1 \to [0,1]$ be a function of bounded variation such that $\int_{\mathbb{T}^1} f(t)dt = \frac{1}{2}$ and $a = \int_{\mathbb{T}^1} f(t)(1 - f(t))dt > 0$. Then, for every $x \in \mathbb{T}^1$ and for every irrational α with continuous fraction expansion $[a_0; \ldots, a_m, \ldots]$ such that the inequality*

$$a_m < m^{1+\epsilon}$$

*is satisfied for any m large enough, with $\epsilon > 0$, the dynamic random walk $(S_n)_{n \in \mathbb{N}}$ is **not** universally representative for $L^p, p \geq 1$. For almost every $\omega \in \Omega$, given any ergodic aperiodic measure-preserving transformation T on $(Y, \mathcal{A}, \mu)$, there exists a function $g \in L^1(Y, \mathcal{A}, \mu)$ such that the averages*

$$A_n^\omega g(y) = \frac{1}{n} \sum_{k=1}^n g(T^{S_k(\omega)} y)$$

diverge a.e. In fact, the sequence has strong sweeping out: given $\epsilon > 0$, we can choose g to be the characteristic function of a set of measure less than ϵ, yet to have

$$\limsup_{n \to \infty} A_n^\omega g(y) = 1 \quad a.e \quad \text{and} \quad \liminf_{n \to \infty} A_n^\omega g(y) = 0 \quad a.e.$$

The above one-dimensional result can be generalized in $r > 1$ under some additional hypotheses on the mutual irrationality of the components of

$\alpha = (\alpha_1, \ldots, \alpha_r)$. Here, we give the main result, valid in $r \geq 1$ (see Appendix B for definitions).

THEOREM 5.5 *Let $f : \mathbb{T}^r \to [0,1]$ be a function of bounded variation in the sense of Hardy and Krause such that $\int_{\mathbb{T}^r} f(t)dt = \frac{1}{2}$ and $a = \int_{\mathbb{T}^r} f(t)(1-f(t))dt > 0$. Then, for every $x \in \mathbb{T}^r$ and for every irrational vector $\alpha = (\alpha_1, \ldots, \alpha_r)$ of type η such that $1 \leq \eta < 1 + \frac{1}{r}$, the dynamic random walk $(S_n)_{n \in \mathbb{N}}$ is **not** universally representative for $L^p, p \geq 1$.*

The set of irrational numbers satisfying the hypotheses of Theorem 5.4 corresponds to irrational numbers badly approximated by rationals and is of full measure. It can also be proved that almost every irrational vector is of type 1 (see [68]).

THEOREM 5.6 *Let Y_n be a sequence of completely independent random variables with a uniformly bounded positive moment. Then setting,*

$$Z_k(\theta, \omega) = \exp\left[2i\pi <\theta, Y_k(\omega)>\right] - \mathbb{E}\exp\left[2i\pi <\theta, Y_k>\right]$$

we have

$$\mathbb{E}\sup_{\theta \in [0,1[^d}\sup_n \left|\frac{1}{\sqrt{n\log n}}\sum_{k=1}^n Z_k(\theta, \omega)\right| < \infty.$$

In order to prove Theorem 5.3, we will need Van der Corput's inequality and Spectral theorem we recall here.

THEOREM 5.7 (VAN DER CORPUT'S INEQUALITY) *Let $(u_k)_{0 \leq k < n}$ be a finite sequence of n points in a Hilbert space $\mathcal{H}$. If H is an integer between 0 and $n-1$, then we have*

$$\left\|\frac{1}{n}\sum_{k=0}^{n-1} u_k\right\|_{\mathcal{H}}^2 \leq \frac{n+H}{n^2(H+1)}\sum_{k=0}^{n-1}\|u_k\|_{\mathcal{H}}^2 + 2\frac{n+H}{n^2(H+1)^2}\sum_{h=1}^H (H+1-h)A(n,h),$$

where $A(n,h) = Re\left(\sum_{k=0}^{n-h-1} <u_{k+h}, u_k>_{\mathcal{H}}\right)$.

This inequality is easily proved by writing down

$$\left\|\frac{1}{n}\sum_{k=0}^{n-1} u_k\right\|_{\mathcal{H}}^2 = \left\|\frac{1}{n}\sum_{k=-H}^{n-1}\left(\frac{1}{H+1}\sum_{h=0}^H u_{k+h}\right)\right\|_{\mathcal{H}}^2,$$

with the convention $u_k = 0$ if $k < 0$ or $k \geq n$ and by using Cauchy-Schwarz's inequality. This Hilbert space version of Van der Corput's lemma is due to Bergelson (see [6]).

THEOREM 5.8 (SPECTRAL THEOREM) *Let T be a contraction of a Hilbert space $\mathcal{H}$ and $p(x)$ be a polynomial defined on $D = [0, 1[^r$. Then for every f in $\mathcal{H}$, there exists a Borelian positive measure which is bounded on D, denoted μ_f such that*

$$\| p(T)f \|_{\mathcal{H}}^2 \leq \int_D | p(x) |^2 \, \mu_f(dx).$$

In the case where T is a measure-preserving transformation, we have an equality in the spectral theorem. It easily follows from Bochner's theorem using the fact that if we define for every $k \in \mathbb{Z}^r$,

$$\gamma_k = < T^k f, f >_{\mathcal{H}}$$

then γ is a non negative definite sequence. The extension to contractions can be obtained by a simple inductive argument on the degree of the trigonometric polynomial p.

3. PROOF OF THEOREMS 5.4 AND 5.5

By Theorem 1.3 in Del Junco and Rosenblatt [36], it is enough to find for a.e. ω, for each $\epsilon > 0$ and $N \in \mathbb{N}$, a set E of measure less than ϵ for which $\mu(\sup_{n \geq N} A_n^\omega 1_E > 1 - \epsilon) > 1 - \epsilon$. Given $\epsilon > 0$, choose $q \in \mathbb{N}$ with $\frac{4}{q} < \epsilon$ and fix n large. Take $\alpha, \beta \in]0, 1[$ such that $\beta(q+1)^{q-1} < 1$. Let $\beta_{-1} = 0$ and for $i = 0, \ldots, q-1$, let $\beta_i = (q+1)^i \beta$. We define a continuous function F on $[0, 1]$ by $F(0) = 0$ and

$$F = (i+1)\alpha \text{ on } [\beta_{i-1} + \beta(q+1)^{i-1}, \beta_i] \text{ for } i = 0, \ldots, q-1,$$

linearly in between and constant elsewhere. We choose α small enough so that $\int_0^1 F'(t)^2 dt \leq 1$. Under the hypotheses of Theorem 5.4 (resp. Theorem 5.5), by Proposition 2 (resp. Proposition 3 in Chapter 1), if $w_n = \sqrt{2na \log \log n}$, for any $\epsilon > 0$ given, for n large enough, we have

$$\frac{1}{w_n} \left| \sum_{i=1}^k (2f(\tau_\alpha^i x) - 1) \right| < \epsilon \text{ for } k = 1, \ldots, n.$$

Then, by Theorem 2.4, for a.e. ω and n large enough,

$$\left| \frac{S_k(\omega)}{w_n} - F\left(\frac{k}{n}\right) \right| < \alpha \text{ for } k = 1, \ldots, n.$$

We fix such n, ω. By standards methods for transferring counterexamples by use of Rokhlin towers, we may work with the translation action of $\mathbb{Z}$ on itself and find $g : \mathbb{Z} \to \{0, 1\}$ taking value 1 on an infrequently visited set but giving large values of the averages for most initial points. For each $i = 1, \ldots, q$, let $I_i = [(i-1)\alpha w_n, i\alpha w_n[$. Let us define the function g by $g(x) = 1$ if $x(\text{mod } q\alpha w_n) \in I_1 \cup I_2 \cup I_q$, 0 otherwise. If g is transferred by means of the Rokhlin lemma to any aperiodic transformation of $\mathbb{Z}$, we have $\mu(g = 1) < \frac{4}{q} < \epsilon$. For each $x \in \mathbb{Z}$, we can choose $i \in \{1, \ldots, q\}$ such that $x(\text{mod } q\alpha w_n) \in I_{q-i+1}$. Then,

$$
\begin{aligned}
A^{\omega}_{n\beta_{i-1}} g(x) & \\
\geq \frac{1}{n\beta_{i-1}} & \sum_{n\beta_{i-2} \leq k \leq n\beta_{i-1}} g(x + S_k(\omega)) \\
= \frac{1}{n\beta_{i-1}} & \sum_{n\beta_{i-2} \leq k \leq n\beta_{i-1}} g(x + w_n F(\frac{k}{n}) + \delta_k w_n) \text{ where } \delta_k \in]-\alpha, \alpha[. \\
\geq \frac{q-1}{qn\beta_{i-1}} & \sum_{n\beta_{i-2} + n\beta(q+1)^{i-2} \leq k \leq n\beta_{i-1}} g(x + i\alpha w_n + \delta_k w_n) \\
\geq \frac{q-1}{q} & \cdot \frac{n\beta_{i-1} - n\beta_{i-2} - n\beta(q+1)^{i-2}}{n\beta_{i-1}} \geq 1 - \frac{3}{q} > 1 - \epsilon
\end{aligned}
$$

since

$$
\begin{aligned}
(x + i\alpha w_n + \delta_k w_n)(\text{mod } q\alpha w_n) & \\
\in & [\delta_k w_n(\text{mod } q\alpha w_n), (\alpha + \delta_k)w_n(\text{mod } q\alpha w_n)[\\
\subset & I_1 \cup I_2 \cup I_q \text{ (since } \delta_k \in]-\alpha, \alpha[).
\end{aligned}
$$

4. PROOF OF THEOREM 5.3

In this section we prove Theorem 5.3 assuming that Theorem 5.6 is satisfied and using Van der Corput's inequality. The proof of Theorem 5.6 will be given in the next section. Van der Corput's inequality permits us to determine sufficient conditions for a sequence of points in a Hilbert space to go to 0 in the Cesàro sense for the Hilbertian norm. Given a finite family of n points $(u_k)_{1 \leq k \leq n}$ in a Hilbert space $\mathcal{H}$, a sufficient condition to get

$$\lim_{n \to \infty} \left\| \frac{1}{n} \sum_{k=1}^{n} u_k \right\|_{\mathcal{H}} = 0 ,$$

is:

(a) $\forall l \geq 0$, $\left| \frac{1}{n} \sum_{k=1}^{n} < u_{k+l}, u_k >_{\mathcal{H}} \right| \leq \psi_{l,n}$ and $\lim_{n \to \infty} \psi_{l,n} = \gamma_l$ exists

and

(b)
$$\lim_{L \to \infty} \frac{1}{L} \sum_{l=0}^{L-1} \gamma_l = 0.$$

For all $l \geq 1$, we denote $S_k^l = X_{k+1} + \ldots + X_{k+l}, \forall k \geq 1$ and we define $\phi_k^{(l)}(\theta), \theta \in [0,1[^d$ the characteristic function of the random variable S_k^l,

$$\phi_k^{(l)}(\theta) = \mathbb{E}(e^{2\pi i <\theta, S_k^l>})$$

The random vectors X_i are independent with independent directions and the law of $X_i^{(j)}$ is known, so

$$\phi_k^{(l)}(\theta) = \prod_{i=1}^{r} \prod_{j=1}^{l} (\cos(2\pi\theta_i) + i(2f_i(\tau_\alpha^{k+j} x) - 1)\sin(2\pi\theta_i)), \theta \in [0,1[^d \quad (5.1)$$

The $\mathbb{Z}^d$-action T can be rewritten as a composition of d commuting automorphisms $T_1, \ldots, T_d$ of the space Y i.e.

$$T = T_1 \circ \ldots \circ T_d.$$

We denote by H the closed space of $L^2(\mu)$ spanned by the functions f such that for every $i = 1, \ldots, d$, $T_i f = \pm f$. Then, the Hilbert space $L^2(\mu)$ can be decomposed as a direct sum of the space H and the orthogonal complement $H^\perp$ of the space H.

LEMMA 16 *Let $g \in H^\perp$ and let μ_g be the spectral measure of T at the point g. Then $\mu_g(\theta) = 0$ for every $\theta \in \{0, \frac{1}{2}\}^d$.*

Proof:
Let $\theta = (\theta_1, \ldots, \theta_d)$ with $\theta_i \in \{0, \frac{1}{2}\}$ for every $i = 1, \ldots, d$. Let us define the new $\mathbb{Z}^d$-action

$$\tilde{T} = \tilde{T}_1 \circ \ldots \circ \tilde{T}_d$$

where $\tilde{T}_i = \exp(-2\pi i \theta_i) T_i, i = 1, \ldots, d$. The $\mathbb{Z}^d$-action $\tilde{T}$ is a contraction of $L^2(\mu)$. Then, by Von Neumann's Theorem for the $\mathbb{Z}^d$-action $\tilde{T}$ (see for instance [149] p.24), the average

$$\frac{1}{n^d} \sum_{k_1, \ldots, k_d=1}^{n} \tilde{T}^k g$$

converges in $L^2(\mu)$ to a function $h \in L^2(\mu)$. Moreover, the function h is fixed for the $\mathbb{Z}^d$-action $\tilde{T}$: for every $i = 1, \ldots, d$, $\tilde{T}_i h = h$ that is to say for every $i = 1, \ldots, d$, $T_i h = \pm h$ so $h \in H$. Using Spectral theorem, we now have

$$\begin{aligned}
<h, g>_{2,\mu} &= \lim_{n \to \infty} \frac{1}{n^d} \sum_{k_1,\ldots,k_d=1}^{n} \exp(-2\pi i <\theta, k>) <T^k g, g>_{2,\mu} \\
&= \int_{[0,1[^d} \lim_{n \to \infty} \frac{1}{n^d} \sum_{k_1,\ldots,k_d=1}^{n} \exp(2\pi i <t-\theta, k>) \mu_g(dt) \\
&= \mu_g(\theta) = 0.
\end{aligned}$$

Let us come back to the proof of Theorem 5.3. Let $g \in H^\perp$. We use Spectral theorem and get the equality

$$\|\frac{1}{n}\sum_{k=1}^{n} g \circ T^{S_k}\|_{2,\mu}^2 = \int_{[0,1[^d} |\frac{1}{n}\sum_{k=1}^{n} \exp(2\pi i <\theta, S_k>)|^2 \mu_g(d\theta).$$

In order to prove that this term goes to 0 as $n \to \infty$, we apply the Van der Corput's inequality to the Hilbert space $L^2([0,1[^d, \mu_g)$ and to the sequence $u_k = \exp(2\pi i <\theta, S_k>), k \geq 1$. Let us begin by verifying the point (a). For every $l \geq 1$,

$$\begin{aligned}
\frac{1}{n}\sum_{k=1}^{n} &<u_{k+l}, u_k>_{2,\mu_g} \\
&= \frac{1}{n}\sum_{k=1}^{n} \int_{[0,1[^d} \exp(2\pi i <\theta, S_k^l>) \mu_g(d\theta) \\
&= \int_{[0,1[^d} \frac{1}{n}\sum_{k=1}^{n} (\exp(2\pi i <\theta, S_k^l>) - \phi_k^{(l)}(\theta)) \mu_g(d\theta) \\
&\quad + \int_{[0,1[^d} \frac{1}{n}\sum_{k=1}^{n} \phi_k^{(l)}(\theta) \mu_g(d\theta).
\end{aligned}$$

By applying Theorem 5.6 to the independent random variables $Y_n = S_{ln+m}^l$, the first integral goes to 0 as $n \to \infty$. We have to estimate the second one. Using (5.1) and the fact that $\log(1-x) \leq -x$ for every $x \in [0,1[$, we get

$$\begin{aligned}
|\frac{1}{n}\sum_{k=1}^{n} \phi_k^{(l)}(\theta)| &\leq \frac{1}{n}\sum_{k=1}^{n} |\phi_k^{(l)}(\theta)| \\
&= \frac{1}{n}\sum_{k=1}^{n} \prod_{i=1}^{d} \exp\left(\frac{1}{2}\sum_{j=1}^{l} \log(1 - 4f_i(\tau_\alpha^{k+j}x)(1-f_i(\tau_\alpha^{k+j}x))\sin^2(2\pi\theta_i))\right) \\
&\leq \max_{k \geq 1} \prod_{i=1}^{d} \exp\left(-\frac{\sin^2(2\pi\theta_i)}{2}\sum_{j=1}^{l} 4f_i(\tau_\alpha^{k+j}x)(1-f_i(\tau_\alpha^{k+j}x))\right) = \psi_l(\theta).
\end{aligned}$$

We denote
$$\gamma_l = \int_{[0,1[^d} \psi_l(\theta)\mu_g(d\theta).$$

Let us now verify that the condition (b) is satisfied so that we need to study the behavior as $L \to \infty$ of the sequence

$$\frac{1}{L}\sum_{l=1}^{L}\gamma_l = \frac{1}{L}\sum_{l=1}^{L}\int_{[0,1[^d}\psi_l(\theta)\mu_g(d\theta).$$

The sequence $\frac{1}{L}\sum_{l=1}^{L}\psi_l(\theta)$ is bounded by 1 for every $\theta \in [0,1[^r$ and we have

$$\frac{1}{L}\sum_{l=1}^{L}\psi_l(\theta)$$

$$\leq \frac{1}{L}\sum_{l=1}^{L}\max_{k\geq 1}\prod_{i=1}^{d}\exp\left(-\frac{\sin^2(2\pi\theta_i)}{2}(\sum_{j=1}^{l}4f_i(\tau_\alpha^{k+j}x)(1-f_i(\tau_\alpha^{k+j}x))-lc_i)\right)$$
$$\times \exp\left(-\frac{lc_i\sin^2(2\pi\theta_i)}{2}\right)$$

where $c_i = \int_{\mathbb{T}^r} f_i(t)(1-f_i(t))dt > 0$. Under the hypotheses of Theorem 5.3, we have for every $i \in \{1,\ldots,d\}$,

$$\sup_{x\in \mathbb{T}^r}\left|\frac{1}{n}\sum_{j=1}^{n}4f_i(\tau_\alpha^j x)(1-f_i(\tau_\alpha^j x))-c_i\right|\underset{n\to\infty}{\to}0$$

(see [149] p.156). Then, $\frac{1}{L}\sum_{l=1}^{L}\gamma_l$ goes to 0 as $L \to \infty$, the spectral measure μ_g having no mass point at the points $(\theta_1,\ldots,\theta_d)$, with $\theta_i \in \{0,\frac{1}{2}\}$ from the previous lemma so that the condition (b) is verified.

We now have to treat the functions belonging to the space H. As these functions are limits of finite linear combinations of the functions spanning the space H, it is enough to treat the latter ones. If g is a function such that for every $i = 1,\ldots,r$, $T_ig = \pm g$, then the sampled ergodic average of g is given by

$$\frac{1}{n}\sum_{k=1}^{n}\prod_{i\in I}(-1)^{S_k^{(i)}}g$$

where $I = \{i \in \{1,..,d\}; T_ig = -g\}$. It is easy to see from the definition of the dynamic random walk that when card(I) is even, this average is

g. Otherwise, it is either 0 when n is even or $-\frac{g}{n}$ for n odd, so that its norm in $L^2(\mu)$ converges to 0 as $n \to \infty$.

5. PROOF OF THEOREM 5.6

Since we assumed that for some positive ϵ, we have $\mathbb{E}(|Y_n|^\epsilon) \leq c < \infty$, there exists $\delta > 0$ so that for almost every ω, the absolute value of the derivative of

$$\sum_{k \leq n} Z_k(\theta, \omega)$$

with respect to θ is less than n^δ if n is large enough. It follows that for each n, for a suitably large $\sigma > 0$, the sup over all θ can be replaced by a sup over the set

$$\Theta_n = \{kn^{-\sigma}; k = 1, \ldots, n^\sigma\}.$$

By Borel-Cantelli lemma, we need to show that for a large enough constant K,

$$\mathbb{P}\left(\sup_{\theta \in \Theta_n} (\log n/n)^{1/2} \left| \sum_{k \leq n} Z_k(\theta, \omega) \right| > K \log n \right) < cn^{-2}.$$

But then we just need to show that for every positive σ, there is a K so that for each θ,

$$\mathbb{P}\left((\log n/n)^{1/2} \left| \sum_{k \leq n} Z_k(\theta, \omega) \right| > K \log n \right) < cn^{-\sigma}.$$

Denoting by R_k the real part of Z_k, we only show

$$\mathbb{P}\left((\log n/n)^{1/2} \sum_{k \leq n} R_k > K \log n \right) < cn^{-\sigma},$$

for large enough K, the remaining cases (imaginary part, $< -K \log n$) being entirely similar. Because of the estimate

$$\mathbb{P}\left((\log n/n)^{1/2} \sum_{k \leq n} R_k > K \log n \right)$$

$$\leq \mathbb{E} \exp\left((\log n/n)^{1/2} \sum_{k \leq n} R_k \right) . e^{-K \log n},$$

we only need to prove that

$$\mathbb{E}\exp\left((\log n/n)^{1/2}\sum_{k\leq n}R_k\right)\leq e^{c\log n}.$$

Denote $s=(\log n/n)^{1/2}$. Since the random variables R_k are uniformly bounded by 2, they have zero mean and $\mathbb{E}(R_k^2)\leq 2$, so that using the power series expansion of the function exp, we obtain the bound

$$\mathbb{E}(e^{sR_k})\leq e^{4s^2}.$$

But then since the R_k are independent,

$$\mathbb{E}\exp\left((\log n/n)^{1/2}\sum_{k\leq n}R_k\right)\leq e^{4ns^2}=e^{4\log n}.$$

References

This chapter is based on [71].

Chapter 6

DYNAMIC RANDOM WALKS ON HEISENBERG GROUPS

1. INTRODUCTION

Random walks on Lie groups have been extensively studied over the last decades ([67, 81, 138]). Among these groups Heisenberg groups play a special rôle. These groups have their origin in quantum mechanics where they can be interpreted as the Lie algebras generated by the location operator, the momentum operator, and the identity operator. They are simply connected nilpotent Lie groups of rank 2 and one-dimensional center. Heisenberg groups are often considered as the simplest non-commutative Lie groups. The geometry of these groups has been investigated by A. Korányi ([102, 103]). Limit theorems for random walks on Heisenberg groups have been proved by P. Crepel, B. Roynette ([30]) and D. Neuenschwander ([138]) in connection with the resolution of Kesten's conjecture on the classification of recurrent and transient groups. A central limit theorem for nilpotent Lie groups has been proved by P. Crepel and A. Raugi ([29]).

The main goal of this chapter is to present some extensions of the theory developed in the previous chapters to Heisenberg groups. The organization of this chapter is as follows: Section 2 provides some introductory material on Heisenberg groups and dynamic random walks. Section 3 is devoted to the proof of a strong law of large numbers and a central limit theorem. The limiting distribution is explicitly calculated.

2. GENERALITIES ON HEISENBERG GROUPS

The continuous Heisenberg group H_d is the group with underlying manifold $\mathbb{R}^{2d+1}$ and group operation

$$X \cdot Y = \left(x_1 + y_1, x_2 + y_2, \ldots, x_{2d} + y_{2d}, z_1 + z_2 + \frac{1}{2}\left(\sum_{i=1}^{2d-1} (x_i y_{i+1} - x_{i+1} y_i) \right) \right)$$

where $X = (x_1, \ldots, x_{2d}, z_1) \in \mathbb{R}^{2d+1}$ and $Y = (y_1, \ldots, y_{2d}, z_2) \in \mathbb{R}^{2d+1}$.

H_d is a nilpotent Lie group of rank 2 and with one-dimensional center. In this paper we identify the Heisenberg group H_d with its Lie algebra $\mathcal{H}_d$.

For simplicity we focus on H_1 but all results of this paper remain true for H_d. Let us remember that another representation H_1 is as a group of upper triangular (3×3)-matrices with 1's on the diagonal:

$$\begin{pmatrix} 1 & x & z \\ 0 & 1 & y \\ 0 & 0 & 1 \end{pmatrix}, x, y, z \in \mathbb{R}$$

If x, y, z are integers (in the ring $\mathbb{Z}$) then we get the discrete Heisenberg group which is known to be the simplest non-abelian nilpotent group. Let $g = (x, y, z) \in H_1$ and

$$|g| = [(x^2 + y^2)^2 + z^2]^{1/4}$$

Let $\delta_r(g) = (rx, ry, r^2 z)$ where $r > 0$. The mapping δ_r is called dilation of ratio r on H_1. The mapping $g \to |g|$ from H_1 into $\mathbb{R}^+$ is an homogeneous norm. This means that:
i) $|g| = 0$ if and only if $g = 0$,
ii) $|g| = |-g|$,
iii) $|\delta_r(g)| = r |g|$.

Remark:
Let V be a compact neighborhood of the identity in H_1. The mapping defined by $|g| = \inf\{n \in \mathbb{N}, g \in V^n\}$ is also an homogeneous norm. It is known [78] that all homogeneous norms on H_1 are equivalent.

$B(0, r) = \{g \in H_1; |g| \leq r\}$ $(r > 0)$ is called a Korányi ball. It replaces the traditional euclidean ball in the geometry developed by A. Korányi ([102, 103]).

We introduce now our model of dynamic random walks on H_1.

Let $(X_n, Y_n, Z_n)_{n \geq 1}$ be a sequence of independent random variables with values in H_1 defined on a probability space $(\Omega, \mathcal{F}, \mathbb{P})$. Let $(E, \mathcal{A}, \mu, T)$ be a dynamical system where $(E, \mathcal{A}, \mu)$ is a probability space and T is a transformation defined on E preserving the measure μ. Let $f_i, i = 1, 2, 3$ be functions defined on E with values in $[0, \frac{1}{3}]$. Let $x \in E$ and $(e_j)_{1 \leq j \leq 3}$ be the unit coordinate vectors of $\mathbb{R}^3$. For each $i \geq 1$, the distribution of the random vector $M_i = (X_i, Y_i, Z_i)$ is given by

$$\mathbb{P}(M_i = z) = \begin{cases} f_j(T^i x) & \text{if } z = e_j \\ \frac{1}{3} - f_j(T^i x) & \text{if } z = -e_j \\ 0 & \text{otherwise} \end{cases}$$

We are interested in the right dynamic random walk

$$S_n = (X_1, Y_1, Z_1) \cdot (X_2, Y_2, Z_2) \cdots (X_n, Y_n, Z_n), n \geq 1.$$

We prove in the next sections a law of large numbers and a central limit theorem for $(S_n)_{n \geq 1}$.

3. LIMIT THEOREMS
3.1 LAW OF LARGE NUMBERS

Y. Guivarc'h [78] proved a strong law of large numbers for random walks on Lie groups. Our formulation of the law of large numbers for dynamic random walks on Heisenberg groups is in the same spirit.
In this section, we assume that the dynamical system $(E, \mathcal{A}, \mu, T)$ is ergodic. The functions $f_i, i = 1, 2, 3$ are all with integral $1/6$. Let $d\left(0, \frac{S_n}{n}\right) = \frac{1}{n} \mid S_n \mid$, d will be called Korányi distance.

THEOREM 6.1 *For μ-almost every $x \in E$, as $n \to +\infty$,*

$$d\left(0, \frac{S_n}{n}\right) \to 0 \quad \mathbb{P} - a.s..$$

Proof:
From the definition of the Korányi distance,

$$d\left(0, \frac{S_n}{n}\right) = \left(\left(\left(\frac{S_n^{(1)}}{n}\right)^2 + \left(\frac{S_n^{(2)}}{n}\right)^2\right)^2 + \left(\frac{S_n^{(3)}}{n^2}\right)^2\right)^{1/4}$$

The two first coordinates of S_n are given by

$$S_n^{(1)} = X_1 + X_2 + \ldots + X_n$$

and

$$S_n^{(2)} = Y_1 + Y_2 + \ldots + Y_n.$$

From Birkhoff's theorem, the sequence

$$\frac{\mathbb{E}(S_n^{(1)})}{n} = \frac{1}{n}\sum_{k=1}^{n}(2f_1(T^k x) - \frac{1}{3})$$

converges for μ-almost every $x \in E$ to $\int_E (2f_1 - \frac{1}{3})\, d\mu = 0$. With the same arguments, the sequence

$$\frac{\mathbb{E}(S_n^{(2)})}{n} = \frac{1}{n}\sum_{k=1}^{n}(2f_2(T^k x) - \frac{1}{3})$$

is proved to converge for μ-almost every $x \in E$ to $\int_E (2f_2 - \frac{1}{3})\, d\mu = 0$. Thus, from Kolmogorov's theorem, the sequences $(S_n^{(1)}/n)_n$ and $(S_n^{(2)}/n)_n$ converge $\mathbb{P}$-almost surely to 0 for μ-almost every $x \in E$. The third component of the dynamic random walk on H_1 is given by

$$\begin{aligned} S_n^{(3)} &= Z_1 + Z_2 + \ldots + Z_n \\ &+ \frac{1}{2}\{X_1 Y_2 + (X_1 + X_2)Y_3 + \ldots + (X_1 + \ldots + X_{n-1})Y_n - Y_1 X_2 \\ &- (Y_1 + Y_2)X_3 - \ldots - (Y_1 + \ldots + Y_{n-1})X_n\} \end{aligned}$$

For μ-almost every $x \in E$, the sequence

$$\frac{1}{n^2}(Z_1 + Z_2 + \ldots + Z_n)$$

converges $\mathbb{P}$-almost surely to 0 so we are only interested in the asymptotic behavior of the sequence

$$\frac{T_n}{n^2} = \frac{1}{2n^2}(T_n^{(1)} - T_n^{(2)})$$

where

$$T_n^{(1)} = X_1 Y_2 + (X_1 + X_2)Y_3 + \ldots + (X_1 + \ldots + X_{n-1})Y_n$$

$$T_n^{(2)} = Y_1 X_2 + (Y_1 + Y_2)X_3 + \ldots + (Y_1 + \ldots + Y_{n-1})X_n.$$

Let us prove that $T_n^{(1)}/n^2$ converges almost surely to 0. Since the Y's can only take values 0 and ± 1, we can bound $|\frac{T_n^{(1)}}{n^2}|$ by

$$\frac{1}{\left(\sum_{i=2}^{n}(i-1)|Y_i|\right)}\sum_{i=2}^{n}(i-1)|Y_i|\frac{1}{i-1}\left|\sum_{j=1}^{i-1}X_j\right|.$$

We conclude by using Toeplitz's Theorem (see for instance Theorem 2.34 in [44]) since the sequence $\sum_{j=1}^{n} X_j/n$ converges $\mathbb{P}$-almost surely to 0 for μ-almost every $x \in E$ as n goes to infinity. By inverting the random variables X_j's and Y_j's, we evidently get that $T_n^{(2)}/n^2$ converges $\mathbb{P}$-almost surely to 0 for μ-almost every $x \in E$ as n goes to infinity. Consequently, T_n/n^2 converges $\mathbb{P}$-almost surely to 0 for μ-almost every $x \in E$ as n goes to infinity as well as the third component of the dynamic random walk. So the theorem is proved.

3.2 CENTRAL LIMIT THEOREM

We still denote by $A = (a_{ij})_{1 \leq i,j \leq 3}$ the matrix with coefficients

$$a_{jj} = \frac{4}{9} \int_E (1 - 9f_j^2(x)) d\mu(x)$$

$$a_{ij} = a_{ji} = \frac{1}{9} \int_E (1 - 36 f_i(x) f_j(x)) d\mu(x).$$

Let us assume that $f_i, f_i f_j \in \mathcal{C}_2(S)$ for every $i, j = 1, 2, 3$ and $\int_E f_i \, d\mu = \frac{1}{6}$. For every $x \in E$, the two first coordinates of S_n converge $\mathbb{P}$-almost surely to 0 (see the proof of Theorem 6.1). In the third component given by

$$Z_1 + Z_2 + \ldots + Z_n + \frac{1}{2} \{X_1 Y_2 + (X_1 + X_2) Y_3 + \ldots + (X_1 + \ldots + X_{n-1}) Y_n$$

$$- Y_1 X_2 - (Y_1 + Y_2) X_3 - \ldots - (Y_1 + \ldots + Y_{n-1}) X_n \},$$

for every $x \in E$, the sequence

$$\frac{1}{n}(Z_1 + Z_2 + \ldots + Z_n)$$

also converges $\mathbb{P}$-almost surely to 0 so we are only interested in the asymptotic distribution of the sequence

$$T_n = \frac{1}{2} \{X_1 Y_2 + (X_1 + X_2) Y_3 + \ldots + (X_1 + \ldots + X_{n-1}) Y_n$$

$$- Y_1 X_2 - (Y_1 + Y_2) X_3 - \ldots - (Y_1 + \ldots + Y_{n-1}) X_n \}.$$

We assume that the matrix C defined by $(a_{ij})_{1 \leq i,j \leq 2}$ is diagonal, namely

$$C = \begin{pmatrix} a_{11} & 0 \\ 0 & a_{22} \end{pmatrix}$$

with $a_{11} = 4[\frac{1}{9} - \int_E f_1^2 \, d\mu]$ and $a_{22} = 4[\frac{1}{9} - \int_E f_2^2 \, d\mu]$.

Since our random variables $(X_i)_{i \geq 1}$ and $(Y_i)_{i \geq 1}$ are not centered, we need to make the following assumption (A):

$$\sup_{x \in E} \left| \sum_{i=2}^{k} \sum_{j=1}^{i-1} (f_2(T^i x) - \frac{1}{6})(f_1(T^j x) - \frac{1}{6}) \right| = o(k).$$

THEOREM 6.2 *For every $x \in E$, for every $t > 0$, the sequence*

$$\left(\frac{T_{[nt]}}{n} \right)_{n \geq 1}$$

converges in distribution to the random variable

$$A(t) = \frac{1}{2} \left\{ \int_0^t B_s^{(1)} dB_s^{(2)} - \int_0^t B_s^{(2)} dB_s^{(1)} \right\}$$

where $B_t = (B_t^{(1)}, B_t^{(2)})$ is a two-dimensional Brownian motion with zero mean and covariance matrix Ct. The density of $A(t)$ is given by

$$d_t(x) = \frac{1}{t\sqrt{a_{11}a_{22}} \cosh\left(\frac{\pi x}{t\sqrt{a_{11}a_{22}}} \right)}.$$

The random variable $A(t)$ is usually called the Lévy stochastic area. In our context $A(t)$ is driven by dynamic random walks, therefore we will call $A(t)$, by analogy, the dynamic Lévy stochastic area.

Remarks:
1- The hypotheses on the functions f_1, f_2 and $f_1 f_2$ imply that the assumption (A) is satisfied as soon as for every $i, j \geq 1$, for every $x \in E$,

$$(f_2(T^i x) - \frac{1}{6})(f_1(T^j x) - \frac{1}{6}) = (f_2(T^j x) - \frac{1}{6})(f_1(T^i x) - \frac{1}{6})$$

i.e., for instance, in the following cases:
a)- $f_1 = f_2$ (but not necessarily equal to $1/6$).
b)- f_1 or f_2 equal to $1/6$.
c)- $f_2 = \frac{1}{3} - f_1$.

2- When T is an irrational rotation on the one-dimensional torus $\mathbb{T}^1$, every function with bounded variation belongs to the class $\mathcal{C}_2(S)$ when the angle is badly approximated by the rationals (see Proposition 2 of Section 4.1 in Chapter 1). For example, we can choose $f_1(x) \equiv 1/6$ and $f_2(x) = 1/3 \cos^2(2\pi x)$ which takes its values in $[0, 1/3]$. The hypothesis (**A**) is clearly satisfied thanks to remark 1. The integral of f_2 is equal to $1/6$ so $a_{12} = a_{21} = 0$. After simple computations, $a_{11} = 1/3$ and $a_{22} = 5/18$. In this particular example, the density of the limit distribution $A(t)$ in Theorem 6.2 is then given by

$$d_t(x) = \frac{1}{t\sqrt{5/54}} \cosh\left(\frac{\pi x}{t\sqrt{5/54}} \right), \quad x \in \mathbb{R}.$$

3.2.1 A PRELIMINARY RESULT

In order to prove Theorem 6.2, we shall need a central limit theorem for the dynamic $\mathbb{Z}^2$-random walk $(S_n^{(1)}, S_n^{(2)})_{n \geq 1}$.

PROPOSITION 10 *For every $x \in E$, the sequence of random vectors $\frac{1}{\sqrt{n}}(S_n^{(1)}, S_n^{(2)})$, $n \geq 1$ converges in distribution, as $n \to +\infty$, to the centered Gaussian random variable G_0 with covariance matrix C.*

Proof:
Let us introduce the characteristic function ϕ_n of

$$\frac{1}{\sqrt{n}} \sum_{k=1}^{n} (X_k, Y_k), \quad n \geq 1.$$

By independence of the random vectors $(X_k, Y_k), k \geq 1$,

$$\phi_n(u) = \prod_{k=1}^{n} \mathbb{E}\left(\exp\left(i\frac{u_1 X_k + u_2 Y_k}{\sqrt{n}}\right)\right)$$

$$= \prod_{k=1}^{n} Q_n^{(k)}(u_1, u_2)$$

where

$$Q_n^{(k)}(u_1, u_2) = \frac{1}{3}\left(1 + \sum_{j=1}^{2} \cos\left(\frac{u_j}{\sqrt{n}}\right)\right) + 2i \sum_{j=1}^{2} (f_j(T^k x) - \frac{1}{6}) \sin\left(\frac{u_j}{\sqrt{n}}\right)$$

A direct calculation gives

$$|Q_n^{(k)}(u)|^2 = \frac{1}{9}\left(1 + \sum_{j=1}^{2}(1 - \frac{u_j^2}{2n}) + \mathcal{O}(n^{-2})\right)^2 + \frac{1}{9}\sum_{j=1}^{2}(6f_j(T^k x) - 1)^2 \frac{u_j^2}{n}$$

$$+ \frac{2}{9}(6f_1(T^k x) - 1)(6f_2(T^k x) - 1)\frac{u_1 u_2}{n} + \mathcal{O}(n^{-2})$$

$$= 1 - \frac{1}{3n}\sum_{j=1}^{2} u_j^2 + \frac{1}{9n}\sum_{j=1}^{2} u_j^2(6f_j(T^k x) - 1)^2$$

$$+ \frac{2}{9n}(6f_1(T^k x) - 1)(6f_2(T^k x) - 1)u_1 u_2 + \mathcal{O}(n^{-2})$$

and then

$$|\phi_n(u)| = \prod_{k=1}^{n} |Q_n^{(k)}(u)|$$

$$= \exp(-\frac{1}{2} <u, Cu> + o(1))$$

The imaginary part of the characteristic function can be rewritten as

$$\prod_{k=1}^{n} \exp\left(i \arctan\left(\frac{\sum_{j=1}^{2}(6f_j(T^k x) - 1)\sin(\frac{u_j}{\sqrt{n}})}{(1 + \sum_{j=1}^{2}\cos(\frac{u_j}{\sqrt{n}}))}\right)\right)$$

$$= \exp\left(\frac{i}{3}\sum_{k=1}^{n}\sum_{j=1}^{2}(6f_j(T^k x) - 1)\frac{u_j}{\sqrt{n}} + o(1)\right) = 1 + o(1)$$

using the fact that f_1, f_2 belong to the class $\mathcal{C}_2(S)$ and for every $j = 1, 2$, the integral of f_j is equal to $1/6$.

3.2.2 PROOF OF THE CENTRAL LIMIT THEOREM 6.2

We only prove Theorem 6.2 for the particular value t equal to 1. The proof can easily be adapted to get the result for every $t > 0$. The proof of the theorem is decomposed into four parts.

a) Straightforward calculations give us

$$\begin{aligned}
2T_{nk} &= [N_0^{(k)} + \ldots + N_{n-1}^{(k)}] - [\tilde{N}_0^{(k)} + \ldots + \tilde{N}_{n-1}^{(k)}] \\
&+ [M_0^{(k)}(Q_1^{(k)} - Q_0^{(k)}) + \ldots + M_{n-2}^{(k)}(Q_{n-1}^{(k)} - Q_{n-2}^{(k)})] \\
&- [Q_0^{(k)}(M_1^{(k)} - M_0^{(k)}) + \ldots + Q_{n-2}^{(k)}(M_{n-1}^{(k)} - M_{n-2}^{(k)})]
\end{aligned}$$

where

$$\begin{aligned}
N_p^{(k)} &= X_{pk+1}Y_{pk+2} + (X_{pk+1} + X_{pk+2})Y_{pk+3} + \ldots + (X_{pk+1} + \ldots \\
&+ X_{(p+1)k-1})Y_{(p+1)k} \\
\tilde{N}_p^{(k)} &= Y_{pk+1}X_{pk+2} + (Y_{pk+1} + Y_{pk+2})X_{pk+3} + \ldots + (Y_{pk+1} \\
&+ \ldots + Y_{(p+1)k-1})X_{(p+1)k} \\
M_p^{(k)} &= X_1 + \ldots + X_{(p+1)k} \\
Q_p^{(k)} &= Y_1 + \ldots + Y_{(p+1)k}.
\end{aligned}$$

The choice of this decomposition can appear artificial but it will be justified in the sequel.

b) Let $x \in E$ fixed. From Proposition 10, for every $p \geq 0$, the sequence of random vectors

$$W_p^{(k)} = \frac{1}{\sqrt{k}} \sum_{j=pk+1}^{(p+1)k} (X_j, Y_j), \quad k \geq 1$$

converges in distribution, as $k \to +\infty$, to the centered Gaussian random vector G_0 with covariance matrix C. Since the random vectors

$(W_p^{(k)})_{p\geq 0}$ are independent, we deduce that $(W_0^{(k)}, W_1^{(k)}, \ldots, W_{n-1}^{(k)})$ converges in distribution, as $k \to +\infty$, to the centered Gaussian random vector $(G_0, G_1, \ldots, G_{n-1})$ where the G_i's are independent copies of G_0. Since C is diagonal, each random vector G_i can be decomposed in two independent centered Gaussian random variables $G_i^{(1)}$ and $G_i^{(2)}$ with variance a_{11} and a_{22} respectively. So it implies that $(W_0^{(k)}, W_1^{(k)}, \ldots, W_{n-1}^{(k)})$ converges in distribution, as $k \to +\infty$, to the centered Gaussian random vector $(G_0^{(1)}, G_0^{(2)}, G_1^{(1)}, G_1^{(2)}, \ldots, G_{n-1}^{(1)}, G_{n-1}^{(2)})$. Now, $W_p^{(k)} = (M_p^{(k)} - M_{p-1}^{(k)}, Q_p^{(k)} - Q_{p-1}^{(k)})/\sqrt{k}$ (with the convention $M_{-1}^{(k)} = Q_{-1}^{(k)} = 0$). The convergence in distribution being preserved by linear transformation we get that

$$\frac{1}{\sqrt{k}}(M_0^{(k)}, M_1^{(k)}, \ldots, M_{n-1}^{(k)}, Q_0^{(k)}, Q_1^{(k)}, \ldots, Q_{n-1}^{(k)})$$

converges in distribution, as $k \to +\infty$, to the centered Gaussian random vector $(G_0^{(1)}, G_0^{(1)}+G_1^{(1)}, \ldots, G_0^{(1)}+\ldots+G_{n-1}^{(1)}, G_0^{(2)}, G_0^{(2)}+G_1^{(2)}, \ldots, G_0^{(2)}+\ldots+G_{n-1}^{(2)})$. Then,

$$\frac{1}{k}[M_0^{(k)}(Q_1^{(k)} - Q_0^{(k)}) + \ldots + M_{n-2}^{(k)}(Q_{n-1}^{(k)} - Q_{n-2}^{(k)})]$$
$$- \frac{1}{k}[Q_0^{(k)}(M_1^{(k)} - M_0^{(k)}) + \ldots + Q_{n-2}^{(k)}(M_{n-1}^{(k)} - M_{n-2}^{(k)})]$$

converges in distribution, as $k \to +\infty$, to the random variable

$$[G_0^{(1)}G_1^{(2)} + (G_0^{(1)} + G_1^{(1)})G_2^{(2)} + \ldots + (G_0^{(1)} + \ldots + G_{n-2}^{(1)})G_{n-1}^{(2)}]$$
$$- [G_0^{(2)}G_1^{(1)} + (G_0^{(2)} + G_1^{(2)})G_2^{(1)} + \ldots + (G_0^{(2)} + \ldots + G_{n-2}^{(2)})G_{n-1}^{(1)}].$$

c) Let us now prove that

$$\operatorname{Var}\left(N_0^{(k)} + \ldots + N_{n-1}^{(k)}\right) = \mathcal{O}(nk^2) \tag{6.1}$$

and

$$\operatorname{Var}\left(\tilde{N}_0^{(k)} + \ldots + \tilde{N}_{n-1}^{(k)}\right) = \mathcal{O}(nk^2). \tag{6.2}$$

The above result is quite evident in the classical case studied in [67] where the random variables X_i and Y_i are centered and uncorrelated but due to the temporal inhomogeneity of our model it is not at all the case and we need the following lemma.

LEMMA 17 *Uniformly in* $p \geq 0$,

$$\operatorname{Var}(N_p^{(k)}) = \mathcal{O}(k^2) \tag{6.3}$$

and
$$\text{Var } (\tilde{N}_p^{(k)}) = \mathcal{O}(k^2). \tag{6.4}$$

Proof:
Let us recall that
$$N_p^{(k)} = \sum_{l=1}^{k-1} Y_{pk+l+1} \Big(\sum_{i=1}^{l} X_{pk+i} \Big).$$

For every random variable X such that $\mathbb{E}(|X|) < +\infty$, we define $\bar{X} = X - \mathbb{E}(X)$. The random variable $N_p^{(k)}$ can be rewritten
$$N_p^{(k)} = \Sigma_1(k) + \Sigma_2(k) + \Sigma_3(k) - \Sigma_4(k)$$

where

$$\Sigma_1(k) = \sum_{l=1}^{k-1} \bar{Y}_{pk+l+1} \Big(\sum_{i=1}^{l} \bar{X}_{pk+i} \Big)$$

$$\Sigma_2(k) = \sum_{l=1}^{k-1} \mathbb{E}(Y_{pk+l+1}) \Big(\sum_{i=1}^{l} X_{pk+i} \Big)$$

$$\Sigma_3(k) = \sum_{l=1}^{k-1} Y_{pk+l+1} \Big(\sum_{i=1}^{l} \mathbb{E}(X_{pk+i}) \Big)$$

$$\Sigma_4(k) = \sum_{l=1}^{k-1} \mathbb{E}(Y_{pk+l+1}) \Big(\sum_{i=1}^{l} \mathbb{E}(X_{pk+i}) \Big).$$

Using that the random variables $\bar{X}_i$ and $\bar{Y}_j$ are independent when $i \neq j$ and centered, a direct computation gives
$$\text{Var } (\Sigma_1(k)) \leq Ck^2$$

where $C > 0$ is a constant independent of p. Furthermore, the random variable $\Sigma_2(k)$ can be rewritten as
$$\Sigma_2(k) = \sum_{l=1}^{k-1} X_{pk+l} \Big(\sum_{i=l}^{k-1} \mathbb{E}(Y_{pk+i+1}) \Big)$$

thus
$$\text{Var } (\Sigma_2(k)) = \sum_{l=1}^{k-1} \Big(\sum_{i=l}^{k-1} \mathbb{E}(Y_{pk+i+1}) \Big)^2 \text{Var } (X_{pk+l})$$
$$= 4 \sum_{l=1}^{k-1} \Big(\sum_{i=l}^{k-1} (f_2(T^{pk+i+1}x) - \frac{1}{6}) \Big)^2 \text{Var } (X_{pk+l})$$

Since the function f_2 belongs to the class $C_2(S)$ and $\int_E f_2\, d\mu = 1/6$, there exists a constant C which does not depend on p and k such that for every $l \geq 1$,

$$\Big|\sum_{i=1}^{k-l}\Big(f_2(T^{pk+l+i}x) - \frac{1}{6}\Big)\Big| \leq C\sqrt{k}.$$

Then,

$$\text{Var}\,(\Sigma_2(k)) \leq C\sum_{l=1}^{k-1}\Big|\sum_{j=1}^{k-l}\Big(f_2(T^{pk+l+i}x) - \frac{1}{6}\Big)\Big|^2 = \mathcal{O}(k^2).$$

The variance of the third sum $\Sigma_3(k)$ is estimated in the same manner using the fact that the function f_1 belongs to the class $C_2(S)$ and $\int_E f_1\, d\mu = 1/6$. So, we get

$$\text{Var}\,(\Sigma_3(k)) = \mathcal{O}(k^2).$$

The variance of the last sum is evidently zero.
The following inequality holds for square integrable random variables X and Y

$$\text{Var}\,(X+Y) \leq 2(\text{Var}\,(X) + \text{Var}\,(Y)).$$

So, by applying twice this inequality, we get

$$\text{Var}\,(N_p^{(k)}) \leq 4\sum_{i=1}^{4}\text{Var}\,(\Sigma_i(k)) = \mathcal{O}(k^2)$$

uniformly in p. The estimation (6.4) is obtained in the same manner by inverting the X's and the Y's.

From Lemma 17 we can now establish (6.1) as follows: the random variables $N_p^{(k)}$ and $N_{p'}^{(k)}$ are independent when $p \neq p'$, so we have

$$\text{Var}\,\Big(N_0^{(k)} + \ldots + N_{n-1}^{(k)}\Big) = \sum_{p=0}^{n-1}\text{Var}\,\Big(N_p^{(k)}\Big) \leq Cnk^2$$

using Lemma 17, so (6.1) is proved and (6.2) can be obtained by inverting the X's and the Y's.

Let $\varepsilon > 0$,

$$\mathbb{P}\Big(\frac{1}{2nk}\Big|\sum_{p=0}^{n-1}[N_p^{(k)} - \mathbb{E}(N_p^{(k)})] - \sum_{p=0}^{n-1}[\tilde{N}_p^{(k)} - \mathbb{E}(\tilde{N}_p^{(k)})]\Big| \geq \varepsilon\Big)$$

$$\leq \mathbb{P}\Big(\frac{1}{nk}\Big|\sum_{p=0}^{n-1}[N_p^{(k)} - \mathbb{E}(N_p^{(k)})]\Big| \geq \varepsilon\Big) + \mathbb{P}\Big(\frac{1}{nk}\Big|\sum_{p=0}^{n-1}[\tilde{N}_p^{(k)} - \mathbb{E}(\tilde{N}_p^{(k)})]\Big| \geq \varepsilon\Big)$$

$$\leq \frac{Cnk^2}{\varepsilon^2 n^2 k^2} = \frac{C}{\varepsilon^2 n}$$

Consequently,
$$\mathbb{P}\Big(\frac{1}{2nk}\Big|\sum_{p=0}^{n-1}[N_p^{(k)} - \mathbb{E}(N_p^{(k)})] - \sum_{p=0}^{n-1}[\tilde{N}_p^{(k)} - \mathbb{E}(\tilde{N}_p^{(k)})]\Big| \geq \varepsilon\Big) \to 0$$

as n goes to infinity, uniformly in k. Using that $|e^{i\theta} - 1| \leq |\theta| \wedge 2$, we obtain

$$\mathbb{E}\Big(\Big|\exp\Big\{i\frac{\theta}{2nk}\Big(\sum_{p=0}^{n-1}[N_p^{(k)} - \mathbb{E}(N_p^{(k)})] \tag{6.5}$$

$$- \sum_{p=0}^{n-1}[\tilde{N}_p^{(k)} - \mathbb{E}(\tilde{N}_p^{(k)})]\Big)\Big\} - 1\Big|\Big) \tag{6.6}$$

$$\leq \varepsilon + 2\mathbb{P}\Big(\frac{1}{2nk}\Big|\sum_{p=0}^{n-1}[N_p^{(k)} - \mathbb{E}(N_p^{(k)})] - \sum_{p=0}^{n-1}[\tilde{N}_p^{(k)} - \mathbb{E}(\tilde{N}_p^{(k)})]\Big| \geq \varepsilon\Big) \to 0$$

as n goes to infinity, uniformly in k.
Let us remark that for every $\theta \in \mathbb{R}$,

$$\Big|\mathbb{E}\Big(e^{i\theta \frac{T_{nk}}{nk}}\Big) - \mathbb{E}\Big(\exp\Big[i\frac{\theta}{2nk}\Big\{M_0^{(k)}(Q_1^{(k)} - Q_0^{(k)}) + \ldots$$
$$+ M_{n-2}^{(k)}(Q_{n-1}^{(k)} - Q_{n-2}^{(k)})$$
$$- Q_0^{(k)}(M_1^{(k)} - M_0^{(k)}) - \ldots - Q_{n-2}^{(k)}(M_{n-1}^{(k)} - M_{n-2}^{(k)})$$
$$+ \sum_{p=0}^{n-1}[\mathbb{E}(N_p^{(k)}) - \mathbb{E}(\tilde{N}_p^{(k)})]\Big\}\Big]\Big)\Big|$$

$$\leq \mathbb{E}\Big(\Big|\exp\Big\{i\frac{\theta}{2nk}\Big(\sum_{p=0}^{n-1}[N_p^{(k)} - \mathbb{E}(N_p^{(k)})] - \sum_{p=0}^{n-1}[\tilde{N}_p^{(k)} - \mathbb{E}(\tilde{N}_p^{(k)})]\Big)\Big\} - 1\Big|\Big)$$

converging to 0 as n goes to infinity, uniformly in k.
Let $\varepsilon > 0$, we can choose n_0 such that for every k,

$$\Big|\mathbb{E}\Big(e^{i\theta \frac{T_{n_0 k}}{n_0 k}}\Big) - \mathbb{E}\Big(\exp\Big[i\frac{\theta}{2n_0 k}\Big\{M_0^{(k)}(Q_1^{(k)} - Q_0^{(k)}) + \ldots$$
$$+ M_{n_0-2}^{(k)}(Q_{n_0-1}^{(k)} - Q_{n_0-2}^{(k)}) - Q_0^{(k)}(M_1^{(k)} - M_0^{(k)}) - \ldots$$
$$- Q_{n_0-2}^{(k)}(M_{n_0-1}^{(k)} - M_{n_0-2}^{(k)}) + \sum_{p=0}^{n_0-1}[\mathbb{E}(N_p^{(k)}) - \mathbb{E}(\tilde{N}_p^{(k)})]\Big\}\Big]\Big)\Big|$$
$$\leq \varepsilon.$$

The characteristic function of the random variable

$$[G_0^{(1)} G_1^{(2)} + (G_0^{(1)} + G_1^{(1)}) G_2^{(2)} + \ldots + (G_0^{(1)} + \ldots + G_{n-2}^{(1)}) G_{n-1}^{(2)}]$$

$$- [G_0^{(2)}G_1^{(1)} + (G_0^{(2)} + G_1^{(2)})G_2^{(1)} + \ldots + (G_0^{(2)} + \ldots + G_{n-2}^{(2)})G_{n-1}^{(1)}]$$

can be written as

$$I_n(\theta) = \frac{1}{(2\pi\sqrt{a_{11}a_{22}})^n} \int_{\mathbb{R}^{2n}} e^{g_n(x,y)} \, dx_1 \ldots dx_n dy_1 \ldots dy_n$$

where g_n is the function defined for every $x=(x_1,\ldots,x_n), y=(y_1,\ldots,y_n)\in \mathbb{R}^n$ by

$$\begin{aligned}g_n(x,y) &= \frac{i\theta}{2n}[x_1y_2 + (x_1+x_2)y_3 + \ldots + (x_1+\ldots+x_{n-1})y_n \\ &\quad - y_1x_2 - (y_1+y_2)x_3 - \ldots - (y_1+\ldots+y_{n-1})x_n] \\ &\quad - \frac{1}{2a_{11}}(x_1^2 + \ldots + x_n^2) - \frac{1}{2a_{22}}(y_1^2 + \ldots + y_n^2).\end{aligned}$$

Now from Lemma 37 in [67], we know that

$$\lim_{n\to +\infty} \frac{1}{(2\pi)^n} \int_{\mathbb{R}^{2n}} e^{g_n(x\sqrt{a_{11}},y\sqrt{a_{22}})} \, dxdy = \frac{1}{\cosh(\frac{\theta\sqrt{a_{11}a_{22}}}{2})}$$

uniformly in θ on a compact set of $\mathbb{R}$. So we also impose n_0 to be large enough so as to have

$$\left| I_{n_0}(\theta) - \frac{1}{\cosh(\frac{\theta\sqrt{a_{11}a_{22}}}{2})} \right| \leq \varepsilon$$

uniformly in θ on a compact set.

Under the assumption (A), it is possible to control the drift coming from the non-centered random variables $N_p^{(k)}$ and $\tilde{N}_p^{(k)}$.

LEMMA 18 *Under hypothesis (A),*

$$\frac{1}{n_0 k} \sum_{p=0}^{n_0-1} [\mathbb{E}(N_p^{(k)}) - \mathbb{E}(\tilde{N}_p^{(k)})] = o(k). \tag{6.7}$$

Proof:
From the definition of the law of the random variables X_i and Y_i, the above sum can be explicitly calculated and bounded as follows:

$$\begin{aligned}&\left| \frac{1}{n_0 k} \sum_{p=0}^{n_0-1} [\mathbb{E}(N_p^{(k)}) - \mathbb{E}(\tilde{N}_p^{(k)})] \right| \\ &\leq \frac{1}{k} \sup_{x\in E} \Big| \sum_{l=1}^{k} \sum_{i=1}^{l} [(2f_2(T^l x) - \frac{1}{3})(2f_1(T^i x) - \frac{1}{3}) \\ &\quad - (2f_1(T^l x) - \frac{1}{3})(2f_2(T^i x) - \frac{1}{3})] \Big|\end{aligned}$$

Using Abel's summation by parts,

$$\sum_{l=1}^{k}\sum_{i=1}^{l}(2f_1(T^l x) - \frac{1}{3})(2f_2(T^i x) - \frac{1}{3})$$

$$= \sum_{l=1}^{k}(2f_1(T^l x) - \frac{1}{3})\sum_{l=1}^{k}(2f_2(T^l x) - \frac{1}{3})$$

$$- \sum_{l=2}^{k}(2f_2(T^l x) - \frac{1}{3})\sum_{i=1}^{l-1}(2f_1(T^i x) - \frac{1}{3})$$

Then,

$$\sum_{l=1}^{k}\sum_{i=1}^{l}[(2f_2(T^l x) - \frac{1}{3})(2f_1(T^i x) - \frac{1}{3})$$

$$- (2f_1(T^l x) - \frac{1}{3})(2f_2(T^i x) - \frac{1}{3})]$$

$$= 2\sum_{l=2}^{k}(2f_2(T^l x) - \frac{1}{3})\sum_{i=1}^{l-1}(2f_1(T^i x) - \frac{1}{3})$$

$$- \sum_{l=1}^{k}(2f_1(T^l x) - \frac{1}{3})\sum_{l=1}^{k}(2f_2(T^l x) - \frac{1}{3})$$

$$+ \sum_{l=1}^{k}(2f_1(T^l x) - \frac{1}{3})(2f_2(T^l x) - \frac{1}{3})$$

In the right hand side the second sum divided by k goes to 0 uniformly in x since f_1 and f_2 belong to the class $C_2(S)$ and $\int_E f_i\, d\mu = 1/6, i = 1,2$. The third one divided by k also goes to 0 uniformly in $x \in E$ since $\int_E f_1 f_2\, d\mu = 1/36$. So, as soon as (A) is satisfied, (6.7) holds.

From item b) and the previous lemma, the characteristic function of the random variable

$$\frac{1}{2n_0 k}\{M_0^{(k)}(Q_1^{(k)} - Q_0^{(k)}) + \ldots + M_{n_0-2}^{(k)}(Q_{n_0-1}^{(k)} - Q_{n_0-2}^{(k)})$$

$$- Q_0^{(k)}(M_1^{(k)} - M_0^{(k)}) - \ldots - Q_{n_0-2}^{(k)}(M_{n_0-1}^{(k)} - M_{n_0-2}^{(k)})$$

$$+ \sum_{p=0}^{n_0-1}[\mathbb{E}(N_p^{(k)}) - \mathbb{E}(\tilde{N}_p^{(k)})]\}$$

converges, as k goes to infinity, to $I_{n_0}(\theta)$. Hence there exists k_0 such that for every $k \geq k_0$,

$$\left|\mathbb{E}\left(e^{i\theta \frac{T_{n_0 k}}{n_0 k}}\right) - \frac{1}{\cosh(\frac{\theta \sqrt{a_{11} a_{22}}}{2})}\right| \leq 3\varepsilon. \tag{6.8}$$

d) Let $p = n_0 k + q, 0 \leq q < n_0, k \geq k_0$ (remark that $p \geq n_0 k_0$) and define

$$\begin{aligned}V_{n_0}^{(k)} &= T_{n_0 k + q} - T_{n_0 k} \\ &= \frac{1}{2}\{(X_1 + \ldots + X_{n_0 k})Y_{n_0 k + 1} + \ldots + (X_1 + \ldots + X_{n_0 k + q - 1})Y_{n_0 k + q} \\ &\quad - (Y_1 + \ldots + Y_{n_0 k})X_{n_0 k + 1} - \ldots - (Y_1 + \ldots + Y_{n_0 k + q - 1})X_{n_0 k + q}\}\end{aligned}$$

From Tchebychev's inequality, for every $\varepsilon > 0$,

$$\mathbb{P}\Big(\Big|\frac{V_{n_0}^{(k)} - \mathbb{E}(V_{n_0}^{(k)})}{n_0 k}\Big| \geq \varepsilon\Big) \leq \frac{\mathrm{Var}(V_{n_0}^{(k)})}{\varepsilon^2 n_0^2 k^2}.$$

Now, using the same techniques than the ones developed in the proof of Lemma 17, it can be proved that

$$\mathrm{Var}(V_{n_0}^{(k)}) = \mathcal{O}(n_0^2 k).$$

Assumption (A) implies (see the proof of Lemma 18) that

$$\lim_{k \to +\infty} \frac{\mathbb{E}(V_{n_0}^{(k)})}{n_0 k} = 0.$$

Then, we deduce that

$$\lim_{k \to +\infty} \mathbb{P}\Big(\Big|\frac{V_{n_0}^{(k)}}{n_0 k}\Big| \geq \varepsilon\Big) = 0.$$

Straightforward calculations lead to

$$\Big|\mathbb{E}\big(e^{i\theta \frac{T_p}{p}}\big) - \mathbb{E}\big(e^{i\theta \frac{T_{n_0 k}}{n_0 k}}\big)\Big|$$

$$\leq \Big|\mathbb{E}\big(e^{i\theta \frac{n_0 k}{n_0 k + q}\frac{T_{n_0 k}}{n_0 k} + \frac{V_{n_0}^{(k)}}{n_0 k + q}}\big) - \mathbb{E}\big(e^{i\theta \frac{T_{n_0 k}}{n_0 k}}\big)\Big|$$

$$\leq \Big|\mathbb{E}\big(e^{i\theta \frac{n_0 k}{n_0 k + q}\frac{T_{n_0 k}}{n_0 k} + \frac{V_{n_0}^{(k)}}{n_0 k + q}}\big) - \mathbb{E}\big(e^{i\theta \frac{T_{n_0 k}}{n_0 k} + \frac{V_{n_0}^{(k)}}{n_0 k + q}}\big)\Big|$$

$$+ \Big|\mathbb{E}\big(e^{i\theta \frac{T_{n_0 k}}{n_0 k} + \frac{V_{n_0}^{(k)}}{n_0 k + q}}\big) - \mathbb{E}\big(e^{i\theta \frac{T_{n_0 k}}{n_0 k}}\big)\Big|$$

$$\leq \mathbb{E}\Big(\Big|e^{i\theta \frac{q}{n_0 k + q}\frac{T_{n_0 k}}{n_0 k}} - 1\Big|\Big) + \mathbb{E}\Big(\Big|e^{i\theta \frac{V_{n_0}^{(k)}}{n_0 k + q}} - 1\Big|\Big).$$

Now, we have (since $|e^{i\theta} - 1| \leq 2 \wedge |\theta|$) for k large enough

$$\mathbb{E}\Big(\Big|e^{i\theta \frac{V_{n_0}^{(k)}}{n_0 k + q}} - 1\Big|\Big) \leq \varepsilon + 2\mathbb{P}\Big(\Big|\frac{\theta V_{n_0}^{(k)}}{n_0 k + q}\Big| \geq \varepsilon\Big) \leq 3\varepsilon.$$

From (6.8), for k large enough,

$$\left|\mathbb{E}\left(e^{i\theta \frac{q}{n_0 k+q} \frac{T_{n_0 k}}{n_0 k}}\right) - 1\right| \leq 3\varepsilon. \tag{6.9}$$

On a compact set of $\mathbb{R}$ (for θ), for p large enough,

$$\begin{aligned}
\left|\mathbb{E}\left(e^{i\theta \frac{T_p}{p}}\right) - \frac{1}{\cosh(\frac{\theta \sqrt{a_{11} a_{22}}}{2})}\right| &\leq \left|\mathbb{E}\left(e^{i\theta \frac{T_p}{p}}\right) - \mathbb{E}\left(e^{i\theta \frac{T_{n_0 k}}{n_0 k}}\right)\right| \\
&+ \left|\mathbb{E}\left(e^{i\theta \frac{T_{n_0 k}}{n_0 k}}\right) - \frac{1}{\cosh(\frac{\theta \sqrt{a_{11} a_{22}}}{2})}\right| \\
&\leq 9\varepsilon,
\end{aligned}$$

so Theorem 6.2 is proved.

Chapter 7

DYNAMIC QUANTUM BERNOULLI RANDOM WALKS

1. INTRODUCTION

Welcome to the world of quantum probability! Quantum probability is a very active research area motivated by applications in physics, information theory and biology. In this chapter we touch lightly on this topic. The recent books of Biane [13], Franz-Schott [59], Meyer [135] and Parthasaraty [148] are interesting introductions to this field. Appendix E provides some basic definitions concerning Hilbert spaces, representation theory, *-algebras and von Neumann algebras which may be helpful for some readers. The organization of this chapter is as follows: Section 2 introduces some notions of quantum probability. In Section 3 we recall some facts on quantum Bernoulli random walks (the simplest model in quantum probability). These random walks can be realized on a space called the dual of $SU(2)$ which is just the set of natural numbers equipped with some non-commutative law. Section 4 provides more insight into this space while Section 5 explains the above mentioned realization. Section 6 describes the model of dynamic random walks on the dual of $SU(2)$ which we will use, focuses on some limit theorems (local limit theorem, central limit theorem, law of large numbers), provides a large deviation principle and gives a characterization of a large class of transient dynamic random walks.

2. QUANTUM PROBABILISTIC NOTIONS

This introductory material is based on [166].
A quantum probability space is a pair (A, Φ) consisting of a *-algebra A and a state (i.e. a normalized positive linear functional) Φ on A.
Let $(\Omega, \mathcal{F}, \mathbb{P})$ be a classical probability space. A *-algebra A of complex-

valued functions on Ω which is a linear subspace of $L^1(\Omega, \mathcal{F}, \mathbb{P})$ is a quantum probability space if we take Φ to be the restriction $\Phi_\mathbb{P}$ to A of the linear functional $f \to \int_\Omega f \, d\mathbb{P}$ on $L^1(\Omega, \mathcal{F}, \mathbb{P})$. One possible choice is $A = L^\infty(\Omega, \mathcal{F}, \mathbb{P})$. Motivated by this example, a quantum probability space is, sometimes, defined to be a pair (N, Φ) consisting of a von Neumann algebra N and a normal state Φ on N. This has the advantage that commutative quantum probability spaces, i.e. spaces with commutative N, reduce to spaces of the form $L^\infty(\Omega, \mathcal{F}, \mathbb{P}), \Phi_\mathbb{P})$ that is to an algebra $\mathcal{F}$ of events and a classical probability measure $\mathbb{P}$.

DEFINITION 7.1 *A quantum random variable over a quantum probability space (A, Φ) on a *-algebra B is a *-homomorphism*

$$j : B \to A.$$

A classical random variable $X : \Omega \to E$ over a probability space $(\Omega, \mathcal{F}, \mathbb{P})$, taking values in a measurable space (E, μ), gives rise to a quantum random variable j_X if we take for B an algebra of complex-valued functions contained in $L^1(E, \mu, \mathbb{P}_X)$ with $\mathbb{P}_X$ the distribution of X and define $j_X(f)$ to be the function $f \circ X$ on Ω. On can take any A that contains $j_X(B)$. In particular $B = L^\infty(E, \mu)$ and $A = L^\infty(\Omega, \mathcal{F}, \mathbb{P})$ give a quantum random variable.

DEFINITION 7.2 *The distribution of a quantum random variable j is the state $\phi \circ j$ on B.*

DEFINITION 7.3 *A quantum stochastic process over a quantum probability space (A, Φ) on a *-algebra B, indexed by the set T, is a family $(j_t)_{t \in T}$ of quantum random variables*

$$j_t : B \to A$$

The states $\phi_t = \Phi \circ j_t$ are called the 1-dimensional distributions of (j_t).

3. QUANTUM BERNOULLI RANDOM WALKS

The presentation of this introductory material is based [11].
Let $M_2(\mathbb{C})$ be the set of 2×2 matrices with complex coefficients. The set of 2×2 self-adjoint matrices forms a four dimensional real vector subspace of $M_2(\mathbb{C})$. A convenient basis is given by the following matrices

$$I = \begin{pmatrix} 1 & 0 \\ 0 & 1 \end{pmatrix} \quad \sigma_x = \begin{pmatrix} 1 & 0 \\ 0 & -1 \end{pmatrix} \quad \sigma_y = \begin{pmatrix} 0 & 1 \\ 1 & 0 \end{pmatrix} \quad \sigma_z = \begin{pmatrix} 0 & -i \\ i & 0 \end{pmatrix}$$

$\sigma_x, \sigma_y, \sigma_z$ are the traditional Pauli matrices, they satisfy the commutation relations: $[\sigma_x, \sigma_y] = 2i\sigma_z$, and those obtained by cyclic permutations of σ_x, σ_y, σ_z. Every unitary matrix U gives rise to a *-automorphism of $M_2(\mathbb{C})$

$$\tau_U : M \to UMU^*.$$

These automorphisms leave the hermitian product $<M, N> = \frac{1}{2}\text{tr}(MN^*)$ invariant and $M_2(\mathbb{C})$ decomposes into two invariant subspaces for this action: $\mathbb{C}I$ and the subspace generated by $\sigma_x, \sigma_y, \sigma_z$. τ_U acts by rotation on $\mathbb{R}\sigma_x + \mathbb{R}\sigma_y + \mathbb{R}\sigma_z$.

A state on $M_2(\mathbb{C})$ is given by a density matrix which we will suppose to be of the form

$$\begin{pmatrix} p & 0 \\ 0 & q \end{pmatrix}$$

with $0 < p \leq 1$, $p + q = 1$, since this can be achieved by conjugation by a unitary matrix. We denote by ρ_p the state given by this density matrix. Given a self adjoint element of $M_2(\mathbb{C})$, we can compute its law in the state ρ_p: let $A = \lambda\sigma_x + \mu\sigma_y + \nu\sigma_z$, and $\xi = (\lambda^2 + \mu^2 + \nu^2)^{1/2}$, the spectrum of A is $\{-\xi, +\xi\}$, its expectation in the state ρ_p is $\lambda(p-q)$ and its law is given by

$$\mathbb{P}(A = \xi) = \frac{1}{2}\left[\frac{\lambda(p-q)}{\xi} + 1\right] \quad (7.1)$$

$$\mathbb{P}(A = -\xi) = \frac{1}{2}\left[\frac{\lambda(q-p)}{\xi} + 1\right] \quad (7.2)$$

In particular, in the state ρ_p, σ_y and σ_z are symmetric Bernoulli random variables, whereas σ_x takes 1 with probability p and -1 with probability q.

Let $M_1, M_2, \ldots, M_n, \ldots$, be infinitely many copies of $M_2(\mathbb{C})$. For each $p \in\,]0,1]$, we consider the algebra

$$\mathcal{M}_p = M_1 \otimes \ldots \otimes M_k \otimes \ldots$$

where the product is taken with respect to the product state

$$w_p = \rho_p \otimes \ldots \otimes \rho_p \otimes \ldots$$

Consider the elements x_k, y_k, z_k of $\mathcal{M}_p$ given by:

$$x_k = I \otimes \ldots \otimes I \otimes \sigma_x \otimes I \otimes I \ldots \quad (7.3)$$
$$y_k = I \otimes \ldots \otimes I \otimes \sigma_y \otimes I \otimes I \ldots \quad (7.4)$$
$$z_k = I \otimes \ldots \otimes I \otimes \sigma_z \otimes I \otimes I \ldots \quad (7.5)$$

where each σ appears at the kth place. Define for $k \geq 1$,

$$X_k = \sum_{i=1}^{k} x_i, \quad Y_k = \sum_{i=1}^{k} y_i, \quad Z_k = \sum_{i=1}^{k} z_i$$

and $X_0 = Y_0 = Z_0 = 0$.
The triple $(X_n, Y_n, Z_n)_{n \in \mathbb{N}}$ is called a quantum random walk.

4. THE DUAL OF $SU(2)$

The presentation in this section follows [67].
$SU(2)$ is the group of matrices

$$g = \begin{pmatrix} a & b \\ -\bar{b} & \bar{a} \end{pmatrix}$$

where a and b are complex numbers such that $|a|^2 + |b|^2 = 1$. It is a compact Lie group. For each $x \in \mathbb{N}$, let H_x be the vector space on $\mathbb{C}$ of polynomials of degree less than or equal to x. If g is as indicated above and if $p \in H_x$, let

$$[\pi_x(g)p](z) = (bz + \bar{a})^x p\left(\frac{az - \bar{b}}{bz + \bar{a}}\right), \quad z \in \mathbb{C}.$$

In this way we define an irreducible, continuous representation of dimension $x+1$ from $SU(2)$ into H_x. It is known that as x goes trough $\mathbb{N}$ then the π_x produce (up to equivalence) the full list of unitary irreducible representations of $SU(2)$. Therefore $\mathbb{N}$ can be identified with the dual group of $SU(2)$.
The character of π_x is given by the formula

$$\xi_x(g) = \frac{1}{x+1} \text{tr}(\pi_x(g)) = \frac{\sin[(x+1)\theta]}{(x+1)\sin\theta}$$

where $e^{i\theta}$, $e^{-i\theta}$ are the eigenvalues of the matrix $g \in SU(2)$. ξ_x obeys $\xi_x(g_1 g_2) = \xi_x(g_1)\xi_x(g_2)$ for all $g_1 \in SU(2)$, $g_2 \in SU(2)$.
Notation: $\phi(\theta)$ will stand for $\phi(M)$ where

$$M = \begin{pmatrix} e^{i\theta} & 0 \\ 0 & e^{-i\theta} \end{pmatrix}$$

For the characters we get the orthogonality relations with respect to the measure

$$d\lambda(\theta) = \frac{2}{\pi} \sin^2\theta \, d\theta$$

More precisely, for $x \in \mathbb{N}$ and $y \in \mathbb{N}$,

$$\int_0^\pi \xi_x(\theta)\xi_y(\theta)\,d\lambda(\theta) = \begin{cases} 0 & \text{if } x \neq y \\ (x+1)^{-2} & \text{if } x = y \end{cases}$$

Finally, we recall Clebsch-Gordan's formula for $SU(2)$.
If $x \leq y$ are in $\mathbb{N}$, then the tensorial product of the representations π_x and π_y splits into the direct sum of a finite number of irreducible representations as follows:

$$\pi_x \otimes \pi_y = \pi_{y-x} \oplus \pi_{y-x+2} \oplus \cdots \oplus \pi_{y+x-2} \oplus \pi_{y+x}$$

This formula leads to the character multiplication formula:

$$\xi_x \xi_y = \frac{|x-y|+1}{(x+1)(y+1)}\xi_{|x-y|} + \frac{|x-y|+3}{(x+1)(y+1)}\xi_{|x-y|+2} + \cdots$$
$$+ \frac{x+y+1}{(x+1)(y+1)}\xi_{x+y}$$

where $x \in \mathbb{N}$, $y \in \mathbb{N}$ and the integers run from $|x-y|$ to $x+y$ by jumps of length two.
Now let $\mathcal{P}(\mathbb{N})$ be the set of probability measures $\mu = \sum_{x \in \mathbb{N}} a_x \delta_x$ on $\mathbb{N}$, where δ_x is the Dirac measure at point x and a_x are coefficients which are non negative and $\sum_{x \geq 0} a_x = 1$. We define a generalized convolution denoted $\star$ as follows:

$$\delta_x \star \delta_y = \frac{|x-y|+1}{(x+1)(y+1)}\delta_{|x-y|} + \frac{|x-y|+3}{(x+1)(y+1)}\delta_{|x-y|+2} + \cdots$$
$$+ \frac{x+y+1}{(x+1)(y+1)}\delta_{x+y}$$

and more generally, if μ, ν are in $\mathcal{P}(\mathbb{N})$:

$$\mu \star \nu = \left(\sum_{x \geq 0} a_x \delta_x\right) \star \left(\sum_{y \geq 0} b_y \delta_y\right) = \sum_{x,y \geq 0} a_x b_y \delta_x \star \delta_y$$

and we denote by μ^n the probability measure $\mu \star \mu \star \cdots \star \mu$ (n times).
The (generalized) Fourier transform of $\mu = \sum_{x \in \mathbb{N}} a_x \delta_x \in \mathcal{P}(\mathbb{N})$ is the function $\hat{\mu}$ defined on $[0, \pi]$ by

$$\hat{\mu}(\theta) = \sum_{x \geq 0} a_x \xi_x(\theta) = \sum_{x \geq 0} \frac{a_x \sin[(x+1)\theta]}{(x+1)\sin\theta}.$$

The coefficient a_x of the measure μ can be obtained from $\hat{\mu}$ by the following formula

$$\begin{aligned} a_x &= (x+1)^2 \int_0^\pi \hat{\mu}(\theta) \xi_x(\theta)\, d\lambda(\theta) \\ &= \frac{2(x+1)}{\pi} \int_0^\pi \hat{\mu}(\theta) \sin[(x+1)\theta] \sin\theta\, d\theta. \end{aligned} \quad (7.6)$$

In particular $\hat{\delta}_x = \xi_x$ and $\widehat{(\delta_x \star \delta_y)} = \hat{\delta}_x \hat{\delta}_y$. More generally,

$$\widehat{(\mu \star \nu)} = \hat{\mu}\, \hat{\nu}.$$

The (generalized) Laplace transform of $\mu = \sum_{x \in \mathbb{N}} a_x \delta_x \in \mathcal{P}(\mathbb{N})$ is the function $\tilde{\mu}$ defined on $\mathbb{R}^+$ by

$$\tilde{\mu}(t) = \sum_{x \geq 0} \frac{a_x \sinh[(x+1)t]}{(x+1)\sinh(t)}$$

For every $x \geq 0$, we denote by ψ_x the function

$$t \to \frac{\sinh[(x+1)t]}{(x+1)\sinh(t)}.$$

The Laplace transform of the Dirac measure at $x \in \mathbb{N}$ is then $\tilde{\delta}_x = \psi_x$. The Clebsch-Gordan's formula can be extended as follows:

$$\begin{aligned} \psi_x \psi_y &= \\ &\frac{|x-y|+1}{(x+1)(y+1)} \psi_{|x-y|} + \frac{|x-y|+3}{(x+1)(y+1)} \psi_{|x-y|+2} + \cdots \\ &+ \frac{x+y+1}{(x+1)(y+1)} \psi_{x+y} \end{aligned} \quad (7.7)$$

From the definition of the generalized convolution, relation (7.7) and the fact that $\tilde{\delta}_x = \psi_x$, we obtain the formula

$$\widetilde{(\mu \star \nu)} = \tilde{\mu}\, \tilde{\nu}. \quad (7.8)$$

Let $\mu \in \mathcal{P}(\mathbb{N})$. For each $x \in \mathbb{N}$ and for each subset A of $\mathbb{N}$, we consider the transition kernel from $\mathbb{N}$ to $\mathbb{N}$:

$$P(x, A) = \delta_x \star \mu(A).$$

Let $(\Omega = \mathbb{N}^{\mathbb{N}}, (X_n)_{n \geq 0}, (P_x)_{x \in \mathbb{N}})$ be the canonical Markov chain associated with the kernel P. This chain will be called the random walk of law μ on $\mathbb{N}$. In other words: the probability $P(x, y)$ to be in the state y at time $n+1$ when departing from state x at time n, is equal to the coefficient on δ_y of the probability measure $\delta_x \star \mu$.

5. QUANTUM BERNOULLI RANDOM WALKS AS RANDOM WALKS ON THE DUAL OF $SU(2)$

P. Biane [11] has proved that QBRW can be related to a random walk of the dual of $SU(2)$. In this section we outline his observation.

Let $\mathcal{A}$ be the von Neumann algebra of $SU(2)$. This is the von Neumann algebra of operators on $L^2(SU(2))$ generated by the left translation operators λ_g: $\lambda_g(f(h)) = f(g^{-1}h)$ (see [40]). It is thought of as an algebra of functions on a non-commutative lattice: the dual of $SU(2)$.

Let $\frac{1}{\sqrt{2}}X$, $\frac{1}{\sqrt{2}}Y$, $\frac{1}{\sqrt{2}}Z$ be an orthonormal basis of the Lie algebra of right invariant vector fields on $SU(2)$ (with respect to the Killing inner product). Then iX, iY, iZ induce self-adjoint operators on $L^2(SU(2))$ which are affiliated to $\mathcal{A}$. The spectrum of these operators is $\mathbb{Z}$, so that one can see them as coordinate functions on the dual of $SU(2)$. The algebra $\mathcal{A}$ is endowed with a structure of cocommutative bialgebra by the morphism of algebras $m : \mathcal{A} \to \mathcal{A}$ determined by $\lambda_g \to \lambda_g \otimes \lambda_g$. In order to define a non commutative random walk on $\mathcal{A}$, we let θ be an irreducible 2-dimensional representation of $SU(2)$ (see Section 3). We get a quantum Markov chain on $\mathcal{A}$ whose generator Q is given by

$$Q_p(\lambda_g) = \rho_p(\theta(g))\lambda_g.$$

Let $\mathcal{N}$ be a von Neumann algebra and τ a morphism: $\mathcal{A} \to \mathcal{A} \otimes \mathcal{N}$. Let $\mathcal{W} = \mathcal{A} \otimes \mathcal{N} \ldots \otimes \mathcal{N} \otimes \ldots$, the tensor product being taken with respect to the product space $w = \rho \otimes \ldots \otimes \rho \otimes \ldots$.

We define $T : \mathcal{W} \to \mathcal{W}$ by $T = \tau \otimes s$ where $s : \mathcal{N}^{[1,\infty[} \to \mathcal{N}^{[2,\infty[}$ is the right shift. We construct morphisms $j_n : \mathcal{A} \to \mathcal{W}$ by putting $j_n = T^n \circ i$ where $i : \mathcal{A} \to \mathcal{W}$ is the canonical injection.

The family of morphisms (j_n) forms a non-commutative process in the sense of Accardi, Frigerio, Lewis [1] and the triple $(j_n(iX), j_n(iY), j_n(iZ))$ forms a quantum Bernoulli random walk.

6. DYNAMIC RANDOM WALKS ON THE DUAL OF $SU(2)$

6.1 DEFINITION

In Section 4 we have considered the set $\mathcal{P}(\mathbb{N})$ of probability measures $\mu = \sum_{x \in \mathbb{N}} a_x \delta_x$ on $\mathbb{N}$, where δ_x is the Dirac measure at point x and a_x are coefficients which are non negative and $\sum_{x \in \mathbb{N}} a_x = 1$.

We consider now a sequence of probability measures $(\mu_i)_{i \geq 1}$ where $\mu_i = \sum_{x \in \mathbb{N}} a_x^{(i)} \delta_x$ on $\mathbb{N}$, where $a_x^{(i)}$ are coefficients which depend on i and x, are non negative and for every $i \geq 1$, $\sum_{x \in \mathbb{N}} a_x^{(i)} = 1$.

For every $i \geq 1$, for each $x \in \mathbb{N}$ and for each subset A of $\mathbb{N}$, we can define the transition kernel from $\mathbb{N}$ to $\mathbb{N}$:

$$P_i(x, A) = \delta_x \star \mu_i(A).$$

The dynamic random walk on the dual of $SU(2)$ denoted by $(S_n)_{n \geq 0}$ is defined as the (inhomogeneous) Markov chain with state space $\mathbb{N}$ and transition kernel at time n given by P_n. The probability to be in a subset A of $\mathbb{N}$ at time n when departing from state x at time 0 is then given by

$$P^{(n)}(x, A) = \delta_x \star \mu_1 \star \ldots \star \mu_n(A) = \delta_x \star \mu^{(n)}(A)$$

with the notation $\mu^{(n)} = \mu_1 \star \ldots \star \mu_n$.

6.2 LIMIT THEOREMS FOR DYNAMIC RANDOM WALKS ON THE DUAL OF $SU(2)$

6.2.1 A LOCAL LIMIT THEOREM

Assume that:

(H_1): $\forall i \geq 1$, μ_i is aperiodic. In particular, for every $r \in]0, \pi[$, there exists $\delta_i = \delta_i(r)$ such that

$$|\hat{\mu}_i(\theta)| \leq 1 - \delta_i, \quad \forall \theta \in [r, \pi].$$

We will assume that

$$n^{3/2} \prod_{i=1}^n (1 - \delta_i) = o(1),$$

(H_2): $\forall i \geq 1$,

$$\sum_{x \in \mathbb{N}} a_x^{(i)} (x+1)^2 < +\infty,$$

(H_3): There exists a sequence of non negative reals $(A_x)_{x \in \mathbb{N}}$ such that $\sum_{x \in \mathbb{N}} A_x = 1$, $\sum_{x \in \mathbb{N}} A_x(x^2 + 2x) < \infty$ and

$$\lim_{n \to +\infty} \sum_{x \in \mathbb{N}} \left| \frac{1}{n} \sum_{i=1}^n a_x^{(i)} - A_x \right| (x+1)^2 = 0,$$

(H_4):

$$\sup_{i \geq 1} \left| \sum_{x \in \mathbb{N}} (a_x^{(i)} - A_x)(x+1)^2 \right| < +\infty.$$

Remark: If the union of the supports of the measures μ_i is reduced to a finite set G, assumptions $(H_i), i = 2, 3, 4$ can be replaced by the

following one:
(H_5): There exists a sequence of reals $(A_x)_{x\in\mathbb{N}}$ such that

$$\lim_{n\to+\infty} \frac{1}{n}\sum_{i=1}^n a_x^{(i)} = A_x.$$

The support of the measure $\mu = \sum_{x\in\mathbb{N}} A_x \delta_x$ is evidently a subset of G.

THEOREM 7.1 *Under the assumptions* (H_1), (H_2), (H_3) *and* (H_4),

$$\lim_{n\to+\infty} \sup_{x\in\mathbb{N}} \left| 2\sqrt{\pi} n^{3/2} P^{(n)}(x,0) - \frac{2n}{(x+1)\sqrt{C}} e^{-\frac{(x+1)^2+1}{4Cn}} \times \sinh\left(\frac{x+1}{2Cn}\right) \right|$$
$$= 0$$

where

$$C = \frac{1}{6}\sum_{x\geq 0} A_x(x^2 + 2x).$$

In particular: as $n \to \infty$,

$$P^{(n)}(x,0) \sim (2\sqrt{\pi})^{-1} C^{-3/2} n^{-3/2}.$$

Proof:
From formula (7.6),

$$2\sqrt{\pi} n^{3/2} P^{(n)}(x,0) = \frac{4n^{3/2}}{\sqrt{\pi}(x+1)} \int_0^\pi \widehat{\mu^{(n)}}(\theta) \sin\left((x+1)\theta\right) \sin(\theta)\, d\theta.$$

Using the change of variables: $\theta = \frac{\alpha}{\sqrt{n}}$, we get

$$2\sqrt{\pi} n^{3/2} P^{(n)}(x,0) =$$
$$\frac{4n}{\sqrt{\pi}(x+1)} \int_0^{\pi\sqrt{n}} \widehat{\mu^{(n)}}\left(\frac{\alpha}{\sqrt{n}}\right) \sin\left((x+1)\frac{\alpha}{\sqrt{n}}\right) \sin\left(\frac{\alpha}{\sqrt{n}}\right) d\alpha$$
$$= I_0(n) + I_1(n,A) + I_2(n,A) + I_3(n,A,r) + I_4(nA,r)$$

where for some $A > 0$ and $0 < r < \pi$,

$$I_0(n) = \frac{4n}{\sqrt{\pi}(x+1)} \int_0^{+\infty} e^{-C\alpha^2} \sin\left((x+1)\frac{\alpha}{\sqrt{n}}\right) \sin\left(\frac{\alpha}{\sqrt{n}}\right) d\alpha$$

$$I_1(n,A) = \frac{4n}{\sqrt{\pi}(x+1)} \int_0^A \left[\widehat{\mu^{(n)}}\left(\frac{\alpha}{\sqrt{n}}\right) - e^{-C\alpha^2}\right] \sin\left((x+1)\frac{\alpha}{\sqrt{n}}\right)$$
$$\times \sin\left(\frac{\alpha}{\sqrt{n}}\right) d\alpha$$

$$I_2(n,A) = -\frac{4n}{\sqrt{\pi}(x+1)} \int_A^{+\infty} e^{-C\alpha^2} \sin\left((x+1)\frac{\alpha}{\sqrt{n}}\right) \sin\left(\frac{\alpha}{\sqrt{n}}\right) d\alpha$$

$$I_3(n,A,r) = \frac{4n}{\sqrt{\pi}(x+1)} \int_A^{r\sqrt{n}} \widehat{\mu^{(n)}}\left(\frac{\alpha}{\sqrt{n}}\right) \sin\left((x+1)\frac{\alpha}{\sqrt{n}}\right) \sin\left(\frac{\alpha}{\sqrt{n}}\right) d\alpha$$

$$I_4(n,A,r) = \frac{4n}{\sqrt{\pi}(x+1)} \int_{r\sqrt{n}}^{\pi\sqrt{n}} \widehat{\mu^{(n)}}\left(\frac{\alpha}{\sqrt{n}}\right) \sin\left((x+1)\frac{\alpha}{\sqrt{n}}\right) \sin\left(\frac{\alpha}{\sqrt{n}}\right) d\alpha.$$

Estimation of $I_0(n)$:
From the well-known equality,

$$\frac{1}{\sqrt{2\pi}} \int_\mathbb{R} \cos(tx) e^{-x^2/2} dx = e^{-t^2/2} \qquad (7.9)$$

we deduce that

$$\int_0^{+\infty} e^{-C\alpha^2} \sin\left((x+1)\frac{\alpha}{\sqrt{n}}\right) \sin\left(\frac{\alpha}{\sqrt{n}}\right) d\alpha$$

$$= \frac{1}{2} \int_0^{+\infty} e^{-C\alpha^2} \cos\left(\frac{x\alpha}{\sqrt{n}}\right) d\alpha - \frac{1}{2} \int_0^{+\infty} e^{-C\alpha^2} \cos\left(\frac{(x+2)\alpha}{\sqrt{n}}\right) d\alpha$$

$$= \frac{1}{2}\sqrt{\frac{\pi}{C}} e^{-\frac{(x+1)^2+1}{4Cn}} \sinh\left(\frac{x+1}{2Cn}\right).$$

Estimation of $I_1(n,A)$:
For every $i \geq 1$, the Fourier transform of the distribution μ_i can be written as

$$\hat{\mu}_i(\theta) = \frac{\phi_i(\theta)}{\sin(\theta)}$$

with

$$\phi_i(\theta) = \sum_{x \in \mathbb{N}} \frac{a_x^{(i)}}{(x+1)} \sin((x+1)\theta).$$

In the same manner, the Fourier transform of the distribution $\mu = \sum_{x \in \mathbb{N}} A_x \delta_x$ is given by

$$\hat{\mu}(\theta) = \frac{\phi(\theta)}{\sin(\theta)}$$

with

$$\phi(\theta) = \sum_{x \in \mathbb{N}} \frac{A_x}{(x+1)} \sin((x+1)\theta).$$

For every x, there exists $\eta = \eta(x) \in \,]0,1[$ such that

$$\sin(x) = x - \frac{x^3}{6} - \frac{x^3}{6}[\cos(\eta x) - 1].$$

Then, from hypotheses (H_2) and (H_3), for every $i \geq 1$,

$$\phi_i(\theta) = \theta - \frac{\theta^3}{6}\sum_{x\in\mathbb{N}} a_x^{(i)}(x+1)^2 - \frac{\theta^3}{6}\sum_{x\in\mathbb{N}} a_x^{(i)}(x+1)^2[\cos(\eta_{x,\theta}(x+1)\theta) - 1] \quad (7.10)$$

and

$$\phi(\theta) = \theta - \frac{\theta^3}{6}\sum_{x\in\mathbb{N}} A_x(x+1)^2 - \frac{\theta^3}{6}\sum_{x\in\mathbb{N}} A_x(x+1)^2[\cos(\eta_{x,\theta}(x+1)\theta) - 1] \quad (7.11)$$

where $\eta_{x,\theta}$ are real numbers in $]0,1[$ depending on x and θ.
Thus, for every $\alpha \in [0, A]$,

$$\left|\widehat{\mu^{(n)}}\left(\frac{\alpha}{\sqrt{n}}\right) - \hat{\mu}\left(\frac{\alpha}{\sqrt{n}}\right)^n\right| = \left|\prod_{i=1}^n \hat{\mu}_i\left(\frac{\alpha}{\sqrt{n}}\right) - \hat{\mu}\left(\frac{\alpha}{\sqrt{n}}\right)^n\right|$$

$$= \frac{1}{|\sin(\frac{\alpha}{\sqrt{n}})|^n}\left|\prod_{i=1}^n \phi_i\left(\frac{\alpha}{\sqrt{n}}\right) - \phi\left(\frac{\alpha}{\sqrt{n}}\right)^n\right|$$

$$= \frac{|\phi(\frac{\alpha}{\sqrt{n}})|^n}{|\sin(\frac{\alpha}{\sqrt{n}})|^n}\left|\exp\left(\sum_{i=1}^n \log\left(\frac{\phi_i(\frac{\alpha}{\sqrt{n}})}{\phi(\frac{\alpha}{\sqrt{n}})}\right)\right) - 1\right|$$

Using the following inequality,

$$|e^z - 1| \leq |z| \, e^{|z|}, \forall z \in \mathbb{C},$$

we obtain that

$$\left|\widehat{\mu^{(n)}}\left(\frac{\alpha}{\sqrt{n}}\right) - \hat{\mu}\left(\frac{\alpha}{\sqrt{n}}\right)^n\right| \leq$$

$$\frac{|\phi(\frac{\alpha}{\sqrt{n}})|^n}{|\sin(\frac{\alpha}{\sqrt{n}})|^n}\left|\sum_{i=1}^n \log\left(\frac{\phi_i(\frac{\alpha}{\sqrt{n}})}{\phi(\frac{\alpha}{\sqrt{n}})}\right)\right|\exp\left|\sum_{i=1}^n \log\left(\frac{\phi_i(\frac{\alpha}{\sqrt{n}})}{\phi(\frac{\alpha}{\sqrt{n}})}\right)\right|$$

Now, thanks to (7.11),

$$\lim_{n\to\infty}\sup_{\alpha\in[0,A]}\left|\frac{\phi(\frac{\alpha}{\sqrt{n}})^n}{\sin(\frac{\alpha}{\sqrt{n}})^n} - \exp(-C\alpha^2)\right| = 0.$$

It remains to prove that

$$\lim_{n\to\infty}\sup_{\alpha\in[0,A]}\left|\sum_{i=1}^n \log\left(\frac{\phi_i(\frac{\alpha}{\sqrt{n}})}{\phi(\frac{\alpha}{\sqrt{n}})}\right)\right| = 0.$$

For n large enough, thanks to (H_4), (7.10) and (7.11),

$$\sum_{i=1}^{n} \log\left(\frac{\phi_i\left(\frac{\alpha}{\sqrt{n}}\right)}{\phi\left(\frac{\alpha}{\sqrt{n}}\right)}\right) = \sum_{i=1}^{n} \log\left(1 + \frac{\phi_i\left(\frac{\alpha}{\sqrt{n}}\right) - \phi\left(\frac{\alpha}{\sqrt{n}}\right)}{\phi\left(\frac{\alpha}{\sqrt{n}}\right)}\right)$$

$$= -\frac{\alpha^2}{6n} \sum_{i=1}^{n} \sum_{x \in \mathbb{N}} [a_x^{(i)} - A_x](x+1)^2$$

$$\quad - \frac{\alpha^2}{6n} \sum_{i=1}^{n} \sum_{x \in \mathbb{N}} [a_x^{(i)} - A_x](x+1)^2 [\cos(\eta_{x,\theta}(x+1)\theta) - 1]$$

$$\quad + o(1)$$

$$= o(1)$$

using hypothesis (H_3).
From the inequality: $|\sin(x)| \leq x$ for $x \geq 0$, we get that

$$|I_1(n,A)| \leq \frac{4A^3}{3\sqrt{\pi}} \sup_{0 \leq \alpha \leq A} |\widehat{\mu^{(n)}}(\frac{\alpha}{\sqrt{n}}) - \exp(-C\,\alpha^2)| = o(1)$$

uniformly in x.

Estimation of $I_2(n, A)$:
Since $|\sin(x)| \leq x$ for $x \geq 0$, we have

$$|I_2(n,A)| \leq \frac{4}{\sqrt{\pi}} \int_A^{+\infty} \alpha^2 \exp(-C\,\alpha^2)\, d\alpha \to 0,$$

when A goes to infinity, uniformly in x.

Estimation of $I_3(n, A, r)$:

$$I_3(n,A,r) = \frac{4n}{\sqrt{\pi}(x+1)} \int_A^{r\sqrt{n}} \widehat{\mu^{(n)}}(\frac{\alpha}{\sqrt{n}}) \sin\left((x+1)\frac{\alpha}{\sqrt{n}}\right) \sin(\frac{\alpha}{\sqrt{n}})\, d\alpha$$

We can choose r small enough to have

$$|\widehat{\mu^{(n)}}(\frac{\alpha}{\sqrt{n}})| \leq \exp(-\frac{\alpha^2}{6n} \sum_{i=1}^{n} \sum_{x \in \mathbb{N}} a_x^{(i)}(x^2 + 2x))$$

Then, from hypothesis (H_3), there exists a constant C' strictly positive such that

$$|\widehat{\mu^{(n)}}(\frac{\alpha}{\sqrt{n}})| \leq \exp(-C'\alpha^2).$$

So,
$$|I_3(n,A,r)| \leq \frac{4}{\sqrt{\pi}} \int_A^{+\infty} \alpha^2 \exp(-C'\alpha^2) \, d\alpha \to 0$$

when A goes to infinity, uniformly in x.

Estimation of $I_4(n,A,r)$:

$$\begin{aligned} I_4(n,A,r) &= \frac{4n}{\sqrt{\pi}(x+1)} \int_{r\sqrt{n}}^{\pi\sqrt{n}} \widehat{\mu^{(n)}}(\frac{\alpha}{\sqrt{n}}) \sin\left((x+1)\frac{\alpha}{\sqrt{n}}\right) \sin(\frac{\alpha}{\sqrt{n}}) \, d\alpha \\ &= \frac{4n^{3/2}}{\sqrt{\pi}(x+1)} \int_r^{\pi} \widehat{\mu^{(n)}}(\theta) \sin((x+1)\theta) \sin(\theta) \, d\theta \end{aligned}$$

For every $i \geq 1$, the probability measure μ_i being aperiodic, there exists $\delta_i = \delta_i(r) > 0$ such that
$$|\hat{\mu}_i(\theta)| \leq 1 - \delta_i, \quad \forall \theta \in [r, \pi],$$

so
$$|I_4(n,A,r)| \leq \frac{4\sqrt{\pi} n^{3/2}}{(x+1)} \prod_{i=1}^n (1-\delta_i) = o(1)$$

uniformly in x.

6.2.2 RECURRENCE AND TRANSIENCE

The question of recurrence-transience is natural and can be deduced from a slightly modification of the previous local limit theorem.

DEFINITION 7.4 *The dynamic random walk on the dual of $SU(2)$ is transient if and only if for every point $x \in \mathbb{N}$, $y \in \mathbb{N}$, the walk starting from x visits y only a finite number of times. It is equivalent to say that, for every point $x \in \mathbb{N}$, $y \in \mathbb{N}$, the potential kernel defined by*

$$G(x,y) = \sum_{n \geq 0} P^{(n)}(x,y)$$

is finite.

Under the assumptions (H_1), (H_2), (H_3) and (H_4),

$$\lim_{n \to +\infty} \sup_{x \in \mathbb{N}} \left| 2\sqrt{\pi} n^{3/2} P^{(n)}(x,y) - \frac{2n(y+1)}{\sqrt{C}(x+1)} e^{-\frac{(x+1)^2+(y+1)^2}{4Cn}} \right.$$
$$\left. \times \sinh\left(\frac{(x+1)(y+1)}{2Cn}\right) \right| = 0$$

where
$$C = \frac{1}{6} \sum_{x \geq 0} A_x(x^2 + 2x).$$

In particular: for every point $x \in \mathbb{N}$, $y \in \mathbb{N}$, as $n \to \infty$,
$$P^{(n)}(x,y) \sim (y+1)^2 (2\sqrt{\pi})^{-1} C^{-3/2} n^{-3/2}.$$

From this result, we easily deduce the following theorem

THEOREM 7.2 *Under assumptions* (H_1), (H_2), (H_3) *and* (H_4), *the dynamic random walk on the dual of* $SU(2)$ *is transient.*

6.2.3   A CENTRAL LIMIT THEOREM

From the local limit theorem established in Section 6.2.1., we deduce a central limit theorem for the dynamic random walk on the dual of $SU(2)$.

THEOREM 7.3 *Under the assumptions* (H_1), (H_2), (H_3) *and* (H_4), *the sequence* $Y_n = \frac{S_n}{\sqrt{2Cn}}$ *where*
$$C = \frac{1}{6} \sum_{x \geq 0} A_x(x^2 + 2x).$$

converges in distribution, as $n \to +\infty$, *to the random variable whose the distribution density is*
$$\sqrt{\frac{2}{\pi}} x^2 \exp(-\frac{x^2}{2}).$$

Proof:
For $0 \leq a \leq b$, we have the relation

$$\mathbb{P}(a \leq Y_n \leq b)$$
$$= \mathbb{P}(a\sqrt{2Cn} \leq S_n \leq b\sqrt{2Cn})$$
$$= \sum_{a\sqrt{2Cn} \leq x \leq b\sqrt{2Cn}; x \in \mathbb{N}} \frac{2(x+1)}{\pi} \int_0^\pi \widehat{\mu^{(n)}}(\theta) \sin\big((x+1)\theta\big) \sin(\theta)\, d\theta.$$

Using the changes of variables: $y = \frac{(x+1)}{\sqrt{n}}$ and $\theta = \frac{\alpha}{\sqrt{n}}$, we get

$$\mathbb{P}(a \leq Y_n \leq b)$$
$$= \sum_{a\sqrt{2Cn}+1 \leq y\sqrt{n} \leq b\sqrt{2Cn}+1; y\sqrt{n} \in \mathbb{N}} \frac{2y}{\pi} \int_0^{\pi\sqrt{n}} \widehat{\mu^{(n)}}(\frac{\alpha}{\sqrt{n}}) \sin(y\alpha) \sin(\frac{\alpha}{\sqrt{n}})\, d\alpha$$

Now,

$$\lim_{n\to +\infty} \frac{1}{\sqrt{n}} \sum_{a\sqrt{2Cn}+1 \leq y\sqrt{n} \leq b\sqrt{2Cn}+1; y\sqrt{n}\in\mathbb{N}} y\sin\left(y\alpha\right) = \int_{a\sqrt{2C}}^{b\sqrt{2C}} y\sin(y\alpha)\, dy$$

and

$$\lim_{n\to +\infty} \widehat{\mu^{(n)}}\left(\frac{\alpha}{\sqrt{n}}\right) = \exp(-C\alpha^2).$$

Thus,

$$\lim_{n\to\infty} \mathbb{P}(a \leq Y_n \leq b) = \frac{2}{\pi} \int_{a\sqrt{2C}}^{b\sqrt{2C}} y \left[\int_0^\infty \alpha \sin(y\alpha)\exp(-C\alpha^2)\, d\alpha\right] dy.$$

Integrating by parts and using formula (7.9), we get

$$\int_0^\infty \alpha \sin(y\alpha)\exp(-C\alpha^2)\, d\alpha = \frac{y\sqrt{\pi}}{4C^{3/2}} \exp(-y^2/(4C)).$$

Thus,

$$\lim_{n\to\infty} \mathbb{P}(a \leq Y_n \leq b) = \sqrt{\frac{2}{\pi}} \int_a^b y^2 \exp\left(-\frac{y^2}{2}\right) dy.$$

Example: Let $S = (E, \mathcal{A}, \mu, T)$ be a dynamical system where $(E, \mathcal{A}, \mu)$ is a probability space and T is a measure-preserving transformation defined on E.
For each $e \in E$, let $a_x^{(i)} = a_x(T^i e)$ where a_x are positive functions such that $\sum_{x>0} a_x^{(i)}(e) = 1$. A particular example is $a_x = 0$ if $x \neq 1,2$ and $a_1(e) = \bar{f}(e)$ and $a_2(e) = 1 - f(e)$ where f is a measurable function defined on E with values in $]0,1[$. This is a model similar to the dynamic random walks considered in [68, 69, 70, 71, 72, 73, 74, 75, 77]. Under assumption of uniform ergodicity of the dynamical system, assumptions $(H_i), i = 1, \ldots, 4$ are clearly satisfied and in that case, Theorems 7.1, 7.2 and 7.3 hold with the constant $C = \frac{1}{3}[4 - \frac{5}{2}\int_E f\, d\mu] > 0$.

6.2.4 A WEAK LAW OF LARGE NUMBERS

From the central limit theorem (Theorem 7.3), we deduce the

THEOREM 7.4 *Under the assumptions (H_1), (H_2), (H_3) and (H_4), for every $\epsilon > 0$, the sequence of random variables $(S_n/n^{1/2+\epsilon})_{n\geq 1}$ converges in probability to 0, as n goes to infinity.*

Remark:
$\epsilon = 1/2$ provides the traditional law of large numbers. A strong law of large numbers should hold using the method developed in Section III of [63]. Details are omitted. This will be the object of further investigation.

6.2.5 A LARGE DEVIATION PRINCIPLE

In the previous section, S_n/n was proved to converge in probability to 0 under convenient assumptions, so the sequence of random variables $(S_n/n)_n$ is a good candidate for a large deviation principle. Let us first recall what we mean by Large Deviations Principle: Let Γ be a Polish space endowed with the Borel σ-field $\mathcal{B}(\Gamma)$. A good rate function is a lower semi-continuous function $\Lambda^* : \Gamma \to [0, \infty]$ with compact level sets $\{x; \Lambda^*(x) \leq \alpha\}, \alpha \in [0, \infty[$. Let $v = (v_n)_n \uparrow \infty$ be an increasing sequence of positive reals. A sequence of random variables $(Y_n)_n$ with values in Γ defined on a probability space $(\Omega, \mathcal{F}, \mathbb{P})$ is said to satisfy a large deviation principle (LDP) with speed $v = (v_n)_n$ and the good rate function Λ^* if for every Borel set $B \in \mathcal{B}(\Gamma)$,

$$- \inf_{x \in B^o} \Lambda^*(x) \leq \liminf_n \frac{1}{v_n} \log \mathbb{P}(Y_n \in B)$$
$$\leq \limsup_n \frac{1}{v_n} \log \mathbb{P}(Y_n \in B) \leq - \inf_{x \in \bar{B}} \Lambda^*(x).$$

When the upper inequality holds, $(Y_n)_n$ is said to satisfy an upper large deviation principle. Let us consider now a sequence of probability measures $(\mu_i)_{i \geq 1}$ where $\mu_i = \sum_{x \in \mathbb{N}} a_x^{(i)} \delta_x$ on $\mathbb{N}$ (remember that $a_x^{(i)}$ are coefficients which are non negative and $\sum_{x \in \mathbb{N}} a_x^{(i)} = 1$). We denote by $\tilde{\mu}_i$ the Laplace transform of the measure μ_i.
We make the following assumption:
(H): the function Γ defined by

$$\forall t \in \mathbb{R}^+, \quad \Gamma(t) = \limsup_{n \to +\infty} \frac{1}{n} \sum_{i=1}^n \log \tilde{\mu}_i(t)$$

is finite in a neighbourhood of 0.

THEOREM 7.5 *Under Assumption (H), the sequence of random variables $(S_n/n)_{n \geq 1}$ satisfies an upper Large Deviations Principle: for every $y \in \mathbb{R}^{+,*}$,*

$$\limsup_{n \to +\infty} \frac{1}{n} \log \mathbb{P}\left(\frac{S_n}{n} \geq y\right) \leq -\Lambda^*(y)$$

where

$$\Lambda^*(y) = \sup_{t; \Gamma(t) < \infty} \{ty - \Gamma(t)\}.$$

Assume that the function Λ^* is strictly positive on $\mathbb{R}^{+,*}$, then the sequence of random variables $(S_n/n)_{n \geq 1}$ converges almost surely to 0 as n goes to infinity.

Proof:
Since the function $x \to \frac{\sinh[(x+1)t]}{(x+1)\sinh(t)}$ is increasing on $\mathbb{R}^+$, using the Markov inequality, we get

$$\mathbb{P}\left(\frac{S_n}{n} \geq y\right) = \mathbb{P}\left(\frac{\sinh[(S_n+1)t]}{(S_n+1)\sinh(t)} \geq \frac{\sinh[(ny+1)t]}{(ny+1)\sinh(t)}\right)$$

$$\leq \mathbb{E}\left(\frac{\sinh[(S_n+1)t]}{(S_n+1)\sinh(t)}\right) \frac{(ny+1)\sinh(t)}{\sinh[(ny+1)t]}.$$

Now, from the definition of the Laplace transform and formula (7.8),

$$\mathbb{E}\left(\frac{\sinh[(S_n+1)t]}{(S_n+1)\sinh(t)}\right) = \sum_{x \in \mathbb{N}} \psi_x(t)\, \delta_0 \star \mu^{(n)}(x)$$

$$= (\widetilde{\delta_0 \star \mu^{(n)}})(t)$$

$$= \prod_{i=1}^{n} \tilde{\mu}_i(t)$$

Moreover

$$\frac{\sinh(t)}{\sinh[(ny+1)t]} \leq e^{-nyt}.$$

Hence,

$$\frac{1}{n}\log \mathbb{P}\left(\frac{S_n}{n} \geq y\right) \leq -ty + \frac{1}{n}\log(ny+1) + \frac{1}{n}\sum_{i=1}^{n} \log(\tilde{\mu}_i(t))$$

Then, using assumption (H),

$$\limsup_{n \to +\infty} \frac{1}{n}\log \mathbb{P}\left(\frac{S_n}{n} \geq y\right) \leq -ty + \limsup_{n \to +\infty} \frac{1}{n}\sum_{i=1}^{n} \log(\tilde{\mu}_i(t))$$

$$\leq -ty + \Gamma(t).$$

Taking the infimum in the right hand side over all $t > 0$, we get the result.

Let $\epsilon > 0$. For n large enough,

$$\mathbb{P}\left(\frac{S_n}{n} \geq \epsilon\right) \leq e^{-n\Lambda^*(\epsilon)/2}$$

Since the series $\sum_n e^{-n\alpha}$ is finite for any $\alpha > 0$, from Borel-Cantelli lemma, we deduce that S_n/n converges almost surely to 0 as n goes to infinity.

Example: Consider the example from Section 6.2.3 that is a dynamical system $(E, \mathcal{A}, \mu, T)$ where $(E, \mathcal{A}, \mu)$ is a probability space and T is a transformation on E. Let $e \in E$ be a fixed point, and choose $a_x^{(i)} = a_x(T^i e)$ where $a_x = 0$ if $x \neq 1, 2$, $a_1(e) = f(e)$ and $a_2(e) = 1 - f(e)$ where f is a measurable function defined on E with values in $]0, 1[$. Then, for every $i \geq 1$, the Laplace transform of the measure $\mu_i = \sum_x a_x^{(i)} \delta_x$ is given by

$$\tilde{\mu}_i(t) = \cosh(t) f(T^i e) + (1 - f(T^i e)) \frac{\sinh(3t)}{3 \sinh(t)}.$$

When the dynamical system is uniformly ergodic (e.g. T is an irrational rotation on the torus), assumption (H) is clearly satisfied with the function

$$\Gamma(t) = \int_E \log \left[\cosh(t) f(x) + (1 - f(x)) \frac{\sinh(3t)}{3 \sinh(t)} \right] d\mu(t).$$

It is worth remarking that using Jensen's inequality, the function Γ is less than or equal to $\log(\tilde{\mu}(t))$ where μ is the probability measure $(\int_E f \, d\mu) \delta_1 + (1 - \int_E f \, d\mu) \delta_2$. Consequently,

$$-\sup\{tx - \Gamma(t)\} \leq -\sup\{tx - \log \tilde{\mu}(t)\}.$$

Assumption (H) is weaker than assuming that the average of the $\tilde{\mu}_i$'s converges to $\tilde{\mu}$. Take for instance the irrational rotation on the one-dimensional torus and $f = \mathbf{1}_{[0,1/2]}$, then

$$\Gamma(t) = \frac{1}{2} \left[\log \left(\cosh(t) \right) + \log \left(\frac{\sinh(3t)}{3 \sinh(t)} \right) \right]$$

and

$$\log(\tilde{\mu}(t)) = \log \left[\frac{1}{2} \cosh(t) + \frac{1}{2} \frac{\sinh(3t)}{3 \sinh(t)} \right].$$

We recover the fact that the dynamic random walk is more concentrated around the origin as the random walk with stationary increments, a phenomenon already encountered in Section 5.3 of Chapter 2.

II
APPLICATIONS

Chapter 8

DISTRIBUTED ALGORITHMS WITH DYNAMICAL RANDOM TRANSITIONS

1. COLLIDING STACKS

We consider the evolution of two stacks (for simplicity) inside a shared, contiguous memory area of a fixed size m. This shared storage allocation algorithm lets the stacks grow from the two ends of the memory until the combined size of the stacks exhausts the available storage. This algorithm is to be compared to another strategy, namely allocating separate zones of size $m/2$ to each stack. This problem of Knuth [99] has been investigated by Yao [187], Flajolet [55], Louchard and Schott [121], Louchard [120] and Maier [128]. The natural formulation

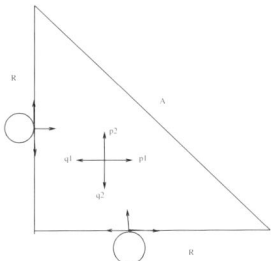

Figure 8.1. Evolution of two stacks

of the two stacks problem is in terms of random walks in a triangle in a two-dimensional lattice space: a state is a pair formed by the size of both stacks. The random walk $Y_m(.)$ has two reflecting barriers (R) along the axes and one absorbing barrier (A) as shown in Figure 8.1.

The distribution of steps (steps of unit length) ΔY is given by:

$$P(\Delta Y = (1,0)) = p_1, \ P(\Delta Y = (-1,0)) = q_1, \ P(\Delta Y = (0,1)) = p_2,$$
$$P(\Delta Y = (0,-1)) = q_2$$

with the boundary conditions of Figure 8.1. In [99, 55, 121, 120, 187] the transition probabilities are supposed to be constant and $p_1 + q_1 + p_2 + q_2 = 1$. In [128] some space dependency is allowed for the transition probabilities. Real-world applications require the transition probabilities to be time and space dependent. Since the analysis of such a model is out of reach of todays mathematics, we will only allow some time dependency (see Section 3). The following questions are of practical interest. With initial condition $Y_m(0) = x_m$, what are asymptotically (as $m \to \infty$):

- the hitting place (Z_m) distribution on the (A) boundary?

- the hitting time (T_m) distribution on the (A) boundary?

We consider the d-dimensional dynamic random walk $(Y_m(.))_m$ as defined in Section 2 of Chapter 1 evolving inside the bounded domain

$$\Omega = \{x = (x_1, \ldots, x_d) \in \mathbb{N}^d \ ; 0 \leq x_i \leq m, i = 1, \ldots, d; \ \sum_{i=1}^{d} x_i \leq m\}$$

with boundary conditions: reflections on the axes $\{0, 1, \ldots, m\}$ and absorption on the sloping side

$$\Gamma_a = \{x = (x_1, \ldots, x_d) \in \Omega \ ; \ \sum_{i=1}^{d} x_i = m\}.$$

THEOREM 8.1 *Let us assume that for every* $j, l \in \{1, \ldots, d\}, f_j \in \mathcal{C}_2(S)$, $f_j f_l \in \mathcal{C}_2(S)$ *and* $\int_E f_j d\mu = \frac{1}{2d}$. *Then, the sequence of processes*

$$(\frac{1}{m} Y_m([m^2 t]))_m$$

converges in $\mathcal{D}$ *to a d-dimensional reflected and absorbed Brownian motion* $(W(t))_t$ *with zero mean and covariance matrix At evolving inside the bounded domain*

$$D = \{x = (x_1, \ldots, x_d) \in \mathbb{R}^d \ ; \ 0 \leq x_i \leq 1, i = 1, \ldots, d; \ \sum_{i=1}^{d} x_i \leq 1\}$$

with boundary conditions: reflections on the axes $[0,1]$ and absorption on the sloping side

$$H_a = \{x = (x_1, \ldots, x_d) \in D \ ; \ \sum_{i=1}^{d} x_i = 1\}.$$

Let us denote $T_m = \inf\{k \geq 1; Y_m(k) \in \Gamma_a\}$ and $Z_m = Y_m(T_m)$. Then,

$$\frac{T_m}{m^2} \xrightarrow{\mathcal{L}} T$$

and

$$\frac{Z_m}{m} \xrightarrow{\mathcal{L}} Z$$

where $T = \inf\{t > 0; W(t) \in H_a\}$ and $Z = W(T)$.

Proof:
The theorem follows from Theorem 2.2 and the continuous mapping theorem (see [15], corollary 1, p. 31) by remarking that before absorption the process $(Y_m(m))_m$ can be written as

$$S_m + \max_{0 \leq i \leq m}(-S_i)$$

and that the hitting time T is a continuous functional of the limiting process $W(.)$.

We give a method to find the transition densities of $(W(t))_t$ before the hitting time T for any dimension, in the case when $A = \frac{1}{d} I_d$ where I_d is the identity matrix. In particular, we show that the characteristics of the process $(W(t))_t$ before the hitting time can be obtained via a geometric transformation from the d-dimensional Brownian motion $(W^R(t))_t$, centered, with covariance matrix $\frac{t}{d}I_d$, evolving inside the hypercube $[0,1]^d$ with normal reflections on the sides.

THEOREM 8.2 *Let us assume that for every* $j, l \in \{1, \ldots, d\}, f_j \in \mathcal{C}_2(S)$, $f_j f_l \in \mathcal{C}_2(S)$, $\int_E f_j d\mu = \frac{1}{2d}$ *and* $\int_E f_j f_l d\mu = \frac{1}{4d^2}$. *The transition densities of the limiting process* $(W(t))_t$ *before the hitting time* T *of* H_a *are given for* $x, y \in D$ *by*

$$q(t, x, y) = p(t, x, y) - \int_{H_a} \int_0^t q^{(R)}(t - \tau, x, v) p(\tau, v, y) \, dv \, d\tau$$

where

$$p(t, x, y) = \prod_{i=1}^{d} \Sigma(t, x_i, y_i)$$

with $\Sigma(t, x_i, y_i) = \sum_{k \in \mathbb{Z}} \exp(-k^2\pi^2 t/(2d)) \cos(k\pi x_i) \cos(k\pi y_i)$ and $q^{(R)}(t, x, v)$ is the density of the first hitting time of the point $v \in H_a$ by the process $(W^R(t))_t$.

Moreover, the distribution of the hitting time and the hitting place of the process $(W(t))_t$ are given by

$$\mathbb{P}_x(T \in dt) = -\frac{1}{2d} \int_{\psi^{-1}(H_a)} \partial_{z_1} q(t, x, \psi(z)) \, dz \, dt \qquad (8.1)$$

and

$$\mathbb{P}_x(Z \in dz_2 dz_3 \ldots dz_d) = -\frac{1}{2d} \int_0^\infty \partial_{z_1} q(t, x, \psi(z)) \Big|_{z_1=0} dt \, dz_2 dz_3 \ldots dz_d \qquad (8.2)$$

where ψ is an orthonormal change of basis in $\mathbb{R}^d$ such that

$$(\psi^{-1}(e_2), \ldots, \psi^{-1}(e_d))$$

is an orthonormal basis of the $(d-1)$-dimensional sub-space H_a and $\psi^{-1}(e_1)$ is outward orthonormal to H_a.

Proof:
The density of the classical one-dimensional driftless Brownian motion $(W_a(t))_t$ with variance at, $a > 0$, with reflecting boundaries at 0 and 1 is well-known and given by

$$\mathbb{P}_{x_1}(W_a(t) \in dy_1) = \Sigma_a(t, x_1, y_1) dy_1$$

where (see Feller [50])

$$\Sigma_a(t, x, y) = \sum_{k \in \mathbb{Z}} \exp(-k^2\pi^2 at/2) \cos(k\pi x) \cos(k\pi y).$$

The matrix A being assumed diagonal, the components of $(W^R(t))_t$ are independent and the transition densities of this process are given for $x, y \in [0, 1]^d$ by

$$p(t, x, y) = \prod_{i=1}^d \Sigma(t, x_i, y_i)$$

where

$$\Sigma(t, x_i, y_i) = \sum_{k \in \mathbb{Z}} \exp(-k^2\pi^2 t/(2d)) \cos(k\pi x_i) \cos(k\pi y_i)$$

For all $x, y \in D$, the probability density of a passage from x to y for the process $(W^R(t))_t$ with an intermediate passage through a point v of the absorbing barrier H_a is the convolution of the density $q^{(R)}(t,x,v)$ of the first passage from x to v and $p(t,v,y)$. Then, the transition densities of $(W(t))_{t\geq 0}$ prior to absorption are given by

$$q(t,x,y) = p(t,x,y) - \int_{H_a} \int_0^t q^{(R)}(t-\tau,x,v) p(\tau,v,y)\, dv\, d\tau.$$

Let us prove (8.1). The probability that during $[0,t]$, the process $W(\tau)$ evolves inside D and that $W(t)$ is in dy is equal to $q(t,x,y)dy$ where q is also the solution of the following PDE's problem

$$\begin{aligned}
\frac{\partial f}{\partial t} &= \frac{1}{2d}\Delta f \text{ in } D\\
f &= 0, \forall\, t \text{ along } H_a\\
\frac{\partial f}{\partial \nu} &= 0, \forall\, t \text{ along } \partial D - H_a\\
f(0,x,y) &= \delta(x-y) \text{ (Dirac distribution)}
\end{aligned}$$

($\frac{\partial}{\partial \nu}$ is the outward normal derivative to ∂D).
The probability that during $[0,t]$, $W(\tau)$ evolves inside D is then given by

$$\mathbb{P}_x(T > t) = \int_D q(t,x,y)\, dy.$$

Differentiating with respect to t, we get

$$\begin{aligned}
\frac{\partial}{\partial t}\mathbb{P}_x(T > t) &= \int_D \frac{\partial}{\partial t} q(t,x,y)\, dy\\
&= \frac{1}{2d}\int_D \Delta q(t,x,y)\, dy\\
&= \frac{1}{2d}\int_{\partial D} \frac{\partial q}{\partial \nu}(t,x,y) dy\\
&= \frac{1}{2d}\int_{H_a} \frac{\partial q}{\partial \nu}(t,x,y) dy
\end{aligned}$$

using Green's identity and the fact that $\frac{\partial q}{\partial \nu} = 0, \forall\, t$ along $\partial D - H_a$. Let ψ be an orthonormal change of basis in $\mathbb{R}^d$ such that $(\psi^{-1}(e_2),\ldots,\psi^{-1}(e_d))$ is an orthonormal basis of the $(d-1)$–dimensional sub-space H_a and $\psi^{-1}(e_1)$ is orthonormal to H_a, then denoting by $(z_1,\ldots,z_d)$ the new co-ordinates in the new basis $(\psi^{-1}(e_1),\ldots,\psi^{-1}(e_d))$, we deduce (8.1). The probability that the process $W(t)$ hits for the first time the absorbing barrier H_a in $dz_2\ldots dz_d$ is equal to

$$\mathbb{P}_x(Z \in dz_2 dz_3\ldots dz_d) = -\frac{1}{2d}\int_0^\infty \partial_{z_1} q(t,x,\psi(z))\Big|_{z_1=0} dt\, dz_2 dz_3\ldots dz_d.$$

By the previous result, we can deduce the density of the hitting time and the hitting place for any diagonal matrix A by remarking that the PDE's problem

$$\begin{aligned}
\frac{\partial f}{\partial t} &= \frac{1}{2}\sum_{i=1}^{d} a_{ii}\frac{\partial^2 f}{\partial x_i^2} \text{ in } D \\
f &= 0, \forall\, t \text{ along } H_a \\
\frac{\partial f}{\partial \nu} &= 0, \forall\, t \text{ along } \partial D - H_a \\
f(0,x,y) &= \delta(x-y) \text{ (Dirac distribution)}
\end{aligned}$$

is equivalent to the simpler PDE's system

$$\begin{aligned}
\frac{\partial \tilde{f}}{\partial t} &= \frac{1}{2d} \Delta \tilde{f} \text{ in } \tilde{D} \\
\tilde{f} &= 0, \forall\, t \text{ along } \tilde{H}_a \\
\frac{\partial \tilde{f}}{\partial \nu} &= 0, \forall\, t \text{ along } \partial \tilde{D} - \tilde{H}_a \\
\tilde{f}(0,x,y) &= \delta(x-y) \text{ (Dirac distribution)}
\end{aligned}$$

where $\tilde{D} = (dA)^{-1/2} D$ and $\tilde{H}_a$ is the hyperplane $(dA)^{-1/2} H_a$. If $\tilde{f}$ is a solution of this system, then $f(t,x,y) = \tilde{f}(t,(dA)^{-1/2}x,(dA)^{-1/2}y)$ is the solution of the first system.

In order to give an explicit formula for the density $f(t,x,y)$ of the limiting process $(W(t))$ before absorption when its components are not necessarily independent we have to solve the following difficult PDE's problem,

$$\begin{aligned}
\frac{\partial f}{\partial t} &= \frac{1}{2} L f \text{ in } D \\
f &= 0, \forall\, t \text{ along } H_a \\
\frac{\partial f}{\partial \nu} &= 0, \forall\, t \text{ along } \partial D - H_a \\
f(0,x,y) &= \delta(x-y)
\end{aligned}$$

where L is the differential operator

$$\left(\frac{\partial}{\partial x_1} \cdots \frac{\partial}{\partial x_d}\right) A \left(\frac{\partial}{\partial x_1} \cdots \frac{\partial}{\partial x_d}\right)^t,$$

$\frac{\partial}{\partial \nu}$ is the outward normal derivative to ∂D, δ is the d-dimensional Dirac distribution. The matrix A is positive definite, therefore there exists a matrix change of basis P and a diagonal matrix U such that $A = PUP^{-1}$. By this change of basis the above problem is equivalent to finding the density before absorption of a Brownian Motion with zero mean and covariance matrix Ut evolving inside the domain D with respect to the new basis, with new boundary conditions in particular with reflections not necessarily normal.

In the particular case for $d = 2$, the distribution of $W(t)$ before the hitting time T and the distributions of T and Z, can be completely determined. We will use the following change of variables $y = \psi(z)$ where

$$\begin{cases} y_1 = \frac{1}{2} + \frac{z_1+z_2}{\sqrt{2}} \\ y_2 = \frac{1}{2} + \frac{z_1-z_2}{\sqrt{2}} \end{cases}$$

THEOREM 8.3 *Let us assume that $f_1, f_2, f_1f_2 \in C_2(S)$, $\int_E f_1 d\mu = \int_E f_2 d\mu = \frac{1}{4}$ and $\int_E f_1 f_2 d\mu = \frac{1}{16}$. The density of $W(.)$ before the hitting time T is given by*

$$\begin{aligned}\mathbb{P}_x(W(t) \in dy, t < T) &= \Sigma_1(t,x_1,y_1)\Sigma_2(t,x_2,y_2) \quad (8.3)\\ &\quad - \Sigma_1(t,x_1,1-y_2)\Sigma_2(t,x_2,1-y_1)dy_1dy_2\end{aligned}$$

where for $i = 1, 2$,

$$\Sigma_i(t,x,y) = \sum_{k \in \mathbb{Z}} \exp(-k^2\pi^2 a_{ii}t/2)\cos(k\pi x)\cos(k\pi y)$$

and

$$a_{ii} = \frac{3}{4} - 4\int_E f_i^2 d\mu$$

Moreover,

$$\mathbb{P}_x(T \in dt) \qquad (8.4)$$
$$= 2\sum_{k>0\ \text{odd}} \left\{\exp(-k^2\pi^2 a_{11}t/2)\cos(k\pi x_1) + \exp(-k^2\pi^2 a_{22}t/2)\cos(k\pi x_2)\right\}$$
$$+ 4\sum_{k>0\ \text{odd}}\sum_{l>0\ \text{even}} \exp(-\pi^2(a_{11}k^2 + a_{22}l^2)t/2)\cos(k\pi x_1)\cos(l\pi x_2)\frac{k^2+l^2}{k^2-l^2}$$
$$- 4\sum_{k>0\ \text{even}}\sum_{l>0\ \text{odd}} \exp(-\pi^2(a_{11}k^2 + a_{22}l^2)t/2)\cos(k\pi x_1)\cos(l\pi x_2)\frac{k^2+l^2}{k^2-l^2}$$

and for $z_2 \in [-\frac{1}{\sqrt{2}}, \frac{1}{\sqrt{2}}]$,

$$\mathbb{P}_x(Z \in dz_2) \qquad (8.5)$$
$$= \frac{\sqrt{2}}{\pi}\sum_{k=1}^{\infty}\frac{1}{k}\left\{\frac{1}{a_{11}}\cos(k\pi x_1)\sin\left[k\pi\left(\frac{1}{2} + \frac{z_2}{\sqrt{2}}\right)\right]\right.$$

$$+ \quad \frac{1}{a_{22}}\cos(k\pi x_2)\sin\left[k\pi\left(\frac{1}{2}-\frac{z_2}{\sqrt{2}}\right)\right]\Big\}$$

$$+ \quad \frac{2\sqrt{2}}{\pi}\sum_{k=1}^{\infty}\sum_{l=1}^{\infty}\frac{1}{a_{11}k^2+a_{22}l^2}\cos(k\pi x_1)\cos(l\pi x_2)$$

$$\times \quad \Big\{k\sin\left[k\pi\left(\frac{1}{2}+\frac{z_2}{\sqrt{2}}\right)\right]\cos\left[l\pi\left(\frac{1}{2}-\frac{z_2}{\sqrt{2}}\right)\right]$$

$$+ \quad l\cos\left[k\pi\left(\frac{1}{2}+\frac{z_2}{\sqrt{2}}\right)\right]\sin\left[l\pi\left(\frac{1}{2}-\frac{z_2}{\sqrt{2}}\right)\right]\Big\}$$

Proof:
Under the hypothesis $\int_E f_1 f_2 d\mu = \frac{1}{16}$, the matrix A is clearly diagonal and the density of $W(t)$ before the hitting time T can be obtained from the previous theorem and remark. As the domain is symmetric the reflection principle across the absorbing barrier can be used to rewrite the density under the form (8.3).

We now apply the results of Theorem 8.2 in order to find explicitly the distributions of the random variables T and Z. Applying (8.1) and (8.2), with the above change of variables ψ, we obtain

$$\mathbb{P}_x(T \in dt) = -\frac{1}{4}\int_{-\frac{1}{\sqrt{2}}}^{\frac{1}{\sqrt{2}}} \partial_{z_1} q(t, x, \psi(z))|_{z_1=0}\, dz_2\, dt$$

and

$$\mathbb{P}_x(Z \in dz_2) = -\frac{1}{4}\int_0^{\infty} \partial_{z_1} q(t, x, \psi(z))|_{z_1=0}\, dt\, dz_2.$$

Then, tedious but simple computations give (8.4) and (8.5).

Example: Let us consider a particular dynamical system $(\mathbb{T}^1, \mathcal{B}(\mathbb{T}^1), \lambda, T_\alpha)$, the rotation on the one-dimensional torus with irrational angle α. As an example of functions satisfying the hypotheses of Theorem 8.3 consider

$$f_1(x) = \frac{1}{2}\cos^2(2\pi x); \quad f_2(x) = \frac{1}{2}(x - [x])$$

where $[x]$ is the integer part of x.

Evidently, $\int_{\mathbb{T}^1} f_1(x)f_2(x)dx = 1/16$ and in this case, (8.3) gives

$$\mathbb{P}_x(W(t) \in dy, t < T)$$
$$= 2\sum_{k=1}^{\infty} \exp(-3k^2\pi^2 t/16)\cos(k\pi x_1)\{\cos(k\pi y_1) - \cos[k\pi(1-y_2)]\}$$
$$+ 2\sum_{l=1}^{\infty} \exp(-5l^2\pi^2 t/24)\cos(l\pi x_2)\{\cos(l\pi y_2) - \cos[l\pi(1-y_1)]\}$$

$$+ \ 4 \sum_{k,l=1}^{\infty} \exp(-(\frac{3}{16}k^2 + \frac{5}{24}l^2)\pi^2 t) \cos(k\pi x_1) \cos(l\pi x_2)$$
$$\times \ \{\cos(k\pi y_1)\cos(l\pi y_2) - \cos[k\pi(1-y_2)]$$
$$\times \ \cos[l\pi(1-y_1)]\} dy_1 dy_2$$

The distributions of the random variables T and Z can be easily obtained from (8.4) and (8.5). Moreover, the expected size of the largest stack when the shared resource algorithm stops is given by the formula

$$\mathbb{E}(\max(Z_1, Z_2)) = \int_0^1 \max(x, 1-x) \mathbb{P}(Z_1 \in dx)$$

where, letting $x = \frac{1}{2} + \frac{z_2}{\sqrt{2}}$,

$$\mathbb{P}(Z_1 \in dx)$$
$$= \frac{2}{\pi a_{11}} \sum_{k=1}^{\infty} \frac{1}{k} \sin(k\pi x) + \frac{2}{\pi a_{22}} \sum_{k=1}^{\infty} \frac{1}{k} \sin(k\pi(1-x))$$
$$+ \ \frac{4}{\pi} \sum_{k=1}^{\infty} \sum_{l=1}^{\infty} \frac{k}{a_{11}k^2 + a_{22}l^2} \sin(k\pi x) \cos(l\pi(1-x))$$
$$+ \ \frac{4}{\pi} \sum_{k=1}^{\infty} \sum_{l=1}^{\infty} \frac{l}{a_{11}k^2 + a_{22}l^2} l \cos(k\pi x) \sin(l\pi(1-x))$$

Now, Fourier calculations (see the proofs in Appendix D) gives us for every $x \in \ ]0, 1[$,

$$\sum_{j=1}^{\infty} \frac{\sin(2\pi j x)}{j} = \pi(\frac{1}{2} - x).$$

and for every $x \in \ ]0, 2\pi[$,

$$\sum_{j=1}^{\infty} \frac{j \sin(jx)}{a^2 + j^2} = \frac{\pi}{2} \frac{\sinh(a(\pi - x))}{\sinh(a\pi)}.$$

These Fourier expansions allows us to estimate numerically (with Maple) $\mathbb{E}(\max(Z_1, Z_2))$ to 0.658. This result has to be compared with Flajolet's result obtained in the equidistributed case (with our notations: $f_1 = f_2 = \frac{1}{4}$) who found the expected value 0.67526 for the highest stack when the algorithm stops. It means that small perturbations in the temporal evolution of the shared resource algorithm have an impact on its own efficiency.

2. THE BANKER ALGORITHM

We consider a simple distributed algorithm which involves only two customers C_1 and C_2 sharing a fixed quantity of a given resource m (money). There are fixed upper bounds m_1 and m_2 on how much of the resource each of the customers is allowed to use at any time. The banker decides to give to the customer C_i, $i = 1, 2$ the required resource units only if the remaining units are sufficient in order to fulfill the requirements of C_j, $j = 1, 2; j \neq i$. This situation is modelled (see Figure 8.2) by a random walk in a rectangle with a broken corner, i.e.,

$$\{(x_1, x_2) : 0 \leq x_1 \leq m_1, 0 \leq x_2 \leq m_2, x_1 + x_2 \leq m\}$$

where the last constraint generates the broken corner. The random walk is reflected on the sides parallel to the axes and is absorbed on the sloping side.

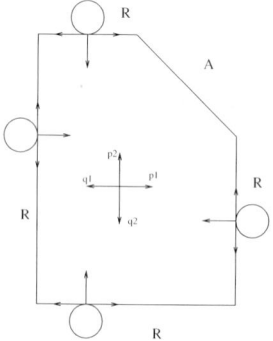

Figure 8.2. Banker algorithm

Again the hitting place and the hitting time of the absorbing boundary are the parameters of interest. This algorithm has been analyzed in [121, 124, 120] in the model where the transitions are constant. The customer's behaviors are, of course, time and space dependent and all comments made for the previous example remain true.

2.1 TIME DEPENDENT TRANSITIONS

Let us now consider the d-dimensional dynamic random walk $(Y_m(.))_m$ as defined in Section 2 of Chapter 1 evolving inside the bounded domain

$$\Omega' = \{x = (x_1, \ldots, x_d) \in \mathbb{N}^d \; ; 0 \leq x_i \leq m, i = 1, \ldots, d; \; \sum_{i=1}^{d} x_i \leq [pm]\}$$

where $p \in]1,2[$, with boundary conditions: reflections on the axes $\{0, 1, \ldots, [pm]\}$ and absorption on the sloping side

$$\Gamma'_a = \{x = (x_1, \ldots, x_d) \in \Omega' \; ; \; \sum_{i=1}^{d} x_i = [pm]\}$$

As in the colliding stacks model, the limiting distributions of the first hitting time and of the hitting place of the absorbing boundary can be found explicitly.

THEOREM 8.4 *Let us assume that for every $j, l \in \{1, \ldots, d\}, f_j \in \mathcal{C}_2(S)$, $f_j f_l \in \mathcal{C}_2(S)$ and $\int_E f_j d\mu = \frac{1}{2d}$. Then, the sequence of processes $(\frac{1}{m} Y_m([m^2 t]))_m$ converges in $\mathcal{D}$ to a d-dimensional reflected and absorbed Brownian motion $(W(t))_t$ with zero mean and covariance matrix At evolving inside the bounded domain*

$$D' = \{x = (x_1, \ldots, x_d) \in \mathbb{R}^d \; ; \; 0 \leq x_i \leq 1, i = 1, \ldots, d; \; \sum_{i=1}^{d} x_i \leq p\}$$

with boundary conditions: reflections on the axes $[0, 1]$ and absorption on the sloping side

$$H'_a = \{x = (x_1, \ldots, x_d) \in D' \; ; \; \sum_{i=1}^{d} x_i = p\}.$$

Let us denote $T_m = \inf\{k \geq 1; Y_m(k) \in \Gamma'_a\}$ and $Z_m = Y_m(T_m)$. Then,

$$\frac{T_m}{m^2} \xrightarrow{\mathcal{L}} T$$

and

$$\frac{Z_m}{m} \xrightarrow{\mathcal{L}} Z$$

where $T = \inf\{t > 0; W(t) \in H'_a\}$ and $Z = W(T)$.

This result holds for any dimension d and the density of the limiting process before absorption, the distribution of the random variables T and Z can also be deduced via a geometric transformation from the Brownian motion evolving inside the hypercube $[0, 1]^d$ with reflections on the sides as in Theorem 8.2. In the two stacks model, we had an explicit expression for the density of the limiting process before absorption, in the case of the Banker algorithm for dimension two, as p is assumed to be strictly greater than one, the symmetry of the domain is lost and no reflection principle is possible. We can give a partial answer in the case for diagonal A, for simplicity let us take $A = I_2$ (the case of any diagonal

130 DYNAMIC RANDOM WALKS

matrix A can be solved using a correct change of variables), we have to solve the PDE's problem,

$$\frac{\partial f}{\partial t} = \frac{1}{2}\Delta f \quad \text{in } D' \tag{8.6}$$

$$f = 0, \forall\, t \text{ along } H'_a \tag{8.7}$$

$$\frac{\partial f}{\partial \nu} = 0, \forall\, t \text{ along } \partial D' - H'_a \tag{8.8}$$

$$f(x, y, 0) = \delta(x - x_0, y - y_0) \tag{8.9}$$

where Δ is the two-dimensional Laplacian operator. A partial answer is obtained using the method developed in [124]. The density of the limiting process $(W(t))_t$ before absorption is given by

$$q(t, x, y) = \sum_{n=1}^{\infty} D_n e^{-\lambda_n t/2} u_n(x, y)$$

where D_n are real coefficients to be chosen in order to satisfy the initial condition (8.9).

The eigenvalues λ_n and the eigenfunctions u_n are solutions of a classical mixed homogeneous boundary value problem for the two-dimensional Helmholtz equation:

$$-\Delta u_n = \lambda_n u_n \quad \text{in } D' \tag{8.10}$$

$$u_n = 0 \quad \text{along } H'_a \tag{8.11}$$

$$\frac{\partial u_n}{\partial \nu} = 0 \quad \text{along } \partial D' - H'_a \tag{8.12}$$

As shown in [38], the eigenfunctions u_n may be written as:

$$u_n = \sum_{p=1}^{\infty} C_{inp} J_{\mu_{ip}}(\sqrt{\lambda_n} r_i) g(\mu_{ip} \theta_i), \quad \text{for } (r_i, \theta_i) \in \bar{D}'_i \tag{8.13}$$

where $\bar{D}'_i, i = 1, \cdots, 5$ is the closure of the i-th subdomain D'_i, $i = 1, \cdots, 5$ that constitute the subdivision of D', $\bar{D}' = \bar{D}'_1 \cup \bar{D}'_2 \cup \cdots \cup \bar{D}'_5$. Every $\bar{D}'_i$ contains at most one vertex of $\partial D'$ and $D'_i \cap D'_j = \phi$ for $i \neq j$. Moreover, $\partial \bar{D}'_i \cap \partial \bar{D}'_j = \Gamma_{ij}$ for $i \neq j$, with $\Gamma_{ij} = \phi$ if it is formed of a finite number of points.

(r_i, θ_i) are polar coordinates centered at a point $P_i \in \partial D' \cap \partial D'_i$.

$C_{inp}, i = 1, \cdots, 5; p = 1, 2, \cdots$ are real coefficients to be properly chosen. More precisely, the C_{inp} should be taken in order to ensure the continuity of u_n and of its normal derivative along $\Gamma_{ij}, i \neq j; i, j = 1, \cdots, 5$.

μ_{ip} are real numbers whose values depend on the particular boundary conditions.

$J_{\mu_{ip}}(\cdot)$ is the Bessel function of the first kind and of order μ_{ip}.
$g(\cdot)$ is a sine or cosine function depending on the boundary conditions.
Finally, taking into account that (for $\lambda_n \neq \lambda_m$)

$$\int\int_{D'} u_n(x,y)u_m(x,y)dx\,dy = \begin{cases} 0 & \text{if } m \neq n \\ \|u_n\|^2 & \text{if } m = n \end{cases} \quad (8.14)$$

and

$$\int\int_{D'} \delta(x-x_0, y-y_0)u_n(x,y)dx\,dy = u_n(x_0, y_0), \quad (8.15)$$

we obtain

$$q(t,x,y) = \sum_{n=1}^{\infty} \frac{u_n(x_0, y_0)u_n(x,y)}{\|u_n\|^2} e^{-\frac{1}{2}\lambda_n t}. \quad (8.16)$$

The separation of D' into subdomains D'_i is done by taking into account the convergence properties of the series (8.13). The decomposition of D' into five subdomains is sketched in Figure 8.3.

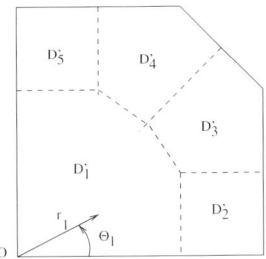

Figure 8.3. Decomposition of domain D' into five subdomains.

As easily verified, the eigenfunctions (8.13) are given by

$$u_n = \begin{cases} \sum_{p=1}^{\infty} C_{inp} J_{2p-2}(\sqrt{\lambda_n} r_i) \cos[(2p-2)\theta_i] & (r_i, \theta_i) \in \bar{D}'_i,\ i=1,2,5 \\ \sum_{p=1}^{\infty} C_{3np} J_{(4p-2)/3}(\sqrt{\lambda_n} r_3) \sin\left[\frac{4p-2}{3}\theta_3\right] & (r_3, \theta_3) \in \bar{D}'_3 \\ \sum_{p=1}^{\infty} C_{4np} J_{(4p-2)/3}(\sqrt{\lambda_n} r_4) \cos\left[\frac{4p-2}{3}\theta_4\right] & (r_4, \theta_4) \in \bar{D}'_4 \end{cases} \quad (8.17)$$

If we take as initial condition (8.9) a Dirac δ-function at the origin, we get

$$D_n = \frac{u_n(0,0)}{\|u_n\|^2} = \frac{C_{1n1}}{\|u_n\|^2} \quad (8.18)$$

and

$$q(t,x,y) = \sum_{n=1}^{\infty} \frac{C_{1n1}}{\|u_n\|^2} u_n(x,y) e^{-\frac{1}{2}\lambda_n t} \qquad (8.19)$$

Unfortunately, the coefficients C_{inp} and the eigenvalues λ_n cannot be computed exactly by an explicit formula. These numbers are to be found by solving a generalized eigenvalue problem for an infinite linear algebraic system (see [38]), which is impossible in the present case. But it is possible to obtain accurate approximations to these values, we refer to [124] where numerical results are given.

2.2 TIME AND SPACE DEPENDENT TRANSITIONS

In this section we present the model developed by Comets-Delarue-Schott [25] which allows the transition probabilities to be (in some sense) time and space dependent.

More precisely we restrict our presentation to some hints since a complete study of this model is out of the scope of this book. See [34] for additional computations and proofs.

Walk without Boundary Conditions.

We define first the transitions of the chain without taking care of the boundary conditions. Recall to this end that the transition matrix of a time-space homogeneous random walk to the nearest neighbours in $\mathbb{Z}^d$, $d \geq 1$, reduces to a probability $p(\cdot)$ on the $2d$ directions of the discrete grid,

$$\mathcal{V} = \{e_1, -e_1, \ldots, e_d, -e_d\},$$

where $(e_i)_{1 \leq i \leq d}$ denotes the canonical basis of $\mathbb{R}^d$. For given $i \in \{1, \ldots, d\}$ and $u \in \mathcal{V}$, $p(u)$ simply denotes the probability of going from the current position x to $x + u$.

We assume for a moment that the transition matrices are space homogeneous but depend on time through some environment (evolving with time) with values in a finite space E, $N = |E|$. We then need to consider, not a single transition probability, but a family $p(1, \cdot), \ldots, p(N, \cdot)$ of N probabilities on $\mathcal{V}$. If the environment at time n is in the state $i \in E$, then the transition of the Markov chain at that time is governed by the probability $p(i, \cdot)$.

When the environment is given by a stochastic process $(\xi_n)_{n \geq 0}$ on E, the jumps $(J_n)_{n \geq 0}$ of the random walk are such that, for every $u \in \mathcal{V}$:

$$\mathbb{P}(J_{n+1} = u | \mathcal{F}_n^{\xi, J}) = p(\xi_n, u), \qquad (8.20)$$

with $\mathcal{F}_n^{\xi,J} = \sigma(\xi_0, \ldots, \xi_n, J_0, \ldots, J_n)$.

From now on, we assume that the environment $(\xi_n)_{n \geq 1}$ is a time-homogeneous Markov chain on E, and we denote by P its transition matrix, $P(k, \ell) = \mathbb{P}(\xi_{n+1} = \ell | \xi_n = k)$ for $k, \ell \in E$. Then, the couple $(\xi_n, J_n)_{n \geq 1}$ is itself a time-homogeneous Markov chain on the product space $E \times \mathbb{Z}^d$ governed by the following transition:

$$\forall k \in E, \; \forall u \in \mathcal{V}, \; \mathbb{P}(\xi_{n+1} = k, J_{n+1} = u | \mathcal{F}_n^{\xi,J}) = P(\xi_n, k)p(\xi_n, u). \tag{8.21}$$

We define the position of the walker in $\mathbb{Z}^d$:

$$S_0 = 0, \quad \forall n \geq 0, \; S_{n+1} = S_n + J_{n+1}. \tag{8.22}$$

In view of the applications mentioned above, the model is not fine enough. For this reason, we also assume that the steps $(J_n)_{n \geq 1}$ depend on the walker position in the following way: for all $k \in E$ and $u \in \mathcal{V}$,

$$\mathbb{P}(\xi_{n+1} = k, J_{n+1} = u | \mathcal{F}_n^{\xi,J}) = P(\xi_n, k)p(\xi_n, S_n/m, u), \tag{8.23}$$

where m denotes a large integer that refers to the size of the box (see Figure 8.2) and, for each $k \in E$ and $y \in \mathbb{R}^d$, $p(k, y, \cdot)$ a probability on $\mathcal{V}$. Note that the random walk $(S_n = S_n^{(m)})_{n \geq 0}$ depends on the parameter m. Nevertheless, for simplicity we will often forget the dependence on m in our notations. In other words, $(\xi_n, S_n)_{n \geq 0}$ defines a Markov chain with rates: $\forall k \in E, \; \forall u \in \mathcal{V}$,

$$\mathbb{P}(\xi_{n+1} = k, S_{n+1} = u + S_n | \mathcal{F}_n^{\xi,S}) = P(\xi_n, k)p(\xi_n, S_n/m, u),$$

where $\mathcal{F}_n^{\xi,S} = \sigma(\xi_0, \ldots, \xi_n, S_0, \ldots, S_n)$.

Walk with Reflection Conditions.

The original problem with reflection on the hyperplanes $y_i = 0$, $i \in \{1, \ldots, d\}$ and $y_i = m$, $i \in \{1, \ldots, d\}$ follows from a slight correction of the former one. The underlying reflected walk $(R_n)_{n \geq 0}$ (also denoted by $(R_n^{(m)})_{n \geq 0}$ to specify the dependence on m) satisfies, with $\mathcal{F}_n^{\xi,R} = \sigma(\xi_0, \ldots, \xi_n, R_0, \ldots, R_n)$:

$$\forall k \in E, \; \forall u \in \mathcal{V}, \; \mathbb{P}(\xi_{n+1} = k, R_{n+1} = u + R_n | \mathcal{F}_n^{\xi,R})$$
$$= P(\xi_n, k)q(\xi_n, R_n/m, u),$$

where q denotes the kernel:

$$\forall k \in E, \; \forall y \in [0,1]^d, \; \forall l \in \{1, \ldots, d\}, \; q(k, y, \pm e_l)$$
$$= p(k, y, \pm e_l) \text{ if } 0 < y_l < 1.$$

If $y_l = 1$, then $q(k,y,e_l) = 0$ and $q(k,y,-e_l) = p(k,y,e_l) + p(k,y,-e_l)$.
If $y_l = 0$, then $q(k,y,-e_l) = 0$ and $q(k,y,e_l) = p(k,y,e_l) + p(k,y,-e_l)$.

The deadlock time of the banker algorithm is then given by $T^{(m)} = \inf\{n \geq 0, \sum_{l=1}^{d} \langle R_n, e_l \rangle \geq \Lambda\}$. This also writes:

$$T^{(m)} = \inf\{n \geq 0,\ R_n \in mF_0\}$$

with

$$F_0 = \{y \in [0,2]^d,\ \sum_{i \leq d} |y_i - 1| \leq (d - \lambda)\} \tag{8.24}$$

(see Figure 8.4 below). This latter form will be useful in the proof of the main results.

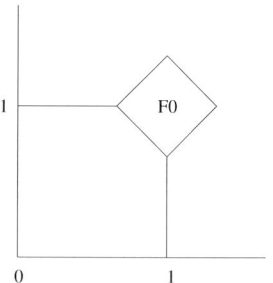

Figure 8.4.   Absorption set.

2.2.1 MAIN ASSUMPTIONS

In formula (8.23), the division by m indicates that the dependence of the transition kernel on the position of the walker takes place at scale m. For large m, the space dependence is mild, since we will assume all through the paper the following smoothness property:

Assumption (A.1). *The function p is twice continuously differentiable with respect to y with bounded derivatives. In particular, there exists a constant $K > 0$ such that:*

$$\forall x \in E,\ u \in \mathcal{V},\ (y,y') \in (\mathbb{R}^d)^2,\ |p(x,y,u) - p(x,y',u)| \leq K|y - y'|. \tag{8.25}$$

It is then readily seen that the transition kernel (8.23) weakly depends on the space position of the walker: a step of the walker modifies only slightly the transition kernel.

We also assume the environment to be ergodic and to fulfill the so-called central limit theorem for Markov chains. We thus impose the following sufficient conditions:

Assumption (A.2). *The matrix P is irreducible on E. It is then well known that P admits a unique invariant probability measure, denoted by μ.*

Assumption (A.3) *The matrix P is aperiodic. In particular, it satisfies the Doeblin condition:*

$$\exists m \geq 0, \ \exists \eta > 0, \ \forall (k,\ell) \in E^2, \ (P^m)(k,\ell) \geq \eta. \tag{8.26}$$

Assumption (A.4) *Define for $k \in E$ and $y \in \mathbb{R}^d$,*

$$g(k,y) = \sum_{u \in \mathcal{V}} [p(k,y,u)u]$$

(i.e., $g(k,y)$ matches the expectation of the measure $p(k,y,\cdot)$) and assume that $g(\cdot,y)$, seen as a function from E to $\mathbb{R}^d$, is centered with respect to the measure μ.

Let us briefly describe the role of each of these conditions:

1. Thanks to Assumption (A.2), the Markov chain $(\xi_n)_{n \geq 0}$ satisfies the ergodic theorem for Markov processes.

2. The general central limit theorem for Markov chains with finite state space follows from the Doeblin condition, given in Assumption (A.3).

3. Assumption (A.4) permits to apply the previous central limit theorem to the function g.

2.2.2   FREQUENTLY USED NOTATIONS

For a square integrable martingale M, $\langle M \rangle$ denotes the bracket of M (do not mix up with the Euclidean scalar product, which is denoted by $\langle x, y \rangle$ for $x, y \in \mathbb{R}^d$).

We also denote by $\mathcal{D}(\mathbb{R}_+, \mathbb{R}^d)$ the path space of right continuous and left limited functions from $\mathbb{R}_+$ into $\mathbb{R}^d$. A function in $\mathcal{D}(\mathbb{R}_+, \mathbb{R}^d)$ is then said to be "càd-làg" for the French acronym "continue à droite-limite à gauche".

2.2.3   MAIN RESULTS
2.2.4   ASYMPTOTIC BEHAVIOUR OF THE WALK WITHOUT REFLECTION

In light of Assumptions (A.1-4), we expect the global effect of the environment process $(\xi_n)_{n \geq 0}$ to reduce for large time to a deterministic one. To this end, we view the process $(\bar{S}_t^{(m)} = m^{-1}S_{\lfloor m^2 t\rfloor}^{(m)})_{t\geq 0}$ as a random element in the space $\mathcal{D}(\mathbb{R}_+, \mathbb{R}^d)$:

THEOREM 8.5 *The process $(\bar{S}_t^{(m)})_{t\geq 0}$ converges in law in $\mathcal{D}(\mathbb{R}_+, \mathbb{R}^d)$ endowed with the Skorohod topology towards the (unique) solution of the martingale problem starting from the origin at time zero and associated to the operator:*

$$\mathbb{L} = \frac{1}{2}\sum_{i,j=1}^{d} \bar{a}_{i,j}(y)\frac{\partial^2}{\partial x_i \partial x_j} + \sum_{i=1}^{d} \bar{b}_i(y)\frac{\partial}{\partial y_i}. \qquad (8.27)$$

The limit coefficients $\bar{a}$ and $\bar{b}$ are given by:

$$\forall y \in \mathbb{R}^d, \quad \bar{a}(y) = \int_E [\alpha + gv^t + vg^t - 2gg^t](i,y)d\mu(i), \qquad (8.28)$$
$$\bar{b}(y) = \int_E [(\nabla_y v - \nabla_y g)g](i,y)d\mu(i), \qquad (8.29)$$

where ∇ stands for the gradient and $\alpha(i,y)$ denotes the second order moment matrix of the measure $p(i,y,\cdot)$:

$$\forall i \in E, \forall y \in \mathbb{R}^d, \quad \alpha(i,y) = \sum_{u \in \mathcal{V}} [uu^t p(i,y,u)], \qquad (8.30)$$

and $v(i,y)$ denotes the well-defined sum:

$$\forall i \in E, \forall y \in \mathbb{R}^d, \quad v(i,y) = \sum_{n\geq 0}\sum_{j\in E}[P^n(i,j)g(j,y)]. \qquad (8.31)$$

Moreover, for every $y \in \mathbb{R}^d$, the lowest eigenvalue of $\bar{a}(y)$ is greater than or equal to the lowest eigenvalue of the matrix:

$$\int_E [\alpha(i,y) - gg^t(i,y)]d\mu(i). \qquad (8.32)$$

In particular, if the covariance matrices of the measures $(p(i,y,\cdot))_{i\in E, y\in \mathbb{R}^d}$ are uniformly elliptic, the matrices $(\bar{a}(y))_{y\in \mathbb{R}^d}$ are also uniformly elliptic.

The existence of v, $\nabla_y v$ and $\nabla_y g$ will be detailed in the sequel of the paper.

We denote by $\bar{\sigma}(x)$, for $x \in \mathbb{R}^d$, the nonnegative symmetric square root of $\bar{a}(x)$. Then, the SDE (stochastic differential equation) associated to $\mathbb{L}$ writes:

$$d\bar{X}_t = \bar{b}(\bar{X}_t)dt + \bar{\sigma}(\bar{X}_t)dB_t, \qquad (8.33)$$

where $(B_t)_{t\geq 0}$ denotes a d-dimensional Brownian motion on a given filtered probability space $(\Omega, (\mathcal{F}_t)_{t\geq 0}, \mathbb{P})$. Under Assumptions (A.1-4),

(8.33) is uniquely strongly solvable.

Of course, if $p(k, y, \cdot)$ does not depend on y, then the drift $\bar{b}$ vanishes in Theorem 8.5 since $b = 0$ and the matrix-valued function $\bar{a}$ is constant so that $\bar{X}$ is a non-standard Brownian motion.

Theorem 8.5 appears actually as an homogenization property: on the long run, the time inhomogeneous rescaled walk $\bar{S}^{(m)}$ behaves like an homogeneous diffusion. Underlying techniques to establish the asymptotic behaviour of $(\bar{S}_t^{(m)})_{t\geq 0}$ are well known in the literature devoted to this subject (see e.g. Bensoussan et al. [5] for a review on homogenization in periodic structures or Jikov et al. [90] for a monograph on stochastic homogenization).

We finally mention that Theorem 8.5 is very close to Theorem 8.4. Indeed, any dynamical system $(E, \mathcal{A}, \mu, T)$ generates an homogeneous Markov chain with degenerate transitions: $\forall k \in E$, $P(k, Tk) = 1$ (but the state space E is then very large). In this framework, the condition $\int_E f_j d\mu = (2d)^{-1}$ in Theorem 8.4 implies that the expectation against μ of the transitions $p(T^i k, e_j) = f_j(T^i x)$ and $p(T^i_k, -e_j) = 1/d - f_j(T^i x)$ vanishes, and thus amounts in some sense to Assumption (A.4) in this section. In the same way, condition (H_2) on the functions f_i, $f_i f_j$ in Theorem 8.4 refers more or less to a "degenerate" central limit theorem for the underlying dynamical system and thus to our standing Assumption (A.3).

However, Theorem 8.5 does not recover exactly the result of Section 2.1: due to the degeneracy of the transitions of a dynamical system, condition (A.3) cannot be satisfied. Actually, the reader must understand that condition (A.3) is a simple but very strong technical condition to ensure the validity of the so-called "central limit theorem" for Markov chains. It thus permits to draw up a clear framework in which the stochastic homogenization theory applies, but is obviously far from being optimal (refer to Olla [142] for a complete overview on central limit theorems for Markov chains).

Theorem 8.5 is proved in [25].

2.2.5 ASYMPTOTIC BEHAVIOUR OF THE REFLECTED WALK

The rescaled walk $(\bar{R}_t^{(m)} = m^{-1} R^{(m)}_{\lfloor m^2 t \rfloor})_{t \geq 0}$ satisfies the following "reflected version" of Theorem 8.5:

THEOREM 8.6 *The process $(\bar{R}_t^{(m)})_{t \geq 0}$ converges in law in $\mathcal{D}(\mathbb{R}_+, \mathbb{R}^d)$ endowed with the Skorohod topology towards the (unique) solution of* *the martingale problem with normal reflection on the boundary of the hypercube* $[0,1]^d$, *with zero as initial condition and with* $\mathbb{L}$ *as underlying operator.*

The reflected SDE associated to the hypercube $[0,1]^d$ and to the operator $\mathbb{L}$ writes:
$$dX_t = \bar{b}(X_t)dt + dH_t - dK_t + \bar{\sigma}(X_t)dB_t. \tag{8.34}$$

We explain now the meaning of the different terms in the r.h.s of (8.34). The quantity $\bar{b}(X_t)dt$ refers to the drift $\bar{b}(\bar{X}_t)dt$ in (8.33) and $\bar{\sigma}(X_t)dB_t$ to the stochastic noise $\bar{\sigma}(\bar{X}_t)dB_t$ in (8.33). The new terms dH and dK stand for the differential elements of two continuous adapted processes with bounded variation that prevent X to leave the hypercube $[0,1]^d$. More precisely, H^1 (the first coordinate of H) is an increasing process that pushes (when necessary) the process X^1 to the right to keep it above zero. In the same way, K^1 is an increasing process that pushes (again when necessary) the process X^1 to the left to keep it below one. For each $l \in \{2,\ldots,d\}$, H^l and K^l act similarly in the direction e_l. Both processes H and K act in a minimal way. In particular, there is no reason to push the process X^ℓ when away from the boundary.

The least action principle for H and K can be summarized as follows. The process H^l, for $l \in \{1,\ldots,d\}$, does not increase when X^l is different from zero, and the process K^l, for $l \in \{1,\ldots,d\}$, does not increase when X^l is different from one, *i.e.*,

$$\int_0^{+\infty} \mathbf{1}_{\{X_t^l > 0\}} dH_t^l = 0, \quad \int_0^{+\infty} \mathbf{1}_{\{X_t^l < 1\}} dK_t^l = 0. \tag{8.35}$$

For a complete review on reflected SDE's, we refer the reader to Tanaka [183], Lions and Sznitman [117] and Saisho [157].

The following corollary describes the asymptotic behaviour of the deadlock time and the deadlock point of the algorithm. We assume the matrix $\alpha - gg^t$ to be uniformly elliptic to ensure $\bar{a}$ to be so. In short, the ellipticity of $\bar{a}$ avoids any singular behaviour (up to a $\mathbb{P}$-null set) of the trajectories of the limit process X.

COROLLARY 1 *Assume that* $\lambda \in [1,d)$ *and that the matrix* $\alpha - gg^t$ *is uniformly elliptic. Denote by* T *the deadlock time of the limit process* X: $T = \inf\{t \geq 0, X_t \in F_0\}$, *where* F_0 *is given by (8.24). Then, the sequence of rescaled deadlock times* $m^{-2}T^{(m)}$ *converges in law towards* T. *In the same way, the sequence of rescaled deadlock points* $m^{-1}R_{T^{(m)}}$ *converges in law towards* X_T.

Theorem 8.6 and Corollary 1 are proved in [25].

2.2.6 BEHAVIOUR OF THE LIMIT PROCESS IN DIMENSION TWO

A crucial question for numerical applications consists in estimating precisely the mean value of the deadlock time of the limit system. When the matrix $\bar{a}$ is constant and diagonal and the drift $\bar{b}$ reduces to zero, several explicit computations in terms of Bessel functions are conceivable (see Section 2).

In our more general framework, the story is quite different. Delarue [34], establishes in the two-dimensional case (*i.e.* $d = 2$) relevant estimates of the expectation of T (now denoted by $T_\lambda = \inf\{t \geq 0, \ X_t^1 + X_t^2 \geq \lambda\}$ to take into account the parameter λ), and in particular to distinguish three different asymptotic regimes as the parameter λ tends to two, each of these regimes depending on the covariance matrix $\bar{a}(1,1)$, and more precisely, on the sign of its off-diagonal components.

Since the matrix $\alpha - gg^t$ is assumed to be uniformly elliptic, the matrix $\bar{a}(1,1)$ is positive and writes:

$$\bar{a}(1,1) = \begin{pmatrix} \rho_1^2 & s\rho_1\rho_2 \\ s\rho_1\rho_2 & \rho_2^2 \end{pmatrix}, \tag{8.36}$$

with $\rho_1, \rho_2 > 0$ and $s \in]-1, 1[$. The matrix $\bar{a}(1,1)$ admits two eigenvalues:

$$\lambda_1 = \frac{1}{2}[\rho_1^2 + \rho_2^2 + \delta], \ \lambda_2 = \frac{1}{2}[\rho_1^2 + \rho_2^2 - \delta], \text{ where } \delta = (\rho_1^4 + \rho_2^4 - 2(1-2s^2)\rho_1^2\rho_2^2)^{1/2}. \tag{8.37}$$

Denote by E_1 and E_2 the associated eigenvectors (up to a multiplicative constant). For $s \neq 0$,

$$E_1 = \begin{pmatrix} 1 \\ (2s\rho_1\rho_2)^{-1}(\delta + \rho_2^2 - \rho_1^2) \end{pmatrix}, \ E_2 = \begin{pmatrix} -(2s\rho_1\rho_2)^{-1}(\delta + \rho_2^2 - \rho_1^2) \\ 1 \end{pmatrix}. \tag{8.38}$$

Since $\delta + \rho_2^2 - \rho_1^2 \geq 0$, the signs of the non-trivial coordinates of E_1 and E_2 are given by the sign of s. The main eigenvector (*i.e.* E_1) has two positive components for $s > 0$, and a positive one and a negative one for $s < 0$. Of course, if s vanishes, E_1 and E_2 reduce to the vectors of the canonical basis.

The three different regimes can be distinguished as follows:

Positive Case.

If $s > 0$, the main eigenvector of $\bar{a}(1,1)$ (*i.e.* E_1) is globally oriented from 0 to the neighbourhood of the corner $(1,1)$ and tends to push the

140   DYNAMIC RANDOM WALKS

limit reflected diffusion towards the border line. The reflection on the boundary cancels most of the effects of the second eigenvalue and keeps on bringing back the diffusion along the main axis. As a consequence, the hitting time of the border line is rather small and the following asymptotic holds for the diffusion starting from 0:

$$\sup_{1<\lambda<2} \mathbb{E}(T_\lambda) < +\infty. \tag{8.39}$$

This phenomenon is illustrated below (see Figure 8.5) when $\bar{b}$ reduces to 0 and $\bar{a}$ is the constant matrix given by $\rho_1 = \rho_2 = 1$ and $s = 0, 9$. We have plotted there a simulated trajectory of the reflected diffusion process, starting from 0 at time 0, and running from time 0 to time 10 in the box $[0, 1]^2$. The algorithm used to simulate the reflected process is given in Słomiński [173]. The eigenvector E_1 exactly matches $(1, 1)^t$.

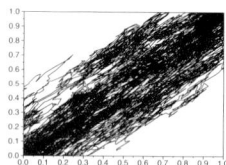

Figure 8.5.   One trajectory of the reflected process over $[0, 10]$ for $s = 0, 9$.

Negative Case.

If $s < 0$, the main eigenvector of $\bar{a}(1, 1)$ is globally oriented from $(1, 0)$ to the neighbourhood of the corner $(0, 1)$ and attracts the diffusion away from the border line. Again, the reflection on the boundary cancels most of the effects of the second eigenvalue, and thus, acts now as a trap: the diffusion stays for a long time along the main axis and hardly touches the boundary. The hitting time satisfies the following asymptotic behaviour when the diffusion starts from 0:

$$\exists c_1, c_2 \geq 1, \ \forall \lambda \in \,]1, 2[, \ c_1^{-1}(2-\lambda)^{-c_2} - c_1 \leq \mathbb{E}(T_\lambda) \leq c_1(2-\lambda)^{-c_2^{-1}} + c_1. \tag{8.40}$$

This point is illustrated by Figure 8.6 below when $\bar{b}$ vanishes and $\bar{a}$ reduces to the constant matrix $\rho_1 = \rho_2 = 1$ and $s = -0, 9$ (again, the initial condition of the process is 0). The eigenvector E_1 is given, in this case, by $(1, -1)^t$.

Null Case.

The case $s = 0$ is intermediate. Eigenvectors are parallel to the axes

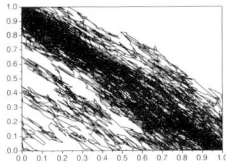

Figure 8.6. One trajectory over $[0, 10]$ for $s = -0, 9$ (the trajectory is stopped before T).

and the behaviour of the diffusion is close to the behaviour of the two-dimensional Brownian motion. For the initial condition 0:

$$\exists c_1 \geq 1, \ \forall \lambda \in \,]1, 2[, \ -c_1^{-1} \ln(2 - \lambda) - c_1 \leq \mathbb{E}(T_\lambda) \leq -c_1 \ln(2 - \lambda) + c_1. \quad (8.41)$$

This is illustrated by Figure 8.7 below when $\bar{b}$ vanishes and $\bar{a}$ reduces to the identity matrix (the initial condition of the process is 0).

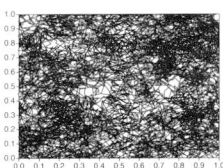

Figure 8.7. One Trajectory over $[0, 10]$ for $s = 0$.

The following theorem sums up these different cases (see Delarue [34] for the proof).

THEOREM 8.7 *Assume that $\alpha - gg^t$ is uniformly elliptic. Then, there exists a constant $C \geq 1$, depending only on known parameters and on the ellipticity constant of $\bar{a}$, such that:*

1. *For $s > 0$,* $\sup\limits_{l \in \,]1,2[} \mathbb{E}(T_l) \leq C$

2. *For $s < 0$, set $\beta_- = -s > 0$, $\beta_+ = s(s-3)(1+s)^{-1} > 0$. Then,*

$$\forall l \in \,]1, 2[, \ C^{-1}(2-l)^{-\beta_-} - C \leq \mathbb{E}(T_l) \leq C(2-l)^{-\beta_+} + C. \quad (8.42)$$

3. *If $s = 0$, $\forall l \in \,]1, 2[, \ -C^{-1} \ln(2-l) - C \leq \mathbb{E}(T_l) \leq -C \ln(2-l) + C$.*

Note that Theorem 8.7 leaves open many questions. For example, we do not know how to compute, for $s < 0$, the exact value of the "true" exponent $\beta = \inf\{c > 0,\ \sup_{l \in]1,2[}[(2-l)^c \mathbb{E}(T_l)] < +\infty\}$. We are even unable to precise the asymptotic behaviour of β as $s \to -1$ (note indeed that $\lim_{s \to -1} \beta_- = 1$, $\lim_{s \to -1} \beta_+ = +\infty$).

We have very few ideas about the extension of Theorem 8.7 to the upper dimensional cases. The only accessible case for us is $\bar{a}(1,\ldots,1) = I_d$, I_d denoting the identity matrix of size d: in this case, the analysis derives from the transience properties of the Brownian motion in dimension $d \geq 3$.

References

This chapter is based on [25, 72].

Chapter 9

DATA STRUCTURES WITH DYNAMICAL RANDOM TRANSITIONS

1. INTRODUCTION

The time and integrated space costs of sequences of operations on list structures have been estimated by Flajolet et al. [56] by combinatorial methods. Louchard [119] and Maier [127] presented two different probabilistic analyzes of these dynamic data structures which led to the same conclusion: the integrated space and time costs of a sequence of n supported operations converge, as n goes to infinity, to Gaussian random variables. All the above mentioned results have been proved under a set of assumptions which constitute the so called Markovian model and assuming uniform ditribution on the set of histories. Following the conclusions of Knuth [100] about deletions that preserve randomness, Louchard et al [123] have shown how to analyze dynamic data structures in the more realistic model proposed by Knuth. The maxima properties (value and position) of these data structures have been analyzed by Louchard, Kenyon and Schott [122]. As discussed by Maier [127], the model of equiprobable histories is unrealistic and necessitates rejection. The main purpose of this chapter is to derive the asymptotic probability distribution on data structure sizes considered as dynamic random walks.

2. PRELIMINARIES

A data type is a specification of the basic operations allowed together with its set of possible restrictions. The following data types are commonly used

Stack: keys are accessed by position, operations are insertion I and deletion D but are restricted to operate on the key positioned first in

the structure (the "top" of the stack).
Linear list: keys are accessed by position, operations are insertion I and deletion D without access restrictions (linear lists make it possible to maintain dynamically changing arrays).
Dictionary: keys belonging to a totally ordered set are accessed by value, all four operations I, D, Q^+, Q^- are allowed without any restriction. Q^+ represents a positive (successful) query (or search). Q^- stands for a negative (unsuccessful) query (or search).
Priority queue: keys belonging to a totally ordered set are accessed by value, the basic operations are I and D, deletion D is performed only on the key of minimal value (of "highest priority").
Symbol table: this type is a particular case of dictionary where deletion always operates on the key last inserted in the structure, only positive queries are performed.
A data organization is a machine implementation of a data type. It consists of a data structure which specifies the way objects are internally represented in the machine, together with a collection of algorithms implementing the operations of the data type.
Stacks are almost always implemented by arrays or linked lists.
Linear lists are often implemented by linked lists and arrays.
Dictionaries are usually implemented by sorted or unsorted lists; binary search trees have a faster execution time and several balancing schemes have been proposed: AVL, 2-3 and red-black trees. Other alternatives are h-tables and digital trees.
Priority queues can be represented by any of the search trees used for dictionaries, more interesting are heaps, P-tournaments, leftist tournaments, binomial tournaments, binary tournaments and pagodas. One can also use sorted lists, and any of the balanced tree structures.
Symbol tables are special cases of dictionaries, all the known implementations of dictionaries are applicable here.

DEFINITION 9.1 *A schema (or path) is a word*

$$\Omega = O_1 O_2 \ldots O_n \in \{I, D, Q^+, Q^-\}^*$$

such that for all j, $1 \leq j \leq n$:

$$|O_1 O_2 \ldots O_j|_I \geq |O_1 O_2 \ldots O_j|_D.$$

$\{I, D, Q^+, Q^-\}^*$ is the free monoid generated by the alphabet

$$\{I, D, Q^+, Q^-\}$$

$|w|$ is the length of the word w.

A schema is to be interpreted as a sequence of n requests (the keys operated on not being represented).
Example: The figure below shows a schema.

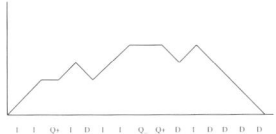

Figure 9.1.   A schema

DEFINITION 9.2 *A structure history is a sequence of the form:*

$$h = O_1(r_1)O_2(r_2)\ldots O_n(r_n)$$

where $\Omega = O_1 O_2 \ldots O_n$ is a schema, and the r_j are integers satisfying: $0 \leq r_j < pos(\alpha_{j-1}(\Omega))$ and $\alpha_j(\Omega) = |O_1 O_2 \ldots O_j|_I - |O_1 O_2 \ldots O_j|_D$ is the size (level) of the structure at step j, pos is a possibility function defined on each request, r_j is the rank (or position) of the key operated upon at step j.

We will only consider schemas and histories with initial and final level 0.

Possibility functions.
Two different models have been considered for defining possibility functions: the markovian model [56, 119] in which possibility functions are linear functions of the size k of the data structure when an allowed operation is performed.
Knuth's model is related to his observation [100] that deletions may not preserve randomness and is more realistic than the markovian model. The following simple example may be helpful to understand Knuth's fundamental remark.
Consider again the sequence of operations $IIDI$ performed, for example, on a linear list which is initially empty. Let $x < y < z$ be the three keys inserted during the sequence III. x, y and z are deleted with equal probability. Let w be the key inserted by the fourth I. Then all four cases $w < x < y < z$, $x < w < y < z$, $x < y < w < z$, $x < y, z < w$ do occur with equal probability, whatever the key deleted. More generally, let us consider a sequence of operations $O_1 O_2 \ldots O_j$ on a dictionary data type, the initial data structure being empty (any data type listed above may be considered). Assume O_j is the ith I or Q^- of the sequence. Let $x_1 < x_2 < \ldots < x_{i-1}$ be the keys inserted and negatively searched

during the sequence $O_1 O_2 \ldots O_{j-1}$, and let w be the ith inserted or negatively searched key. Then all the cases $w < x_1 < x_2 < \ldots < x_{i-1}$, $x_1 < w < x_2 < \ldots < x_{i-1}, \ldots, x_1 < x_2 < \ldots < x_{i-1} < w$ are equally likely, whatever the deleted keys. Put into combinatorial words: after j operations, whose i are I and Q^-'s (thus $j - i$ are D and Q^+'s), the size of the data structure is $k \leq 2i - j$. The keys of the data structure can be considered as a subset of k distinct objects of a set of size i any of the C_i^k possible subsets being equally likely. We say that the number of possibilities of the ith I or Q^- (in a sequence of operations) is equal to i in Knuth's model whatever the size of the data structure when this insertion (or negative query) occurs. We summarize in the two tables below the differences between the markovian and Knuth's models. We consider only a few data structures.

Data type	Npos(I,k)	Npos(D,k)	Npos(Q^+,k)	Npos(Q^-,k)
Dictionary	k+1	k	k	k+1
Priority queue	k+1	1	0	0
Linear list	k+1	k	0	0

Table 9.1.   Possibility functions in the markovian model.

Data type	Npos($i^{theorem}$ I)	Npos(D,k)	Npos(Q^+,k)	Npos($i^{th} Q^-$)
Dictionary	i	k	k	i
Priority queue	i	1	0	0
Linear list	i	k	0	0

Table 9.2.   Possibility functions in Knuth's model

If instead of deleting items from the data structure as described above we wait until there is a new insertion, we get a new operation called lazy deletion (see [122] and the references therein). Batched insertion waits until there is a new deletion. The probabilistic model described below applies also for these operations but we restrict our analysis to dynamic data structures subject to the classical operations: I, D, Q^+, Q^-.

3.   DYNAMIC LINEAR LISTS

Let $(E, \mathcal{A}, \mu, T)$ be a dynamical system where E is a compact metric space, $\mathcal{A}$ the associated Borel σ-field, T a continuous transformation of E and μ a probability measure on $(E, \mathcal{A})$ such that $T\mu = \mu$. It is assumed to be uniquely ergodic. Let f be a continuous function defined on E with values in $[0, 1]$, such that

$$\int_E |\log[f(1-f)]|\, d\mu \; < \; \infty. \tag{9.1}$$

Let $x \in E$ be a fixed point and $(S_k)_{0 \leq k \leq n}$ the associated one-dimensional dynamic random walk defined as in Section 2 of Chapter 1, conditioned to remain nonnegative and such that $S_n = 0$. From Table 9.1, the evolution of dynamic linear lists is modeled by the one-dimensional dynamic random walk $(S_k)_{0 \leq k \leq n}$ each path being assigned relative probability

$$\prod_{i=1}^{n} (S_{i-1} + 1)$$

since we have to take into account the number of places where we can delete or insert an item in the list. We shall denote this weighted random walk by $(S^w_{ll,k})_{0 \leq k \leq n}$. The normalized data structure size as a function of time is then defined by

$$S^w_{ll,n}(t) = \frac{S^w_{ll,[nt]}}{n}, \quad t \in [0, 1].$$

The storage cost function defined as

$$C_{ll,n} = n \sum_{i=1}^{n} S^w_{ll,n}\left(\frac{i}{n}\right)$$

is asymptotically equivalent to

$$n^2 \int_0^1 S^w_{ll,n}(t)\, dt.$$

The asymptotic behavior of the random process $S^w_{ll,n}(t)$ will determine those of $C_{ll,n}$.

Let Ω be the space of functions ϕ defined on $[0, 1]$ vanishing at time 0, with left limit and right continuous at each point $t \in [0, 1]$. The space Ω is endowed with the topology of uniform convergence on the interval $[0, 1]$. The distribution of the random variable $(S^w_{ll,n}(t))_{t \in [0,1]}$ with values in Ω is denoted by $\mathbb{Q}^{(ll)}_{x,n}$.

THEOREM 9.1 *For any $x \in E$, the sequences $(S^w_{ll,n}(t))_{t \in [0,1]}$ converge, as n goes to infinity, to the unique function $(\phi_{ll}(t))_{t \in [0,1]}$, solution of the Euler-Lagrange equation*

$$\frac{d}{dt}\frac{\partial L_{ll}}{\partial y}(\phi, \dot{\phi}) - \frac{\partial L_{ll}}{\partial x}(\phi, \dot{\phi}) = 0$$

where $L_{ll}(x,y) = \Lambda^*(y) - \log(x)$ with boundary conditions: $\phi(0) = \phi(1) = 0$. The convergence is exponential in the sense that for any $\varepsilon > 0$, there exists $\delta = \delta(\varepsilon) > 0$ such that for all sufficiently large n,

$$\mathbb{Q}_{x,n}^{(ll)}(\{\phi \in \Omega \mid \|\phi - \phi_{ll}\|_\infty \geq \varepsilon\}) \leq \exp(-n\delta).$$

Proof:
The proof of the convergence of the sequence $(S_{ll,n}^w(t))_{t \in [0,1]}$ to ϕ_{ll} is essentially based on an adaptation in our particular case of Varadhan's integral lemma (see Section 4.3 in [37]). Let us recall that $\mathbb{P}_{x,n}$ denotes the distribution of the random variable $\left(\frac{S_{[nt]}}{n}\right)_{t \in [0,1]}$. For every $n \geq 1$, we define the functional $F_n : \Omega \longrightarrow [-\infty, +\infty)$ by

$$F_n(\phi(.)) = \begin{cases} \int_0^1 \log(\phi(t) + 1/n) \, dt, & \text{if } \phi(1) = 0. \\ -\infty, & \text{otherwise.} \end{cases}$$

as well as the function

$$F(\phi(.)) = \begin{cases} \int_0^1 \log \phi(t) \, dt, & \text{if } \phi(1) = 0 \\ -\infty, & \text{otherwise.} \end{cases}$$

With these notations, the distribution $\mathbb{Q}_{x,n}^{(ll)}$ of the random variable $(S_{ll,n}^w(t))_{t \in [0,1]}$ with values in Ω can be written as

$$\mathbb{Q}_{x,n}^{(ll)}(A) = \frac{\int_A \exp(nF_n(\phi)) \, d\mathbb{P}_{x,n}}{\int_\Omega \exp(nF_n(\phi)) \, d\mathbb{P}_{x,n}}, \quad \text{for every } A.$$

Theorem 9.1 will be deduced from both following inequalities: For any closed set $A \subseteq \Omega$,

$$\limsup_{n \to \infty} \frac{1}{n} \log \left[\int_A \exp(nF_n(\phi)) \, d\mathbb{P}_{x,n} \right] \leq - \inf_{\phi \in A} (I - F)(\phi) \quad (9.2)$$

and

$$\liminf_{n \to \infty} \frac{1}{n} \log \left[\int_\Omega \exp(nF_n(\phi)) \, d\mathbb{P}_{x,n} \right] \geq - \inf_{\phi \in \Omega} (I-F)(\phi) = F(\phi_{ll}) - I(\phi_{ll}) \quad (9.3)$$

where I is given in Theorem 2.6.

Proof of (9.2): The tail condition:

$$\lim_{L \to \infty} \limsup_{n \to \infty} \frac{1}{n} \log \left[\int_{F_n(\phi) \geq L} \exp(nF_n(\phi)) \, d\mathbb{P}_{x,n} \right] = -\infty$$

is clearly satisfied since for all n and for $\mathbb{P}_{x,n}$–almost every ϕ,
$$F_n(\phi) \leq 1 + ||\phi||_\infty \leq 2.$$
The upper semi-continuity property:
$$\text{whenever } \phi_n \to \phi, \ \limsup_{n \to \infty} F_n(\phi_n) \leq F(\phi)$$
is a consequence of Fatou's Lemma.

Inequality (9.2) is then deduced from the tail condition and from the upper semi-continuity property as in the proof of Lemma 4.3.6 of [37].

Proof of (9.3): The functions F_n and F are not lower semi-continuous, so we are not able to use Lemma 4.3.4 of [37] to establish the lower bound. The expectation with respect to the probability measure $\mathbb{P}_{x,n}$ is denoted by $\mathbb{E}_{x,n}$. Let $\varepsilon > 0$,

$$\mathbb{E}_{x,n}\left(\exp(nF_n)\right) = \int_\Omega \exp(nF_n(\phi)) \, d\mathbb{P}_{x,n}$$

$$\geq \int_\Omega \mathbf{1}_{\{\phi > \phi_{ll} \text{ on } [\varepsilon, 1-\varepsilon]\}} \exp(nF_n(\phi)) \, d\mathbb{P}_{x,n}$$

$$= \int_\Omega \mathbf{1}_{\{\phi > \phi_{ll} \text{ on } [\varepsilon, 1-\varepsilon]\}} \exp\left(n \int_0^1 \log(\phi(t) + 1/n) \, dt\right) d\mathbb{P}_{x,n}$$

$$\geq \exp\left(n \int_\varepsilon^{1-\varepsilon} \log(\phi_{ll}(t) + 1/n) \, dt\right)$$

$$\times \int_\Omega \mathbf{1}_{\{\phi > \phi_{ll} \text{ on } [\varepsilon, 1-\varepsilon]\}} \exp\left(n \int_0^\varepsilon \log(\phi(t) + 1/n) \, dt\right)$$

$$\exp\left(n \int_{1-\varepsilon}^1 \log(\phi(t) + 1/n) \, dt\right) d\mathbb{P}_{x,n}$$

In the last integral, the random variable $\mathbf{1}_{\{\phi > \phi_{ll} \text{ on } [\varepsilon, 1-\varepsilon]\}}$ only depends on $\phi(t)_{\varepsilon \leq t \leq 1-\varepsilon}$, $\exp\left(n \int_0^\varepsilon \log(\phi(t) + 1/n) \, dt\right)$ only depends on $\phi(t)_{0 \leq t \leq \varepsilon}$ and $\exp\left(n \int_{1-\varepsilon}^1 \log(\phi(t) + 1/n) \, dt\right)$ only depends on $\phi(t)_{1-\varepsilon \leq t \leq 1}$. As $\mathbb{P}_{x,n}$ satisfies the Markov property, these three random variables are independent conditionally to $\phi(\varepsilon)$ and $\phi(1-\varepsilon)$. Let us define

$$Y_1 = \mathbb{E}_{x,n}\left(\mathbf{1}_{\{\phi > \phi_{ll} \text{ on } [\varepsilon, 1-\varepsilon]\}} \mid \phi(\varepsilon), \phi(1-\varepsilon)\right),$$

$$Y_2 = \mathbb{E}_{x,n}\left(\exp(n \int_0^\varepsilon \log(\phi(t) + 1/n)dt) \mid \phi(\varepsilon)\right),$$

$$Y_3 = \mathbb{E}_{x,n}\left(\exp(n \int_{1-\varepsilon}^1 \log(\phi(t) + 1/n)dt) \mid \phi(1-\varepsilon)\right).$$

The Markov property yields

$$\int_\Omega \mathbf{1}_{\{\phi > \phi_{ll} \text{ on } [\varepsilon, 1-\varepsilon]\}} \exp\left(n \int_0^\varepsilon \log(\phi(t) + 1/n) \, dt\right)$$

$$\exp\left(n\int_{1-\varepsilon}^{1} \log\left(\phi(t)+1/n\right) dt\right) d\mathbb{P}_{x,n}$$
$$= \mathbb{E}_{x,n}(Y_1 \, Y_2 \, Y_3).$$

If $\phi(\varepsilon) \leq \phi_{ll}(\varepsilon)$ or $\phi(1-\varepsilon) \leq \phi_{ll}(1-\varepsilon)$, then $Y_1 = 0$.
Let us define

$$K_n^\varepsilon = \inf_{y \geq \phi_{ll}(\varepsilon)} \mathbb{E}_{x,n}\left(\exp(n\int_0^\varepsilon \log\left(\phi(t)+1/n\right) dt) \mid \phi(\varepsilon) = y\right)$$

and

$$L_n^\varepsilon = \inf_{y \geq \phi_{ll}(1-\varepsilon)} \mathbb{E}_{x,n}\left(\exp(n\int_{1-\varepsilon}^1 \log\left(\phi(t)+1/n\right) dt) \mid \phi(1-\varepsilon) = y\right).$$

Then, $Y_1 \, Y_2 \, Y_3 \geq Y_1 \, K_n^\varepsilon \, L_n^\varepsilon$ and

$$\mathbb{E}_{x,n}(Y_1 \, Y_2 \, Y_3) \geq \mathbb{E}_{x,n}(Y_1) \, K_n^\varepsilon \, L_n^\varepsilon = \mathbb{P}_{x,n}(\phi > \phi_{ll} \text{ on } [\varepsilon, 1-\varepsilon]) \, K_n^\varepsilon \, L_n^\varepsilon.$$

This yields

$$\liminf_{n \to \infty} \frac{1}{n} \log\left[\int_\Omega \exp(nF_n(\phi)) \, d\mathbb{P}_{x,n}\right]$$
$$\geq \int_\varepsilon^{1-\varepsilon} \log \phi_{ll}(t) \, dt + \liminf_{n \to \infty} \frac{1}{n} \log \mathbb{P}_{x,n}(\phi > \phi_{ll} \text{ on } [\varepsilon, 1-\varepsilon])$$
$$+ \liminf_{n \to \infty} \frac{1}{n} \log K_n^\varepsilon + \liminf_{n \to \infty} \frac{1}{n} \log L_n^\varepsilon.$$

From Theorem 2.6, the sequence of probability measures $(\mathbb{P}_{x,n})_n$ satisfies a LDP in the space Ω with speed n and good rate function I. Moreover, $\{\phi > \phi_{ll} \text{ on } [\varepsilon, 1-\varepsilon]\}$ is an open set of Ω, so

$$\liminf_{n \to \infty} \frac{1}{n} \log \mathbb{P}_{x,n}(\phi > \phi_{ll} \text{ on } [\varepsilon, 1-\varepsilon]) \geq - \inf_{\{\phi > \phi_{ll} \text{ on } [\varepsilon, 1-\varepsilon]\}} I(\phi).$$

Now, thanks to the regularity of I,

$$\inf_{\{\phi > \phi_{ll} \text{ on } [\varepsilon, 1-\varepsilon]\}} I(\phi) = \inf_{\{\phi \geq \phi_{ll} \text{ on } [\varepsilon, 1-\varepsilon]\}} I(\phi) \leq I(\phi_{ll}).$$

Let $\varepsilon \to 0$, this yields

$$\liminf_{n \to \infty} \frac{1}{n} \log\left[\int_\Omega \exp(nF_n(\phi)) \, d\mathbb{P}_{x,n}\right]$$
$$\geq F(\phi_{ll}) - I(\phi_{ll}) + \limsup_{\varepsilon \to 0} \liminf_{n \to \infty} \frac{1}{n} \log K_n^\varepsilon + \limsup_{\varepsilon \to 0} \liminf_{n \to \infty} \frac{1}{n} \log L_n^\varepsilon.$$

Let $y > \phi_{ll}(\varepsilon)$ such that $\mathbb{P}_{x,n}(\phi(\epsilon) = y) \neq 0$. The conditional expectation

$$\mathbb{E}_n\left(\exp(n\int_0^\varepsilon \log\left(\phi(t) + 1/n\right) dt) \mid \phi(\varepsilon) = y\right)$$

is a finite sum over the paths ϕ such that $\phi(\varepsilon) = y$ of positive terms

$$\exp(n\int_0^\varepsilon \log\left(\phi(t) + 1/n\right) dt)\, \mathbb{P}_{x,n}(\phi \mid \phi(\varepsilon) = y).$$

We underestimate this sum by the contribution of the path having increments $+1$ between times 0 and $([n\varepsilon]+ny)/2$ and increments -1 between times $([n\varepsilon]+ny)/2$ and $[n\varepsilon]$. For this path, $\phi(t) + 1/n > \phi_{ll}(\varepsilon)t$ on $[0,\varepsilon]$, hence

$$\mathbb{E}_{x,n}\left(\exp(n\int_0^\varepsilon \log\left(\phi(t)+1/n\right) dt) \mid \phi(\varepsilon) = y\right)$$

$$\geq \exp(n\int_0^\varepsilon \log\left(\phi_{ll}(\varepsilon)t\right) dt)\prod_{i=1}^{([n\varepsilon]+ny)/2} f(T^i x) \prod_{i=([n\varepsilon]+ny)/2+1}^{[n\varepsilon]} (1-f(T^i x))$$

Thus,

$$\frac{1}{n}\log K_n^\varepsilon \geq \int_0^\varepsilon \log\left(\phi_{ll}(\varepsilon)t\right) dt + \frac{1}{n}\sum_{i=1}^{[n\varepsilon]} \log f(T^i x) + \frac{1}{n}\sum_{i=1}^{[n\varepsilon]} \log\left(1 - f(T^i x)\right).$$

Then, the unique ergodicity of the dynamical system and hypothesis (9.1) imply that

$$\liminf_{n\to\infty} \frac{1}{n}\log K_n^\varepsilon \geq \int_0^\varepsilon \log\left(\phi_{ll}(\varepsilon)t\right) dt + \varepsilon \int_E \log\left[f(1-f)\right] d\mu.$$

Finally, since $\phi_{ll}(0) = 0$ and $\dot\phi_{ll}(0) > 0$ (this will be proved later on),

$$\limsup_{\varepsilon \to 0} \liminf_{n\to\infty} \frac{1}{n}\log(K_n^\varepsilon) = 0.$$

In the same way, we prove that

$$\limsup_{\varepsilon \to 0} \liminf_{n\to\infty} \frac{1}{n}\log(L_n^\varepsilon) = 0.$$

This proves inequality (9.3).

We now prove Theorem 9.1. Inequalities (9.2) and (9.3) imply that $\mathbb{Q}_{x,n}^{(ll)}$ satisfies a large deviation upper bound: for every closed set $A \subset \Omega$,

$$\limsup_{n\to\infty} \frac{1}{n}\log \mathbb{Q}_{x,n}^{(ll)}(A) \leq -\inf_{\phi \in A}\left[I(\phi) - F(\phi) - \inf_{\phi \in \Omega}\left(I(\phi) - F(\phi)\right)\right].$$

Apply this inequality with $A_\varepsilon = \{\phi \in \Omega \mid \|\phi - \phi_{ll}\|_\infty \geq \varepsilon\}$. The function F is upper semi-continuous. On the set where the good rate function I is finite, the function F is bounded above (because $I(\phi) < +\infty$ implies $|\dot\phi| \leq 1$). Thus, by applying the result of Exercise 4.3.10 in [37], we deduce that for any closed set $C \subseteq \Omega$, the infimum $\inf_{\phi \in C} [I(\phi) - F(\phi)]$ is attained on the set C. The uniqueness of the minimizer follows from the strict convexity of the functional $I - F$ (due to the strict concavity of the function $x \to \log(x)$) on its domain. Since ϕ_{ll} does not belong to A_ε, the infimum over A_ε is strictly greater that the infimum over Ω. So there exists $\delta = \delta(\varepsilon) > 0$ such that for all sufficiently large n,

$$\mathbb{Q}^{(ll)}_{x,n}(A_\varepsilon) \leq \exp(-n\delta).$$

We now prove that the function ϕ_{ll} is concave and satisfies: $\dot\phi_{ll}(0) > 0$, $\dot\phi_{ll}(1) < 0$ and $\phi_{ll} > 0$ on $(0,1)$. The function ϕ_{ll} is concave, otherwise the function $\tilde\phi$ defined by

$$\tilde\phi(t) = \sup\{\int_D \dot\phi_{ll}(t) \mid D \subset [0,1], |D| = t\}$$

would verify $I(\phi_{ll}) = I(\tilde\phi)$ and $F(\phi_{ll}) < F(\tilde\phi)$, contradicting the fact that ϕ_{ll} is a minimizer of $I - F$. Thus, ϕ_{ll} is a nonzero concave function such that $\phi_{ll}(0) = \phi_{ll}(1) = 0$, this implies that $\dot\phi_{ll}(0) > 0$, $\dot\phi_{ll}(1) < 0$ and $\phi_{ll} > 0$ on $(0,1)$. The path ϕ_{ll} satisfies the Euler-Lagrange equation

$$\frac{d}{dt}\frac{\partial L_{ll}}{\partial y}(\phi, \dot\phi) - \frac{\partial L_{ll}}{\partial x}(\phi, \dot\phi) = 0 \tag{9.4}$$

where $L_{ll}(x, y) = \Lambda^*(y) - \log(x)$ with boundary conditions $\phi(0) = \phi(1) = 0$.

Remark:
A direct calculation gives for every $\lambda \in \mathbb{R}$,

$$\Lambda'(\lambda) = \int_E \frac{fe^\lambda - (1-f)e^{-\lambda}}{fe^\lambda + (1-f)e^{-\lambda}} d\mu.$$

Under the hypothesis (9.1), we necessarily have that $\int_E f(1-f) \, d\mu > 0$. Since the function Λ' is continuous on $\mathbb{R}$, with values in $]-1, 1[$, and for every $\lambda \in \mathbb{R}$,

$$\Lambda''(\lambda) = 4 \int_E \frac{f(1-f)}{[fe^\lambda + (1-f)e^{-\lambda}]^2} d\mu > 0,$$

we deduce that Λ' is an homeomorphism from $\mathbb{R}$ to $]-1, 1[$ and then, by a direct calculation,

$$(\Lambda^*)''(\phi) = ((\Lambda')^{-1})'(\phi).$$

Equation (9.4) is then equivalent to the following equation

$$\ddot{\phi}(t)((\Lambda')^{-1})'(\dot{\phi}(t)) = -\frac{1}{\phi(t)} \tag{9.5}$$

with boundary conditions $\phi(0) = \phi(1) = 0$. In order to solve Equation (9.5), we firstly have to determine the function $\psi :]-1,1[\to \mathbb{R}$ satisfying the functional equation

$$\phi = \int_E \frac{fe^{\psi(\phi)} - (1-f)e^{-\psi(\phi)}}{fe^{\psi(\phi)} + (1-f)e^{-\psi(\phi)}} d\mu$$

Then, Equation (9.5) is reduced to

$$\frac{\ddot{\phi}(t)}{\Lambda''(\psi(\dot{\phi}(t)))} = -\frac{1}{\phi(t)}.$$

When $f \equiv 1/2$, the function ψ is just $\arg\tanh$ and $\Lambda(\lambda) = \log(\cosh(\lambda))$, hence the unique solution of (9.5) with boundary conditions: $\phi(0) = \phi(1) = 0$ is given by

$$\phi_{ll}(t) = \frac{1}{\pi}\sin(\pi t)$$

(see [119] or [127] for further details). This example is the most simplest one. For others particular functions f, a solution of the above functional equation can perhaps be obtained with numerical methods. This question is hard and is actually under consideration. We will give in Section 6 an example of function f where direct calculations lead to a degenerate non linear partial differential equation.

Consider the storage cost function

$$C_{ll,n} = n\sum_{i=1}^n S_{ll,n}^w\left(\frac{i}{n}\right).$$

The next result is easily derived from Theorem 9.1.

COROLLARY 2 *Under the hypothesis* (9.1), *for any* $x \in E$, *the random variables* $\left(\frac{C_{ll,n}}{n^2}\right)_{n \geq 1}$ *converge exponentially fast to*

$$m_{ll} = \int_0^1 \phi_{ll}(t)\, dt$$

as n goes to infinity.

Remark:
When $f \equiv 1/2$, $\phi_{ll}(t) = \frac{1}{\pi}\sin(\pi t)$ and $m_{ll} = \frac{2}{\pi^2}$.

4. DYNAMIC PRIORITY QUEUES

This section deals more briefly with priority queues driven by the dynamic random walk defined in Section 2 of Chapter 1. From Table 9.1, we see that the difference between dynamic priority queues and dynamic linear lists is the weight we assign to each path. Let $(S_k)_{0 \leq k \leq n}$ be the dynamic random walk defined as in the previous section: starting from 0, conditioned to remain nonnegative and such that $S_n = 0$. From Table 9.1, the dynamic priority queues are modeled by the one-dimensional dynamic random walk $(S_k)_{0 \leq k \leq n}$ each path being assigned relative probability

$$\prod_{i=1}^{n}(S_{i-1}+1)^{1/2}.$$

This weight comes from the fact that in the priority queue case, the number of insertions is equal to the number of deletions; the structure beginning and ending empty. We shall denote this weighted random walk by $(S^w_{pq,n})_{0 \leq k \leq n}$. The normalized data structure size as a function of time is defined by

$$S^w_{pq,n}(t) = \frac{S^w_{pq,[nt]}}{n}, \quad t \in [0,1]$$

The distribution of the random variable $(S^w_{pq,n}(t))_{t \in [0,1]}$ with values in Ω is denoted by $\mathbb{Q}^{(pq)}_{x,n}$.

THEOREM 9.2 *Under the hypothesis (9.1), for any $x \in E$, the sequences $(S^w_{pq,n}(t))_{t \in [0,1]}$ converge, as n goes to infinity, to the unique solution $(\phi_{pq}(t))_{t \in [0,1]}$ of the following differential equation*

$$\ddot{\phi}(t)(\Lambda^*)''(\dot{\phi}(t)) = -\frac{1}{2\phi(t)}$$

with boundary conditions: $\phi(0) = \phi(1) = 0$. The convergence is exponential in the sense that for any $\varepsilon > 0$, there exists $\delta = \delta(\varepsilon) > 0$ such that for all sufficiently large n,

$$\mathbb{Q}^{(pq)}_{x,n}(\{\phi \in \Omega \mid \|\phi - \phi_{pq}\|_\infty \geq \varepsilon\}) \leq \exp(-n\delta).$$

The proof of Theorem 9.1 can easily be adapted in order to prove Theorem 9.2 except that for dynamic priority queues, we have to find the function ϕ_{pq} in Ω minimizing

$$\int_0^1 L_{pq}(\phi,\dot{\phi}) \, dt$$

with $L_{pq}(x,y) = \Lambda^*(y) - \frac{1}{2}\log(x)$. This problem can be solved using Laplace's method, the associated Euler-Lagrange equation is then

$$\ddot{\phi}(t)(\Lambda^*)''(\dot{\phi}(t)) = -\frac{1}{2\phi(t)}.$$

Let us consider the storage cost function defined as

$$C_{pq,n} = n\sum_{i=1}^{n} S^w_{pq,n}\left(\frac{i}{n}\right).$$

Its asymptotic properties are directly obtained from Theorem 9.2.

COROLLARY 3 *Under the hypothesis (9.1), for any $x \in E$, the sequences $\left(\frac{C_{pq,n}}{n^2}\right)_{n\geq 1}$ converge exponentially fast to*

$$m_{pq} = \int_0^1 \phi_{pq}(t)\, dt$$

as n goes to infinity.

Remark:
When $f \equiv 1/2$, $\phi_{pq}(t) = t(1-t)$ and $m_{pq} = \frac{1}{6}$.

5. DYNAMIC DICTIONARIES
5.1 A NEW DYNAMIC RANDOM WALK

Because of the operations supported by dictionaries, we need a different model of dynamic random walks.
Let $(E, \mathcal{A}, \mu, T)$ be a dynamical system where $(E, \mathcal{A}, \mu)$ is a probability space and T is a transformation defined on E preserving the measure μ. Let f be a function defined on E with values in $[0,1]$. Let $(X_i)_{i\geq 1}$ be a sequence of independent random variables with values in $\mathbb{Z}$. Let $x \in E$ and for every $i \geq 1$, the law of the random variable X_i is given by

$$\mathbb{P}_x(X_i = z) = \begin{cases} \frac{1}{2}f(T^i x) & \text{if } z = 1 \\ \frac{1}{2}f(T^i x) & \text{if } z = -1 \\ 1 - f(T^i x) & \text{if } z = 0 \\ 0 & \text{otherwise} \end{cases}$$

We write

$$S_0 = 0, \quad S_n = \sum_{i=1}^{n} X_i \text{ for } n \geq 1$$

for this new dynamic $\mathbb{Z}$-random walk generated by the family $(X_i)_{i\geq 1}$.

5.2 A FUNCTIONAL LARGE DEVIATION PRINCIPLE

THEOREM 9.3 *Let f be a measurable function defined on E with values in $[0,1]$. Then, for μ-almost every point $x \in E$, the sequence $(\frac{S_n}{n})_{n \geq 1}$ satisfies the LDP with speed n and the good rate function*

$$\Lambda_d^*(y) = \sup_{\lambda \in \mathbb{R}}\{\lambda y - \Lambda_d(\lambda)\}$$

where

$$\Lambda_d(\lambda) = \mathbb{E}\Big(\log\big(1 + f(\cosh(\lambda) - 1)\big) \,\Big|\, \mathcal{I}\Big),$$

$\mathcal{I}$ being the σ-algebra generated by the fixed points of the transformation T.

Let us assume E to be a compact metric space, $\mathcal{A}$ the associated Borel field and T a continuous transformation of E. When $(E, \mathcal{A}, \mu, T)$ is uniquely ergodic, Theorem 9.3 holds when the function f is continuous and for every $x \in E$. In that case,

$$\Lambda_d(\lambda) = \int_E \log\big(1 + f(x)(\cosh(\lambda) - 1)\big) d\mu(x).$$

Under these stronger hypotheses on the dynamical system, we can extend Theorem 9.3 as follows.

Let us define for every $n \geq 1$,

$$S_n^*(t) = \frac{S_{[nt]}}{n}, \quad t \in [0,1]$$

The linear interpolation of $S_n^*(t)$, $t \in [0,1]$, is then defined by

$$\bar{S}_n(t) = S_n^*(t) + \Big(t - \frac{[nt]}{n}\Big) X_{[nt]+1}.$$

THEOREM 9.4 *Let $(E, \mathcal{A}, \mu, T)$ be an uniquely ergodic dynamical system. Let f be a continuous function defined on E with values in $[0,1]$, then for every $x \in E$, $(\bar{S}_n(.))_{n \geq 1}$ satisfies in $\mathcal{C}([0,1])$ the LDP with the good rate function*

$$I_d(x(.)) = \begin{cases} \int_0^1 \Lambda_d^*(\dot{\phi}(t))\, dt, & \text{if } \phi(.) \in \mathcal{AC}, \phi(0) = 0 \\ +\infty, & \text{otherwise.} \end{cases}$$

The proofs of Theorems 2.5 and 2.6 from Section 5 of Chapter 2 can easily be adapted to prove Theorem 9.3 and 9.4.

Let $(E, \mathcal{A}, \mu, T)$ be an uniquely ergodic dynamical system, $x \in E$ fixed and $(S_k)_{0 \leq k \leq n}$ the associated dynamic random walk defined as in

Section 5.1 conditioned to remain nonnegative and such that $S_n = 0$. From Table 9.1, the dynamic dictionaries are here modelled by the one-dimensional dynamic random walk $(S_k)_{0 \leq k \leq n}$ each path being assigned relative probability

$$\prod_{i=1}^{n}(S_{i-1}+1).$$

We shall denote this weighted random walk by $(S^w_{d,n})_{0 \leq k \leq n}$. The normalized data structure size as a function of time is then defined by

$$S^w_{d,n}(t) = \frac{S^w_{d,[nt]}}{n}, \quad t \in [0,1].$$

The distribution of the random variable $(S^w_{d,n}(t))_{t \in [0,1]}$ with values in Ω is denoted by $\mathbb{Q}^{(d)}_{x,n}$. For every $y \in \mathbb{R}$,

$$\Lambda^*_d(y) = \sup_{\lambda \in \mathbb{R}}\{\lambda y - \Lambda_d(\lambda)\}$$

where

$$\Lambda_d(\lambda) = \int_E \log\left(1 + f(x)(\cosh(\lambda) - 1)\right) d\mu(x).$$

THEOREM 9.5 *Under the hypothesis* (9.1), *for any* $x \in E$, *the sequences* $(S^w_{d,n}(t))_{t \in [0,1]}$ *converge, as n goes to infinity, to the unique solution* $(\phi_d(t))_{t \in [0,1]}$ *of the following differential equation*

$$\ddot{\phi}(t)(\Lambda^*_d)''(\dot{\phi}(t)) = -\frac{1}{\phi(t)} \tag{9.6}$$

with boundary conditions: $\phi(0) = \phi(1) = 0$. *The convergence is exponential in the sense that for any* $\varepsilon > 0$, *there exists* $\delta = \delta(\varepsilon) > 0$ *such that for all sufficiently large n,*

$$\mathbb{Q}^{(d)}_{x,n}(\{\phi \in \Omega \mid \|\phi - \phi_d\|_\infty \geq \varepsilon\}) \leq \exp(-n\delta).$$

The proof of Theorem 9.1 can be adapted in order to prove Theorem 9.5. For dynamic dictionaries, we now have to find the function ϕ_d in Ω minimizing

$$\int_0^1 L_d(\phi, \dot{\phi}) \, dt$$

with $L_d(x, y) = \Lambda^*_d(y) - \log(x)$. The associated Euler-Lagrange equation is given by

$$\ddot{\phi}(t)(\Lambda^*_d)''(\dot{\phi}(t)) = -\frac{1}{\phi(t)}. \tag{9.7}$$

Finally we consider the storage cost function for dynamic dictionaries defined by

$$C_{d,n} = n \sum_{i=1}^{n} S_{d,n}^w(\frac{i}{n}).$$

COROLLARY 4 *Under the hypothesis (9.1), for any $x \in E$, the random variables $\left(\frac{C_{d,n}}{n^2}\right)_{n \geq 1}$ converges exponentially fast to*

$$m_d = \int_0^1 \phi_d(t)\, dt$$

as n goes to infinity.

Remark:
When $f \equiv 1/2$, we have the relation $\Lambda_d^* = 2\Lambda^*$ (see [119] or [127]) and thus, the solution ϕ_d of equation (9.6) is the same as in the case of priority queues, namely

$$\phi_d(t) = t(1-t), t \in [0,1]$$

and $m_{pq} = m_d = \frac{1}{6}$.

6. AN EXAMPLE: LINEAR LISTS AND ROTATION ON THE TORUS

Let $(\mathbb{T}^1, \mathcal{B}(\mathbb{T}^1), \lambda, T_\alpha)$ be the dynamical system where T_α is defined by $x \to x + \alpha \mod 1$, with α a given real and λ is the Lebesgue measure on $\mathbb{T}^1$. This particular dynamical system is the so-called rotation on the one-dimensional torus. Twofold motivations are related to this example:
1. Explicit calculations are possible in this case,
2. When α is rational, we get a periodic dynamical system which models a periodic behaviour of the operations on the data structures.
Irrational rotations are uniquely ergodic, i.e. the ergodic average of a continuous function uniformly converges in $x \in \mathbb{T}^1$ to the integral of f. The uniform convergence of the ergodic averages even holds for any function with bounded variation (see [108]). Consequently, if we consider a dynamic random walk associated to an irrational rotation on the torus, Theorem 9.1 holds for every function f with bounded variation.

6.1 CHOICE OF A FUNCTION AND DERIVATION OF THE CORRESPONDING DIFFERENTIAL EQUATION

When the hypothesis (9.1) is not satisfied by the function f, the methods presented in the paper do not apply. For instance, let us choose the function $f = \mathbf{1}_{[0, \frac{1}{2}]}$, then $\Lambda = 0$ and

$$\Lambda^*(y) = \begin{cases} +\infty & \text{if } y \neq 0 \\ 0 & \text{otherwise.} \end{cases}$$

In that case, no large deviation occurs as the dynamic random walk is deterministic in the sense that

$$S_n = \sum_{i=1}^{n} (2\mathbf{1}_{[0,\frac{1}{2}]}(T_\alpha^i x) - 1)$$

for some $x \in [0, 1]$ fixed.

The computation of the function Λ^* for a general function f is difficult (see the remark following Theorem 9.1). However, we are able to compute it for the very particular function $f = \frac{3}{4}\mathbf{1}_{[0,\frac{2}{3}]}$. Unfortunately, this function does not satisfy hypothesis (9.1). We think that the hypothesis (9.1) is only technical and should be dropped out but we don't have any proof of this point yet. In order to illustrate how our method is efficient to determine the asymptotic behavior of the storage cost function associated to a dynamic linear list, we now compute the function ϕ_{ll} for this particular function f. Straightforward computations give us

$$\Lambda(\lambda) = \log(\cosh(\lambda)) + \frac{2}{3}\log\left(1 + \frac{\tanh(\lambda)}{2}\right) + \frac{1}{3}\log\left(1 - \tanh(\lambda)\right).$$

When $y \in\ ]-1, \frac{1}{3}[$,

$$\Lambda^*(y) = \frac{1}{2}(1+y)\log(1+y) + \frac{(1-3y)}{6}\log(1-3y)$$

and $+\infty$ otherwise.

We consider dynamic linear lists driven by this particular dynamic random walk and determine the path ϕ_{ll} satisfying the following Euler-Lagrange equation:

$$\ddot{\phi}(t)(\Lambda^*)''(\dot{\phi}(t)) = -\frac{1}{\phi(t)}. \tag{9.8}$$

In this particular case, on the interval $]-1, 1/3[$,

$$(\Lambda^*)''(y) = ((\Lambda')^{-1})'(y) = \frac{1}{2(1+y)} + \frac{3}{2(1-3y)}.$$

After straightforward computations, the equation (9.8) becomes

$$2\phi\ddot{\phi} = 3\dot{\phi}^2 + 2\dot{\phi} - 1. \tag{9.9}$$

Rewriting the equation like a system of two differential equations, with $\theta = \dot{\phi}$ gives:

$$\begin{cases} \dot{\phi} = \theta \\ \dot{\theta} = \frac{3}{2\phi}(\theta - \frac{1}{3})(\theta + 1) \end{cases} \iff \dot{z} = F(z) \text{ with } z(t) = \begin{bmatrix} \phi(t) \\ \theta(t) \end{bmatrix}$$

which shows that a solution orbit $(\phi(t), \theta(t))$ may cross the $\phi = 0$ line only when $\theta = \frac{1}{3}$ or $\theta = -1$. We note also that the curves $(\phi(t) = \frac{1}{3}(t-t_0)+\phi_0, \theta(t) = \frac{1}{3})$ and $(\phi(t) = -(t-t_0)+\phi_0, \theta(t) = -1)$ are solutions of the equation, moreover they are the only polynomial solutions.

A first overview of the behavior of the differential equation may be suggested by plotting the (normalized) vector field F on a grid in the phase space, see Figure 9.2 (on the figure the letters r,l stand for "right" and "left" and the letters u,m and d for "upper", "middle" and "down"; also x and y are used in place of (respectively) ϕ and θ).

The vector field suggests a symmetry of the trajectories (see Figure 9.2). Furthermore we have also a similarity principle between some solution orbits. More precisely we have:

LEMMA 19 *if $(\phi(t), \theta(t))$, $t \in [0, T]$ is a trajectory of the differential equation then:*

- *(symmetry)* $(\hat{\phi}(t), \hat{\theta}(t))$, $t \in [c, c+T]$ *defined by:*

$$\begin{cases} \hat{\phi}(t) = -\phi(c+T-t) \\ \hat{\theta}(t) = \theta(c+T-t)) \end{cases}$$

- *and (similarity)* $(\tilde{\phi}(\tau), \tilde{\theta}(\tau))$, $\tau \in [c, c+T/\alpha]$, *with $\tau = t/\alpha + c$ defined by:*

$$\begin{cases} \tilde{\phi}(\tau) = \frac{1}{\alpha}\phi(t) \\ \tilde{\theta}(\tau) = \theta(t) \end{cases}$$

where $\alpha > 0$ and c are two constants, are also two trajectories of the differential equation.

Proof:
Straightforward computations show that $(\hat{\phi}(t), \hat{\theta}(t))$, $\forall t \in [c, c+T]$ and $(\tilde{\phi}(\tau), \tilde{\theta}(\tau))$, $\forall \tau \in [c, c+T/\alpha]$ verify the differential equation since $(\phi(t), \theta(t))$, $\forall t \in [0, T]$ verifies it.

Thanks to the symmetry property, it is enough to study the dynamic on the half plane $\phi \geq 0$. This property (together with the similarity property) will also be very useful for the computation of solutions: it will be enough to compute accurately very few trajectories (in fact 3) to get all the others.

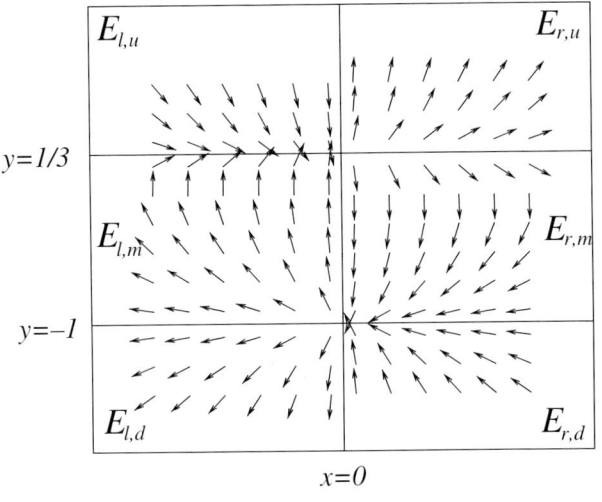

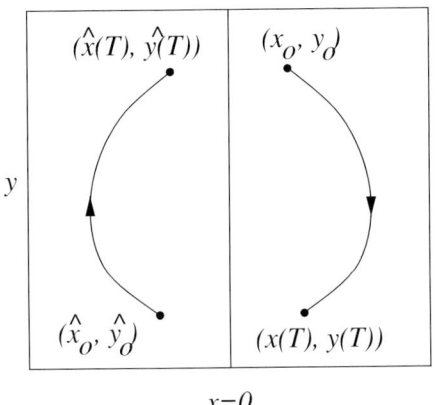

Figure 9.2. Normalized vector field and symmetry of trajectories

PROPOSITION 11 *Starting from the initial condition (ϕ_0, θ_0), with $\phi_0 > 0$, the dynamical behavior may be summarized as follows:*

- *if $(\phi_0, \theta_0) \in E_{r,u} = \{\phi > 0\} \times \{\frac{1}{3} < \theta\}$, then the trajectory stays in the set $E_{r,u}$, $\phi(t)$ and $\theta(t)$ are increasing on $[0, +\infty)$ and:*

$$\lim_{t \to +\infty} \phi(t) = \lim_{t \to +\infty} \theta(t) = +\infty.$$

- if $(\phi_0, \theta_0) \in E_{r,m} = \{\phi > 0\} \times \{-1 < \theta < \frac{1}{3}\}$, then the trajectory stays in the set $E_{r,m}$ until a time $\bar{t}$ where $(\phi(\bar{t}), \theta(\bar{t})) = (0, -1)$ with:

$$\bar{t} = \int_0^{+\infty} 2\phi_0 e^{\frac{2\xi}{3}} \left| \frac{C-1}{Ce^{4\xi} - 1} \right|^{\frac{2}{3}} d\xi, \quad \text{and } C = C(\theta_0) = \frac{\theta_0 - \frac{1}{3}}{\theta_0 + 1}$$

Moreover $\theta(t)$ is decreasing on $[0, \bar{t}]$ and starting from $\theta_0 > 0$, $\phi(t)$ first increases until a time $\hat{t}$ (which corresponds to $\phi(\hat{t}) = 0$) where $\phi(t)$ becomes to decrease.

- if $(\phi_0, \theta_0) \in E_{r,d} = \{\phi > 0\} \times \{\theta < -1\}$, then the trajectory stays in the set $E_{r,d}$ until the time $\bar{t}$ where $(\phi(\bar{t}), \theta(\bar{t})) = (0, -1)$.

- the points $(0, \frac{1}{3})$ and $(0, 1)$ are singular for the dynamic: there are an infinity number of trajectories across them. Any trajectory starting from (ϕ_0, θ_0) in $E_{r,m}$, $E_{r,d}$, $E_{l,m}$, or $E_{l,u}$ reaches one of the 2 singular points but may be extended (after the singularity) in the previous mentioned symmetric way (which lets to have the maximum regularity for these completed trajectories). In particular a trajectory starting at $(\phi_0, \theta_0) \in E_{r,m}$ may be extended at time $\bar{t}$ in $E_{l,m}$ (by the $x = 0$ axis symmetry). Arriving at $(0, \frac{1}{3})$ this second part of the trajectory may also be extended in $E_{r,m}$ and finally reaches the point (ϕ_0, θ_0) at time T (depending on (ϕ_0, θ_0)) giving rise to a periodic trajectory. For these periodic orbits, which cut the $\theta = 0$ axis, say for $\phi = \Phi$, the period T is a linear function of Φ, more precisely:

$$T = \frac{2^{4/3}}{\sqrt{3}} \beta(\frac{1}{6}, \frac{1}{2}) \Phi \simeq 10.6 \Phi.$$

- The integral curves are given by:

$$\phi = \phi_0 \left| \frac{\theta - 1/3}{\theta_0 - 1/3} \right|^{\frac{1}{6}} \left| \frac{\theta + 1}{\theta_0 + 1} \right|^{\frac{1}{2}} \tag{9.10}$$

Proof:
The main tool consists in applying a transformation on the time t ($t \mapsto \tau$) defined by:

$$\begin{cases} \frac{d\tau}{dt} = \frac{1}{2\phi} \\ \tau(t=0) = 0 \end{cases} \text{ while } \phi > 0,$$

because we can get the trajectories analytically as function of τ. θ (as a function of τ) verifies the differential equation:

$$\frac{d\theta}{d\tau} = 3(\theta - \frac{1}{3})(\theta + 1), \text{ while } \phi > 0,$$

which is easily solved by variable separation. After straightforward computations we get:

$$\theta(\tau) = \frac{\frac{1}{3} + Ce^{4\tau}}{1 - Ce^{4\tau}}, \text{ while } \phi > 0 \qquad (9.11)$$

with:

$$C = C(\theta_0) = \frac{\theta_0 - \frac{1}{3}}{\theta_0 + 1} \text{ and } \begin{cases} C > 1 \text{ when } \theta_0 \in (-\infty, -1) \\ C < 0 \text{ when } \theta_0 \in (-1, \frac{1}{3}) \\ C \in (0,1) \text{ when } \theta_0 \in (\frac{1}{3}, +\infty) \end{cases}$$

So we see that, starting from $(\phi_0, \theta_0) \in E_{r,u}$, we have:

$$\lim_{\tau \to \bar{\tau}-} \theta(\tau) = +\infty, \quad \bar{\tau} = -\log(C)/4$$

But we will see later on that in this case the time t goes also to $+\infty$ (as a function of τ) and so there is no blow up at some finite time $\bar{t}$.

The expression of ϕ in terms of τ can be obtained via:

$$\frac{d\phi}{d\tau} = \frac{d\phi}{dt}\frac{dt}{d\tau} = 2\phi\theta, \text{ while } \phi > 0,$$

so that, we have to integrate:

$$\frac{d\phi}{\phi} = 2\theta d\tau$$

but fortunately the antiderivative of θ (seen as a function of τ) can be expressed analytically ! We get after some computations:

$$\phi(\tau) = \phi_0 e^{\frac{2\tau}{3}} \left| \frac{C-1}{Ce^{4\tau}-1} \right|^{\frac{2}{3}}, \text{ while } \phi > 0 \qquad (9.12)$$

Like for $\theta(\tau)$ we see that, starting from $(\phi_0, \theta_0) \in E_{r,u}$ we have:

$$\lim_{\tau \to \bar{\tau}-} \phi(\tau) = +\infty, \quad \bar{\tau} = -\log(C)/4.$$

Finally, from $dt = 2\phi d\tau$ we have the following relation giving the time t in term of τ:

$$t(\tau) = \int_0^\tau 2\phi_0 e^{\frac{2\xi}{3}} \left| \frac{C-1}{Ce^{4\xi}-1} \right|^{\frac{2}{3}} d\xi, \text{ while } \phi > 0. \qquad (9.13)$$

We can now prove the stated behavior:

- if $(\phi_0, \theta_0) \in E_{r,u}$ then the time transformation $t \mapsto \tau$ is a diffeomorphism from $[0, +\infty)$ to $[0, \bar{\tau})$ ($\bar{\tau} = \log(C)/4$). Combined with (9.11) and (9.12) we have then:

$$\lim_{t \to +\infty} \phi(t) = \lim_{t \to +\infty} \theta(t) = +\infty,$$

- if $(\phi_0, \theta_0) \in E_{r,m}$, we get $C < 0$ and the time transformation $t \mapsto \tau$ is a diffeomorphism from $[0, \bar{t})$ to $[0, +\infty)$ with:

$$\bar{t} = \int_0^{+\infty} 2\phi_0 e^{\frac{2\xi}{3}} \left| \frac{C-1}{Ce^{4\xi} - 1} \right|^{\frac{2}{3}} d\xi$$

and (9.11) and (9.12) show that:

$$\lim_{t \to \bar{t}-} \theta(t) = -1, \quad \lim_{t \to \bar{t}-} \phi(t) = 0$$

Furthermore, by computing the derivative of $\phi(\tau)$ and $\theta(\tau)$, we see that $\theta(t)$ is strictly decreasing on $[0, \bar{t})$ and:

  - if $\theta_0 > 0$ then $\phi(t)$ is increasing on $[0, t(\hat{\tau}))$, and decreasing on $(t(\hat{\tau}), \bar{t}))$, with:

$$\hat{\tau} = -\frac{1}{4} \log(-3C)$$

  corresponding to the time when $\theta = 0$.
  - if $\theta_0 \leq 0$ $\phi(t)$ is decreasing on $[0, \bar{t})$.

It is obvious that a trajectory starting from $(\phi_0, \theta_0) \in E_{r,u}$ which reached the point $(0, -1)$ at the time $\bar{t}$ defined previously, may be completed by a first part starting from $(0, 1/3)$ which joins (ϕ_0, θ_0) with a time of $-t(-\infty)$. The family of all these trajectories going from $(0, 1/3)$ to $(0, -1)$ may be parametrized in a unique way by $\Phi > 0$, Φ selecting the only one passing at the point $(\Phi, 0)$. The time to go from $(0, 1/3)$ to $(0, -1)$ is:

$$\int_{-\infty}^{+\infty} 2\phi_0 e^{\frac{2\xi}{3}} \left| \frac{C-1}{Ce^{4\xi} - 1} \right|^{\frac{2}{3}} d\xi = \frac{2^{1/3}}{\sqrt{3}} \beta(\frac{1}{6}, \frac{1}{2}) \Phi \qquad (9.14)$$

If we complete such a trajectory by the symmetrized trajectory in $E_{l,m}$ we have then a periodic orbit of period $T = 2^{4/3}/\sqrt{3} \, \beta(1/6, 1/2)\Phi$.

- if $(\phi_0, \theta_0) \in E_{r,d}$, we have $C > 1$, the transformation $t \mapsto \tau$ is a diffeomorphism from $[0, \bar{t})$ to $[0, +\infty)$, $\bar{t}$ given by the same formula and the same limit for ϕ and θ as in the previous case and here ϕ is decreasing while θ is increasing.

- From Equation (9.11) we can write $e^{4\tau}$ (and so $e^{2\tau/3}$) in terms of $\theta(\tau)$, then, substituting these quantities in Equation (9.12) we get Equation (9.10) giving the integral curves independently of t.

Like for all analytical expressions, we can compute safely the trajectories in terms of the time t. In fact the only numerical work involved consists in approximating the integral (9.13) which is an easy task. By the way, we have also experienced the use of a (good) numerical differential equation solver (at the very beginning of this study) and we note that different results may be obtained when crossing (near) the singular points, where the solver may continue the trajectory to various ones (depending on many things). This is not surprising!

Figure (9.3) shows two periodic trajectories, one as unbroken line passing at $(1,0)$ the other as dotted line passing at $(0.5, 0)$ and computed thanks to the similarity property.

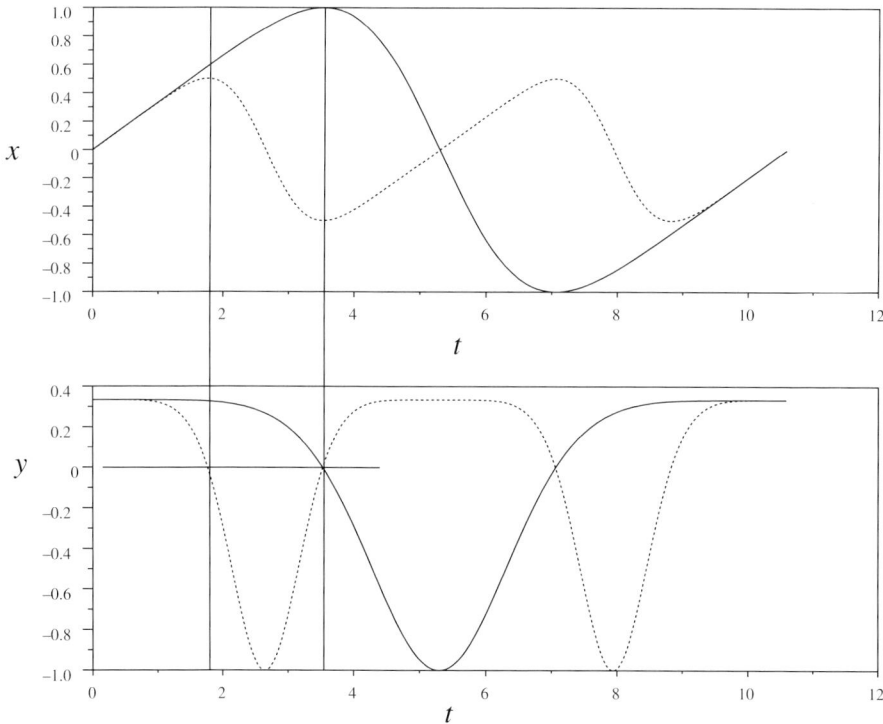

Figure 9.3. Two trajectories of the differential equation

DYNAMIC RANDOM WALKS

Finally we are interested by a solution $\phi_{ll}(t)$, $t \in [0,1]$ of the differential equation such that $\phi_{ll}(0) = \phi_{ll}(1) = 0$. If we constrain ϕ_{ll} to be positive ($t \in (0,1)$), then this solution is unique and corresponds to the trajectory in $E_{r,m}$ which goes from $(0, 1/3)$ to $(0, -1)$ in a unit time, and so, the one parametrized (see Equation (9.14)) by:

$$\Phi = \frac{\sqrt{3}}{2^{1/3}\beta(\frac{1}{6}, \frac{1}{2})}.$$

Like the period T, the area (i.e. m_{ll}) can be computed in term of the beta function, we get:

$$\begin{aligned} m_{ll} &= \int_0^1 \phi_{ll}(t) dt \\ &= \int_{-\infty}^{+\infty} \phi_{ll}(t(\tau)) \frac{dt}{d\tau} d\tau \\ &= \int_{-\infty}^{+\infty} 2\phi_{ll}(\tau)^2 d\tau \\ &= \frac{6}{\beta(\frac{1}{6}, \frac{1}{2})^2} \\ &\simeq 0.113. \end{aligned}$$

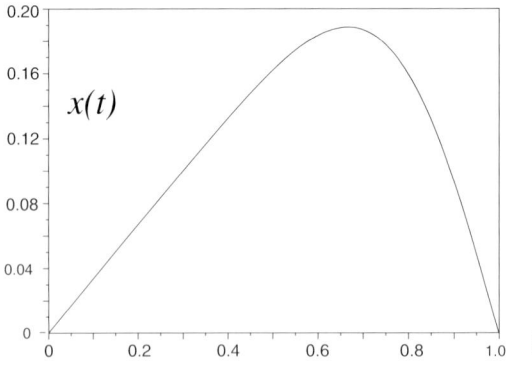

Figure 9.4. The positive solution such that $\phi_{ll}(0) = \phi_{ll}(1) = 0$

References
This chapter is based on [42].

Chapter 10

TRANSIENT RANDOM WALKS ON DYNAMICALLY ORIENTED LATTICES

1. INTRODUCTION

The use of random walks as a tool in mathematical physics is now well established and they have been for example widely used in classical statistical mechanics to study critical phenomena (see [51]). It has been recently observed that analogous methods in quantum statistical mechanics require the study of random walks on oriented lattices, due to the intrinsic non commutative character of the (quantum) world (see e.g. [24, 112]). Although random walks in random and non-random environments have been intensively studied for many years, only a few results on random walks on oriented lattices are known. The recurrence versus transience properties of simple random walks on oriented versions of $\mathbb{Z}^2$ are studied in [23] when the horizontal lines are unidirectional towards a random or deterministic direction. An interesting behavior of this model is that, depending on the orientation, the walk could be either recurrent or transient. In the deterministic "alternate" case, for which the orientations of horizontal lines are alternated, i.e. oriented rightwards at one level and leftwards at the following level, the recurrence of the simple random walk is proved, whereas the transience naturally arises when the orientations are all identical in infinite regions. More surprisingly, it is also proved that the recurrent character of the simple random walk on $\mathbb{Z}^2$ is lost when the orientations are i.i.d. with zero mean.

In this chapter, we study more general models and focus on spatially inhomogeneous or dependent distributions of the orientations. To do so, we introduce lattices for which the distribution of the horizontal orientation is generated by a dynamical system, namely the orientations are increments of a dynamic random walk. We prove that the transience of the simple random walk still holds under some smoothness conditions on this generation and detail examples and counterexamples for various standard dynamical systems. For ergodic dynamical systems, we also prove a strong law of large numbers and, in the case of i.i.d. orientations, a functional limit theorem with an unconventional normalization due to the random character of the environment of the walk, solving an open question of [23].

2. MODEL AND RESULTS
2.1 DYNAMICALLY ORIENTED LATTICES

Let $S = (E, \mathcal{A}, \mu, T)$ be a dynamical system where $(E, \mathcal{A}, \mu)$ is a probability space and T is an invertible transformation of E preserving the measure μ. This dynamical system is used to introduce inhomogeneity or dependencies in the distribution of the random orientations. To do so, we use a function f defined on E with values in $[0, 1]$ and we ask $\int_E f d\mu = 1/2$ to avoid trivialities. By *orientations*, we mean a random field $X = (X_y)_{y \in \mathbb{Z}} \in \{-1, +1\}^{\mathbb{Z}}$, or equivalently a family X of $\{-1, +1\}$-valued random variables X_y, $y \in \mathbb{Z}$, and we distinguish two different approaches to introduce its distribution.

2.1.1 QUENCHED CASE

It describes spatially inhomogeneous distributions of the orientation. We define for fixed $x \in E$ the quenched law $\mathbb{P}_T^{(x)}$ to be the product probability measure on $(\{-1, +1\}^{\mathbb{Z}}, \mathcal{F} = \mathcal{P}(\{-1, +1\})^{\otimes \mathbb{Z}})$ with one-dimensional marginals given by:

$$\mathbb{P}_T^{(x)}(X_y = +1) = f(T^y x)$$
$$= 1 - \mathbb{P}_T^{(x)}(X_y = -1).$$

To simplify, we have used the same notation for the quenched law and its marginals, which should be written $\mathbb{P}_{T,y}^{(x)}$ with $\mathbb{P}_T^{(x)} = \otimes_y \mathbb{P}_{T,y}^{(x)}$. This quenched case is thus an extension of the i.i.d. case of [23], with independent but not necessarily identically distributed random variables: when $f \equiv \frac{1}{2}$, these orientations are the Rademacher random variables of [23] but if the function f is not constant, the sequence of random variables $(X_y)_{y \in \mathbb{Z}}$ is not stationary under $\mathbb{P}_T^{(x)}$. These random variables are viewed as the increments of a dynamic random walk.

2.1.2 ANNEALED CASE

We consider now μ-averages over all $x \in E$ and the distribution of X to be the probability law $\mathbb{P}_\mu$ on $(\{-1,+1\}^{\mathbb{Z}}, \mathcal{P}(\{-1,+1\})^{\otimes \mathbb{Z}})$ defined for all $A \in \mathcal{F}$ by

$$\mathbb{P}_\mu(X \in A) = \int_E \mathbb{P}_T^{(x)}(X \in A) d\mu(x).$$

The one dimensional marginals are thus given for all $y \in \mathbb{Z}$ by

$$\mathbb{P}_\mu(X_y = +1) = \int_E f(T^y x) d\mu(x) = \int_E f(x) d\mu(x) = \frac{1}{2}$$

and the hypothesis $\int_E f d\mu = \frac{1}{2}$ has been taken to get $\mathbb{E}_\mu(X_y) = 0$. The T-invariance of μ implies the translation-invariance of $\mathbb{P}_\mu$ but this latter is not a product measure in general. For example, one compute easily, for $y \neq y' \in \mathbb{Z}$,

$$\mathrm{Cov}_\mu[X_y, X_{y'}] = 4 \int_E f(x) f(T^{y'-y} x) d\mu(x) - 1 \tag{10.1}$$

providing thus an explicit relation between the correlations of the orientation and those of μ, defined when it exists for the function f and for all $y \in \mathbb{Z}$ by

$$\begin{aligned} C_\mu^f(y) &= \int_E f \, f \circ T^y d\mu - \int_E f d\mu \int_E f \circ T^y d\mu \tag{10.2} \\ &= \int_E f \, f \circ T^y d\mu - \frac{1}{4}. \end{aligned}$$

Thus we get

$$\mathrm{Cov}_\mu(X_0, X_y) = 4\, C_\mu^f(y). \tag{10.3}$$

This annealed case leads in Section 4 to another extension of the transience property of the i.i.d. case where this time, on the contrary to the quenched case, independence is dropped but translation-invariance is kept.

2.1.3 LATTICES

We use these dynamic random variables to build our *dynamically oriented* lattices. These lattices are oriented versions of $\mathbb{Z}^2$: the vertical lines are not oriented and the horizontal ones are unidirectional, the orientation at a level $y \in \mathbb{Z}$ being given by the random variable X_y (say right if the value is $+1$ and left if it is -1). More formally we give the

DEFINITION 10.1 (DYNAMICALLY ORIENTED LATTICES) *Let* $X = (X_y)_{y \in \mathbb{Z}}$ *be a sequence of random variables defined as previously. The* dynamically oriented lattice $\mathbb{L}^X = (\mathbb{V}, \mathbb{A}^X)$ *is the (random) directed graph with (deterministic) vertex set* $\mathbb{V} = \mathbb{Z}^2$ *and (random) edge set* $\mathbb{A}^X$ *defined by the condition that for* $u = (u_1, u_2), v = (v_1, v_2) \in \mathbb{Z}^2$, $(u, v) \in \mathbb{A}^X$ *if and only if*

1. *either* $v_1 = u_1$ *and* $v_2 = u_2 \pm 1$
2. *or* $v_2 = u_2$ *and* $v_1 = u_1 + X_{u_2}$.

2.2 SIMPLE RANDOM WALK ON $\mathbb{L}^X$

We consider the usual simple random walk $M = (M_n)_{n \in \mathbb{N}}$ on $\mathbb{L}^X$. For every realization X, it is a $\mathbb{Z}^2$-valued Markov chain M defined on a probability space $(\Omega, \mathcal{B}, \mathbb{P})$, whose ($X$-dependent) transition probabilities are defined for all $(u, v) \in \mathbb{V} \times \mathbb{V}$ by

$$\mathbb{P}(M_{n+1} = v | M_n = u) = \begin{cases} \frac{1}{3} & \text{if } (u, v) \in \mathbb{A}^X \\ 0 & \text{otherwise.} \end{cases}$$

Its transience is proved in [23] for almost every orientation X when they are given by a sequence of independent Rademacher random variables $(X_y)_{y \in \mathbb{Z}}$. This result can be generalized in this dynamical context when the orientations are either *annealed* or *quenched*.

THEOREM 10.1 *Assume that*

$$\int_E \frac{1}{\sqrt{f(1-f)}} \, d\mu < \infty \qquad (10.4)$$

then

1. *in the annealed case, for* $\mathbb{P}_\mu$-*a.e. orientation* X, *the simple random walk on dynamically oriented lattice* $\mathbb{L}^X$ *is transient.*

2. *in the quenched case, for* μ-*a.e.* $x \in E$, *for* $\mathbb{P}_T^{(x)}$-*a.e. realization of the orientation* X, *the simple random walk on the dynamically oriented lattice* $\mathbb{L}^X$ *is transient.*

Let us mention that non-invertible transformations T of the space E can also be considered in the following

THEOREM 10.2 *Assume that T is not invertible but that the distribution of the orientations $(X_y)_{y \in \mathbb{Z}}$ has one-dimensional marginals defined by*

$$\mathbb{P}_T^{(x)}(X_y = +1) = f(T^{|y|}x) \qquad (10.5)$$
$$= 1 - \mathbb{P}_T^{(x)}(X_y = -1). \qquad (10.6)$$

Then conclusions of Theorem 10.1 still hold.

It is worth noting that in the annealed case the measure $\mathbb{P}_\mu$ is not stationary anymore, and we illustrate this result by the example of Manneville-Pomeau maps of the interval in Section 4.

2.3 LIMIT THEOREMS IN THE ERGODIC CASE

Let us assume that the dynamical system $S = (E, \mathcal{A}, \mu, T)$ defined in Section 2.1 is ergodic. Then, it is not difficult to prove that in this particular setting, the annealed measure $\mathbb{P}_\mu$ is also ergodic. We denote $\tilde{\mathbb{P}}_\mu$ for the joint measure $\mathbb{P} \otimes \mathbb{P}_\mu$ and establish the following theorem:

THEOREM 10.3 (STRONG LAW OF LARGE NUMBERS) *The random walk on the lattice $\mathbb{L}^X$ has $\tilde{\mathbb{P}}_\mu$-almost surely zero speed, i.e.*

$$\lim_{n \to +\infty} \frac{M_n}{n} = (0,0) \quad \tilde{\mathbb{P}}_\mu\text{-almost surely.} \tag{10.7}$$

We introduce a real constant $m = \frac{1}{2}$, defined later as the mean of some geometric random variables related to the behavior of the walk in the horizontal direction. The study of the simple random walk on dynamically oriented graph $\mathbb{L}^X$ is closely related to the simple random walks in random sceneries introduced in Chapter 4. Let us consider a standard Brownian motion $(B_t)_{t \geq 0}$ and denote by $(L_t(x))_{t \geq 0}$ its corresponding local time at $x \in \mathbb{R}$. Moreover, we introduce a pair of independent Brownian motions $Z_+(x), Z_-(x), x \geq 0$. We assume these processes to be defined on one probability space and to be independent of each other so that the following process is well-defined for all $t \geq 0$:

$$\Delta_t = \int_0^\infty L_t(x) dZ_+(x) + \int_0^\infty L_t(-x) dZ_-(x). \tag{10.8}$$

Using Theorem 4.3, we shall prove

THEOREM 10.4 (FUNCTIONAL LIMIT THEOREM)

$$\left(\frac{1}{n^{3/4}} M_{[nt]} \right)_{t \geq 0} \overset{\mathcal{D}}{\Longrightarrow} \frac{m}{(1+m)^{3/4}} (\Delta_t, 0)_{t \geq 0} \tag{10.9}$$

where $\overset{\mathcal{D}}{\Longrightarrow}$ stands for convergence in the space of càdlàg functions $\mathcal{D}([0,\infty), \mathbb{R}^2)$ endowed with the Skorohod topology.

The following local limit theorem for the random walk M_n should hold: There exists a constant $C > 0$ such that as $n \to +\infty$,

$$\tilde{\mathbb{P}}_\mu(M_n = (0,0)) \sim C n^{-5/4}. \tag{10.10}$$

3. PROOFS
3.1 VERTICAL AND HORIZONTAL EMBEDDINGS OF THE SIMPLE RANDOM WALK

The simple random walk M defined on $(\Omega, \mathcal{B}, \mathbb{P})$ can be decomposed into vertical and horizontal embeddings by projection to the corresponding axis. These embeddings will carry the interesting asymptotic properties of the walk. The vertical one is a simple random walk $S' = (S'_n)_{n \in \mathbb{N}}$ on the line and we define for all $n \in \mathbb{N}$ and $y \in \mathbb{Z}$ its *local time* at level y to be

$$N_n(y) = \sum_{k=0}^{n} \mathbf{1}_{\{S'_k = y\}}.$$

The horizontal embedding is a random walk with $\mathbb{N}$-valued geometric jumps. More formally, a doubly infinite family $(\xi_i^{(y)})_{i \in \mathbb{N}^*, y \in \mathbb{Z}}$ of independent geometric random variables of parameter $p = \frac{1}{3}$ (and mean $m = \frac{1}{2}$) is given and one defines the embedded horizontal random walk $S = (S_n)_{n \in \mathbb{N}}$ by $S_0 = 0$ and for $n \geq 1$,

$$S_n = \sum_{y \in \mathbb{Z}} X_y \sum_{i=1}^{N_{n-1}(y)} \xi_i^{(y)}$$

with the convention that the last sum is zero when $N_{n-1}(y) = 0$. Of course, the walk M_n does not coincide with (S_n, S'_n) but these objects are closely related: define for all $n \in \mathbb{N}$

$$T_n = n + \sum_{y \in \mathbb{Z}} \sum_{i=1}^{N_{n-1}(y)} \xi_i^{(y)}$$

to be the instant just after the random walk M has performed its nth vertical move. A direct and useful consequence of this decomposition is the following lemma ([23]). We denote the (independent) σ-algebras generated by the vertical walk S' and the orientation X by

$$\mathcal{G} = \sigma(X_y, \, y \in \mathbb{Z}) \quad \text{and} \quad \mathcal{F}_n = \sigma(S'_k, \, k = 1, \ldots, n),$$

with $\mathcal{F} = \sigma(S'_k, \, k \geq 1)$.

LEMMA 20
- $M_{T_n} = (S_n, S'_n), \, \forall n \in \mathbb{N}$.

- For a given orientation X, the transience of $(M_{T_n})_{n \in \mathbb{N}}$ implies the transience of $(M_n)_{n \in \mathbb{N}}$.

Proof:
The first assertion is evident by construction. Let us prove the second one. Denote $(\sigma_n)_{n \geq 0}$ the sequence of return times to 0 of the random walk $(S'_n)_n$, namely $\sigma_0 = 0$ and for $n = 1, 2, \ldots, \sigma_n = \inf\{k \geq \sigma_{n-1}; S'_k = 0\}$. Since $S'_{\sigma_n} = 0$ and only then $M_{T_{\sigma_n}} = (S_{\sigma_n}, 0)$. For $k = T_{\sigma_n}, \ldots, T_{\sigma_{n+1}-1}$, the process M_k can vanish only when k is in the first part of this discrete time interval, before the process performs any vertical move, namely if either $S_{\sigma_n} = 0$ or if the points S_{σ_n} and $S_{\sigma_{n+1}}$ straddle the point 0.

$$S_{\sigma_{n+1}} - S_{\sigma_n} = \sum_{y \in \mathbb{Z}} X_y \left(\sum_{i=1}^{N_{\sigma_n}(y)} \xi_i^{(y)} - \sum_{i=1}^{N_{\sigma_n-1}(y)} \xi_i^{(y)} \right)$$
$$= X_0 \xi_{N_{\sigma_n}(0)}^{(0)}$$

We use the notation: $I(y, Xz)$ is the set $\{y, \ldots, y+z\}$ when $X = 1$ and the set $\{y-z, \ldots, y\}$ when $X = -1$. Let Z be a random variable having the same distribution as $\xi_1^{(0)}$. With these definitions, the process $(M_n)_n$ can vanish for $n \in \{T_{\sigma_n}, \ldots, T_{\sigma_{n+1}} - 1\}$ if and only if 0 belongs to the set $I(S_{\sigma_n}, X_0 Z)$. Consequently,

$$\sum_{n \in \mathbb{N}} \mathbb{P}(M_n = (0,0) | \mathcal{F} \vee \mathcal{G}) = \sum_{n \in \mathbb{N}} \mathbb{P}(0 \in I(S_{\sigma_n}, X_0 Z) | \mathcal{F} \vee \mathcal{G})$$

So we only have to estimate the following probabilities

$$\mathbb{P}(0 \in I(S_{\sigma_n}, X_0 Z) | \mathcal{F} \vee \mathcal{G}) = \frac{1}{3} \mathbb{P}(S_{\sigma_n} = 0 | \mathcal{F} \vee \mathcal{G})$$
$$+ \mathbf{1}_{\{X_0 = -1\}} \mathbb{P}(\bigcup_{y \in \mathbb{N}^*} \{S_{\sigma_n} = y; Z \geq y\} | \mathcal{F} \vee \mathcal{G})$$
$$+ \mathbf{1}_{\{X_0 = 1\}} \mathbb{P}(\bigcup_{y \in \mathbb{N}^*} \{S_{\sigma_n} = -y; Z \geq y\} | \mathcal{F} \vee \mathcal{G}).$$

When the process $(S_{\sigma_n})_{n \in \mathbb{N}}$ is transient, there exists a constant $C > 0$ such that, uniformly in $y \in \mathbb{Z}$, we have

$$\sum_{n \in \mathbb{N}} \mathbb{P}(S_{\sigma_n} = y | \mathcal{F} \vee \mathcal{G}) \leq C < \infty.$$

Consequently,

$$\sum_{n \in \mathbb{N}} \mathbb{P}(\bigcup_{y \in \mathbb{N}^*} \{S_{\sigma_n} = y; Z \geq y\} | \mathcal{F} \vee \mathcal{G}) = \sum_{n \in \mathbb{N}} \sum_{y \geq 1} \left(\frac{2}{3}\right)^y \mathbb{P}(S_{\sigma_n} = y | \mathcal{F} \vee \mathcal{G}) \leq 2C,$$

then the transience of $(M_n)_{n \in \mathbb{N}}$ follows.

3.2 PROOF OF THE TRANSIENCE OF THE SIMPLE RANDOM WALK

Independent of the orientation X, the vertical walk S'_n is known to be recurrent and its asymptotic behavior is rather well controlled. The transience is due to the behavior of the embedded horizontal random walk S_n and to exploit it and prove Theorem 10.1, we introduce a partition of Ω between typical or untypical paths of S'_n.

In all this proof, for any $i \in \mathbb{N}$, δ_i is a strictly positive real number and we write $d_{n,i} = n^{\frac{1}{2}+\delta_i}$. Define the sets

$$A_n = \{\omega \in \Omega; \max_{0 \leq k \leq 2n} |S'_k| < d_{n,1}\} \cap \{\omega \in \Omega; \max_{y \in \mathbb{Z}} N_{2n-1}(y) < d_{n,2}\}$$

and

$$B_n = \{\omega \in A_n; \Big|\sum_{y \in \mathbb{Z}} X_y N_{2n-1}(y)\Big| > d_{n,3}\}.$$

We prove the transience in the annealed case, but the proof in fact contains this of the quenched case. We denote the joint measure $\mathbb{P} \otimes \mathbb{P}_\mu$ by $\tilde{\mathbb{P}}_\mu$ (annealed case) and $\mathbb{P} \otimes \mathbb{P}_T^{(x)}$ by $\tilde{\mathbb{P}}_T^{(x)}$ (quenched case). We establish that

$$\tilde{\mathbb{P}}_\mu(M_{T_n} = (0,0) \text{ i.o.}) = 0 \tag{10.11}$$

or, using Lemma 20,

$$\sum_{n \in \mathbb{N}} \tilde{\mathbb{P}}_\mu(S_{2n} = 0; S'_{2n} = 0) < \infty. \tag{10.12}$$

By definition

$$\sum_{n \in \mathbb{N}} \tilde{\mathbb{P}}_\mu(S_{2n} = 0; S'_{2n} = 0) = \int_E \sum_n \mathbb{P}(\mathbb{P}_T^{(x)}(S_{2n} = 0; S'_{2n} = 0)) d\mu(x)$$

and we first decompose $\tilde{\mathbb{P}}_T^{(x)}(S_{2n} = 0; S'_{2n} = 0)$ into

$$\tilde{\mathbb{P}}_T^{(x)}(S_{2n} = 0; S'_{2n} = 0; A_n^c) + \tilde{\mathbb{P}}_T^{(x)}(S_{2n} = 0; S'_{2n} = 0; B_n) \tag{10.13}$$
$$+ \tilde{\mathbb{P}}_T^{(x)}(S_{2n} = 0; S'_{2n} = 0; A_n \setminus B_n).$$

Some results of the i.i.d. case of [23] still hold uniformly in x and in particular we can prove using standard techniques the following

LEMMA 21 *For every* $x \in E$,

$$\sum_{n \in \mathbb{N}} \tilde{\mathbb{P}}_T^{(x)}(S_{2n} = 0; S'_{2n} = 0; A_n^c) < \infty.$$

Proof:
We have $A_n^c = A_{n,1}^c \cup A_{n,2}^c$ with

$$A_{n,1}^c = \{\omega \in \Omega; \max_{0 \leq k \leq 2n} |S_k'| \geq d_{n,1}\}$$

and

$$A_{n,2}^c = \{\omega \in \Omega; \max_{y \in \mathbb{Z}} N_{2n-1}(y) \geq d_{n,2}\}.$$

Let $R_n = \{a_n, a_n + 1, \ldots, n\}$ with $a_n = [d_{n,1}]$. With this notation,

$$\begin{aligned}
\mathbb{P}(A_{n,1}^c | S_{2n}' = 0) &= \sum_{y \in R_n} \mathbb{P}(\max_{0 \leq k \leq 2n} |S_k'| = y | S_{2n}' = 0) \\
&\leq 2 \sum_{y \in R_n} \mathbb{P}(S_{2n}' = 2y | S_0' = 0) \\
&= 2\mathbb{P}(S_{2n}' \geq 2a_n | S_0' = 0) \\
&\leq 2 \inf_{t>0} \mathbb{P}(\exp(tS_{2n}') \geq \exp(2ta_n) | S_0' = 0) \\
&\leq 2 \inf_{t>0} \frac{(\cosh t)^n}{\exp(2ta_n)} \\
&= 2\exp(-\frac{a_n^2}{n}) \\
&\leq 2\exp(-n^{2\delta_1})
\end{aligned}$$

We have

$$\begin{aligned}
\mathbb{P}(A_{n,2}^c | S_{2n}' = 0) &= \mathbb{P}(\max_{y \in \mathbb{Z}} N_{2n-1}(y) \geq d_{n,2} | S_{2n}' = 0) \\
&\leq \sum_{y \in \mathbb{Z}} \frac{\mathbb{P}(N_{2n-1}(y) \geq d_{n,2} | S_0' = 0)}{\mathbb{P}(S_{2n}' = 0)}
\end{aligned}$$

Let $\sigma_{y,k}$ be the kth return time to the point y for the simple random walk $(S_k')_k$. Remark that the local time $N_{2n-1}(y)$ can exceed the threshold $[d_{n,2}]$ whenever the random walk (S_k'), starting at the origin, attains y before time $2n-1$ and then returns to y at least $[d_{n,2}]$ times before time $2n - 1$. Then,

$$\begin{aligned}
\mathbb{P}(N_{2n-1}(y) \geq d_{n,2} | S_0' = 0) &\leq \mathbb{P}(\sigma_{y,[d_{n,2}]} \leq 2n | S_0' = y) \\
&\leq \exp(2nt)\mathbb{E}(\exp(-t\sigma_{0,[d_{n,2}]})) \\
&\leq \inf_{t>0} \exp(2nt)(1 - \sqrt{1 - \exp(-2t)})^{[d_{n,2}]} \\
&= \exp(-cn^{\delta_2})
\end{aligned}$$

for some positive constant c, uniformly in y. Using that $\mathbb{P}(S'_{2n} = 0) \sim Cn^{-1/2}$, we get the bound

$$\mathbb{P}(A^c_{n,2}|S'_{2n} = 0) \leq Cnn^{1/2}\exp(-cn^{\delta_2}).$$

Choosing $0 < \delta'_2 < \delta_2$ and $\delta = \min(2\delta_1, \delta'_2) > 0$, we conclude that

$$\mathbb{P}(A^c_n|S'_{2n} = 0) \leq 2\exp(-n^{2\delta_1}) + C\exp(-cn^{\delta'_2}) = \mathcal{O}(\exp(-cn^{\delta})).$$

The second term of (10.13) is also a generic term of convergent series due to the untypical character of the paths in B_n. Again from [23] with standard techniques, we have the

LEMMA 22 *For every* $x \in E$,

$$\sum_{n \in \mathbb{N}} \tilde{\mathbb{P}}_T^{(x)}(S_{2n} = 0; S'_{2n} = 0; B_n) < \infty.$$

Proof:
Introduce the random variables

$$N_+ = \sum_{k=1}^{2n} 1_{\{X_{S'_k} = 1\}}$$

$$N_- = \sum_{k=1}^{2n} 1_{\{X_{S'_k} = -1\}}$$

$$\Delta_n = N_+ - N_- = \sum_{y \in \mathbb{Z}} X_y N_{2n-1}(y)$$

Remark that N_+, N_- and Δ_n are $\mathcal{F}_{2n} \vee \mathcal{G}$-measurable and $N_+ + N_- = 2n$. Denoting $m_1 = \mathbb{E}(\xi_1^{(1)})$, $m_2 = \mathbb{E}((\xi_1^{(1)})^2)$ and $s^2 = m_2 - m_1^2$, we have

$$\begin{aligned}
\mathbb{E}(S_{2n}|\mathcal{F}_{2n} \vee \mathcal{G}) &= m_1 \Delta_n \\
\mathbb{E}(S_{2n}^2|\mathcal{F}_{2n} \vee \mathcal{G}) &= 2ns^2 + m_1^2 \Delta_n^2 \\
\text{Var}(S_{2n}|\mathcal{F}_{2n} \vee \mathcal{G}) &= 2ns^2
\end{aligned}$$

For $t \in]-\infty, -\log(q)[$, we define the generating function of the random variable $\xi_1^{(1)}$

$$\phi(t) = \mathbb{E}(\exp(t\xi_1^{(1)})).$$

For $|t|$ small enough, $\phi(t) = \exp(tm_1 + t^2s^2/2 + \mathcal{O}(t^3))$. Then,

$$\begin{aligned}
\mathbb{E}(\exp(tS_{2n})|\mathcal{F}_{2n} \vee \mathcal{G}) &= \phi(t)^{N_+}\phi(-t)^{N_-} \\
&= \exp(tm_1\Delta_n + t^2s^2n + \mathcal{O}(t^3n))
\end{aligned}$$

Assume for the moment that $\Delta_n > d_{n,3}$. Using Markov's inequality, we have for $t < 0$,

$$\begin{aligned}\tilde{\mathbb{P}}_T^{(x)}(S_{2n} = 0|\mathcal{F}_{2n} \vee \mathcal{G}) &\leq \tilde{\mathbb{P}}_T^{(x)}(S_{2n} \leq 0|\mathcal{F}_{2n} \vee \mathcal{G}) \\ &\leq \mathbb{E}(\exp(tS_{2n})|\mathcal{F}_{2n} \vee \mathcal{G}) \\ &= \exp(tm_1\Delta_n + t^2 s^2 n + \mathcal{O}(t^3 n))\end{aligned}$$

Now, choose $t = -\frac{m_1 n^{\delta_3 - 1/2}}{2s^2}$. Hence, on $\{\Delta_n > d_{n,3}\}$, we have

$$\tilde{\mathbb{P}}_T^{(x)}(S_{2n} = 0|\mathcal{F}_{2n} \vee \mathcal{G}) \leq \exp(-\frac{m_1^2}{4s^2} n^{2\delta_3} + \mathcal{O}(n^{3\delta_3 - 1/2})).$$

Similarly, on $\{\Delta_n < -d_{n,3}\}$, we have

$$\tilde{\mathbb{P}}_T^{(x)}(S_{2n} = 0|\mathcal{F}_{2n} \vee \mathcal{G}) \leq \exp(tm_1\Delta_n + t^2 s^2 n + \mathcal{O}(t^3 n))$$

for every $t \in]0, -\log(q)[$, and choosing for large n, $t = \frac{m_1 n^{\delta_3 - 1/2}}{2s^2}$.

Let us write

$$\begin{aligned}p_n^{(x)} &= \tilde{\mathbb{P}}_T^{(x)}(S_{2n} = 0; S'_{2n} = 0; A_n \setminus B_n) \\ &= \mathbb{E}\Big(1_{\{S'_{2n}=0\}} \mathbb{E}(1_{\{A_n \setminus B_n\}} \mathbb{E}(1_{\{S_{2n}=0\}}|\mathcal{F} \vee \mathcal{G})|\mathcal{F})\Big)\end{aligned}$$

where $\mathbb{E}$ stands for the expectation under the measure $\tilde{\mathbb{P}}_T^{(x)}$. To prove the theorem, it remains to show that

$$\int_E \left(\sum_{n \in \mathbb{N}} p_n^{(x)}\right) d\mu(x) < \infty. \tag{10.14}$$

It is well known that for the simple random walk S', there exists $C > 0$ s.t.
$$\mathbb{P}(S'_{2n} = 0) \sim Cn^{-\frac{1}{2}}, \ n \to +\infty \tag{10.15}$$

and we can prove as in [23] the

LEMMA 23 *On the set $A_n \setminus B_n$, we have,*

$$\tilde{\mathbb{P}}_T^{(x)}(S_{2n} = 0|\mathcal{F} \vee \mathcal{G}) = \mathcal{O}\left(\sqrt{\frac{\ln n}{n}}\right) \tag{10.16}$$

uniformly in $x \in E$.

Proof:
Since the random variables $(X_y)_{y \in \mathbb{Z}}$ and $(N_n(y))_{y \in \mathbb{Z}, n \in \mathbb{N}}$ are $\mathcal{F} \vee \mathcal{G}$-measurable, we can rewrite the conditional characteristic function of

the random variable S_{2n} as follows:

$$\chi_1(\theta) = \mathbb{E}(\exp(i\theta S_{2n})|\mathcal{F} \vee \mathcal{G}) = \prod_{y \in \mathbb{Z}} \chi(\theta X_y)^{N_{2n-1}(y)}$$

where $\chi(.)$ is the characteristic function of the ξ's. Hence,

$$\tilde{\mathbb{P}}_T^{(x)}(S_{2n} = 0|\mathcal{F} \vee \mathcal{G}) = \frac{1}{2\pi} \int_{-\pi}^{\pi} \chi_1(\theta) \, d\theta.$$

Now use the decomposition of $\chi(\theta)$ into the modulus part $r(\theta)$ (it is an even function of θ) and its angular part $\alpha(\theta)$

$$\chi(\theta) = r(\theta) \exp(i\alpha(\theta)), \quad \theta \in [-\pi, \pi]$$

and the fact that there is a constant $K < 1$ such that for $\theta \in [-\pi, -\pi/2] \cup [\pi/2, \pi]$, we can bound $r(\theta) < K$ to majorise

$$\tilde{\mathbb{P}}_T^{(x)}(S_{2n} = 0|\mathcal{F} \vee \mathcal{G}) \leq \frac{1}{\pi} \int_0^{\pi/2} r(\theta)^{2n} \, d\theta + \mathcal{O}(K^n).$$

Fix $a_n = \sqrt{\frac{\log(n)}{n}}$ and split the above integral over $[0, \pi/2] = [0, a_n] \cup [a_n, \pi/2]$. For the first part, we can majorise the integrand by 1, so that

$$\int_0^{a_n} r(\theta)^{2n} \, d\theta \leq a_n.$$

For the second part, use the majorisation $r(\theta) \leq \exp(-\frac{3}{8}\theta^2)$ valid for every $\theta \in \,]0, \pi/2]$ to estimate

$$\int_{a_n}^{\pi/2} r(\theta)^{2n} \, d\theta = \mathcal{O}(n^{-3/4}).$$

Since the first part dominates, the result follows.

Hence, the transience of the simple random walk is a direct consequence of the following

PROPOSITION 12 *It is possible to choose $\delta_1, \delta_2, \delta_3 > 0$ such that there exists $\delta > 0$ and*

$$\int_E \tilde{\mathbb{P}}_T^{(x)}(A_n \setminus B_n|\mathcal{F}) \, d\mu(x) = \mathcal{O}(n^{-\delta}). \quad (10.17)$$

Proof:
We have to estimate, on the event A_n, the conditional probability

$$\tilde{\mathbb{P}}_T^{(x)}(|\sum_{y \in \mathbb{Z}} \zeta_y| \leq d_{n,3}|\mathcal{F})$$

where $\zeta_y = X_y N_{2n-1}(y), y \in \mathbb{Z}$. Let G be a centered Gaussian random variable with variance $d_{n,3}^2$, (conditionally on $\mathcal{F}$) independent of the random variables ζ_y's. Clearly,

$$\tilde{\mathbb{P}}_T^{(x)}(\sum_y \zeta_y \in [0, d_{n,3}] | \mathcal{F}) = \frac{\tilde{\mathbb{P}}_T^{(x)}(\sum_y \zeta_y \in [0, d_{n,3}]; 0 \leq G \leq d_{n,3} | \mathcal{F})}{\tilde{\mathbb{P}}_T^{(x)}(0 \leq G \leq d_{n,3} | \mathcal{F})}$$

where $\tilde{\mathbb{P}}_T^{(x)}(0 \leq G \leq d_{n,3} | \mathcal{F}) = c$ is a strictly positive real number independent of n. Since G is independent of the random variables ζ_y's and using the symmetry of the Gaussian distribution, we have

$$\tilde{\mathbb{P}}_T^{(x)}(\sum_y \zeta_y \in [0, d_{n,3}]; 0 \leq G \leq d_{n,3} | \mathcal{F}) \quad (10.18)$$

$$= \tilde{\mathbb{P}}_T^{(x)}(\sum_y \zeta_y \in [0, d_{n,3}]; -d_{n,3} \leq G \leq 0 | \mathcal{F}). \quad (10.19)$$

Consequently, we obtain

$$\tilde{\mathbb{P}}_T^{(x)}(\sum_y \zeta_y \in [0, d_{n,3}] | \mathcal{F}) \leq \frac{1}{c} \tilde{\mathbb{P}}_T^{(x)}(|\sum_y \zeta_y + G| \leq d_{n,3} | \mathcal{F}).$$

In the same way, we get

$$\tilde{\mathbb{P}}_T^{(x)}(\sum_y \zeta_y \in [-d_{n,3}, 0] | \mathcal{F}) \leq \frac{1}{c} \tilde{\mathbb{P}}_T^{(x)}(|\sum_y \zeta_y + G| \leq d_{n,3} | \mathcal{F})$$

and then, we have the following inequality

$$\tilde{\mathbb{P}}_T^{(x)}(|\sum_y \zeta_y| \leq d_{n,3} | \mathcal{F}) \leq \frac{2}{c} \tilde{\mathbb{P}}_T^{(x)}(|\sum_y \zeta_y + G| \leq d_{n,3} | \mathcal{F}).$$

From Plancherel's formula, we deduce that there exists a constant $C > 0$ such that

$$\tilde{\mathbb{P}}_T^{(x)}(|\sum_y \zeta_y + G| \leq d_{n,3} | \mathcal{F}) \leq C \, d_{n,3} \, I_n(x) \quad (10.20)$$

where

$$I_n(x) = \int_{-\pi}^{\pi} \mathbb{E}(e^{it \sum_{y \in \mathbb{Z}} X_y N_{2n-1}(y)} | \mathcal{F}) e^{-t^2 d_{n,3}^2/2} dt.$$

To use that for $td_{n,3}$ small enough, $e^{-t^2 d_{n,3}^2/2}$ dominates the term under the expectation, we split the integral in two parts. For $b_n = \frac{n^{\delta_2}}{d_{n,3}}$, we write

$$I_n(x) = I_n^1(x) + I_n^2(x)$$

with

$$I_n^1(x) = \int_{|t|\leq b_n} \mathbb{E}(e^{it\sum_{y\in\mathbb{Z}} X_y N_{2n-1}(y)}|\mathcal{F})e^{-t^2 d_{n,3}^2/2} dt$$

$$I_n^2(x) = \int_{|t|> b_n} \mathbb{E}(e^{it\sum_{y\in\mathbb{Z}} X_y N_{2n-1}(y)}|\mathcal{F})e^{-t^2 d_{n,3}^2/2} dt.$$

To control the integral $I_n^2(x)$, we write

$$|I_n^2(x)| \leq C\int_{|t|>b_n} e^{-t^2 d_{n,3}^2/2} dt = \frac{C}{d_{n,3}} \int_{|s|>n^{\delta_2}} e^{-s^2/2} ds$$

$$\leq 2\frac{C}{d_{n,3}} n^{-\delta_2} e^{-n^{2\delta_2}/2}$$

to get

$$|I_n^2(x)| = \mathcal{O}(e^{-n^{2\delta_2}/2})$$

uniformly in $x \in E$.

LEMMA 24 For $\delta_3 > 2\delta_2$,

$$\int_E |I_n^1(x)|\, d\mu(x) = \mathcal{O}(n^{-\frac{3}{4}+\frac{\delta_1}{2}}).$$

Proof:
From the definition of the orientations $(X_y)_y$, an explicit formula for the characteristic function $\phi_{X_y}^{(x)}$ of the random variable X_y can be given and we deduce that

$$|\phi_{X_y}^{(x)}(u)|^2 = \cos^2(u) + (2f(T^y x) - 1)^2 \sin^2(u)$$
$$= 1 - 4f(T^y x)(1 - f(T^y x)) \sin^2(u)$$

and by independence of the X's we get

$$|I_n^1(x)| \leq \int_{|t|\leq b_n} |\prod_y \phi_{X_y}^{(x)}(N_{2n-1}(y)t)|\, dt.$$

Denote for all $y \in \mathbb{Z}, n \in \mathbb{N}$, $p_{n,y} = \frac{N_{2n-1}(y)}{2n}$, $C_n = \{y : N_{2n-1}(y) \neq 0\}$ and use Hölder's inequality to get

$$|I_n^1(x)| \leq \prod_y \left[\left(\int_{|t|\leq b_n} |\phi_{X_y}^{(x)}(N_{2n-1}(y)t)|^{1/p_{n,y}} dt\right)^{p_{n,y}}\right]. \qquad (10.21)$$

Now, using the fact that we work on A_n, we choose $\delta_3 > 2\delta_2$ in order to have $b_n N_{2n-1}(y) \to 0$ uniformly in y when n goes to infinity. Using that $\sin(x) \geq \frac{2}{\pi}x$ for $x \in [0, \frac{\pi}{2}]$ and $\exp(-x) \geq 1-x$, $|I_n^1(x)|$ is bounded by

$$\prod_{y\in C_n} \left(\frac{1}{N_{2n-1}(y)} \int_{|v|\leq b_n N_{2n-1}(y)} \exp\left(-\frac{16}{p_{n,y}\pi^2} f(T^y x)(1-f(T^y x))v^2\right) dv\right)^{p_{n,y}}$$

$$\leq \prod_{y\in C_n}\left(\frac{c\mathbf{1}_{\{f(T^y x)(1-f(T^y x))>0\}}}{\sqrt{2nN_{2n-1}(y)f(T^y x)(1-f(T^y x))}}\right)^{p_{n,y}} \quad (c=\pi^{3/2}/4)$$

$$= c\exp\left[-\frac{1}{2}\sum_{y\in C_n} p_{n,y}\log(2nN_{2n-1}(y))\right]\prod_{y\in C_n}\left(\frac{\mathbf{1}_{\{f(T^y x)(1-f(T^y x))>0\}}}{\sqrt{f(T^y x)(1-f(T^y x))}}\right)^{p_{n,y}}.$$

The vector $\mathbf{p}=(p_{n,y})_{y\in C_n}$ defines a probability measure on C_n and we have

$$-\frac{1}{2}\sum_{y\in C_n}p_{n,y}\log(2nN_{2n-1}(y)) = -\log 2n - \frac{1}{2}\sum_{y\in C_n}p_{n,y}\log p_{n,y}$$
$$= -\log 2n + \frac{1}{2}H(\mathbf{p})$$

where $H(\cdot)$ is the entropy of the probability vector $\mathbf{p}$, always bounded by $\log(\mathrm{card}(C_n))$. We thus have on the set A_n,

$$|I_n^1(x)|\leq c\exp\left[-\log 2n + \frac{1}{2}\log(2d_{n,1})\right]\prod_{y\in C_n}\left(\frac{\mathbf{1}_{\{f(T^y x)(1-f(T^y x))>0\}}}{\sqrt{f(T^y x)(1-f(T^y x))}}\right)^{p_{n,y}}.$$

By applying Hölder's inequality and the fact that T preserves the measure μ, we get

$$\int_E |I_n^1(x)|\,d\mu(x)$$
$$\leq Cn^{-\frac{3}{4}+\frac{\delta_1}{2}}\int_E\prod_{y\in C_n}\left(\frac{\mathbf{1}_{\{f(T^y x)(1-f(T^y x))>0\}}}{\sqrt{f(T^y x)(1-f(T^y x))}}\right)^{p_{n,y}}d\mu(x)$$
$$\leq Cn^{-\frac{3}{4}+\frac{\delta_1}{2}}\prod_{y\in C_n}\left[\int_E\left(\frac{\mathbf{1}_{\{f(T^y x)(1-f(T^y x))>0\}}}{\sqrt{f(T^y x)(1-f(T^y x))}}\right)d\mu(x)\right]^{p_{n,y}}$$
$$= Cn^{-\frac{3}{4}+\frac{\delta_1}{2}}\int_E\frac{1}{\sqrt{f(x)(1-f(x))}}\,d\mu(x).$$

Now, using (10.20), write with the usual notation $d_{n,3}=n^{\frac{1}{2}+\delta_3}$:

$$\int_E \tilde{\mathbb{P}}_T^{(x)}(A_n\setminus B_n|\mathcal{F})\,d\mu(x)\leq Cd_{n,3}\int_E\left(|I_n^1(x)|+|I_n^2(x)|\right)d\mu(x)$$

and consider $\delta_3>2\delta_2$. By the previous lemmata, we have

$$d_{n,3}\int_E |I_n^1(x)|\,d\mu(x) = \mathcal{O}(n^{-\frac{1}{4}+\delta_3+\frac{\delta_1}{2}})$$

and

$$d_{n,3}\int_E |I_n^2(x)|\,d\mu(x) = \mathcal{O}(e^{-n^{2\delta_2}/2})$$

and the proposition follows by choosing $\delta_i, i = 1, 2, 3$ small enough.

Combining Equations (10.15), (10.16) and (10.17), we obtain (10.14) and then (10.12). By Borel-Cantelli Lemma, we get (10.11):

$$\tilde{\mathbb{P}}_\mu(M_{T_n} = (0,0) \text{ i.o.}) = \mathbb{P}_\mu(\mathbb{P}(M_{T_n} = (0,0) \text{ i.o.})) = 0$$

and thus for $\mathbb{P}_\mu$-almost every orientation X,

$$\mathbb{P}(M_{T_n} = (0,0) \text{ i.o.}) = 0.$$

This proves that $(M_{T_n})_{n \in \mathbb{N}}$ is transient for $\mathbb{P}_\mu$-almost every orientation X, and by Lemma 20, the $\mathbb{P}_\mu$-almost sure transience of the simple random walk on the annealed oriented lattice. The proof of point 2. of the theorem (transience in the quenched case) is contained in the proof of point 1. but can also be recovered from it as follows. By Fubini-Tonnelli's theorem,

$$\sum_n \tilde{\mathbb{P}}_\mu(M_{T_n} = (0,0)) < \infty \implies \sum_n \int_E \tilde{\mathbb{P}}_T^{(x)}(M_{T_n} = (0,0)) d\mu(x) < \infty$$

$$\implies \int_E \sum_n \tilde{\mathbb{P}}_T^{(x)}(M_{T_n} = (0,0)) d\mu(x) < \infty$$

$$\implies \text{For } \mu\text{-a.e. } x, \sum_n \tilde{\mathbb{P}}_T^{(x)}(M_{T_n} = (0,0)) < \infty$$

$$\implies \text{For } \mu - \text{a.e. } x, \tilde{\mathbb{P}}_T^{(x)}(\{M_{T_n} = (0,0) \text{ i.o.}\}) = 0$$

$$\implies \text{For } \mu\text{-a.e. } x, \mathbb{P}_T^{(x)}(\mathbb{P}(\{M_{T_n} = (0,0) \text{ i.o.}\})) = 0$$

and thus the $\mathbb{P}_T^{(x)}$-almost sure transience on the quenched dynamically oriented lattice holds:

For μ-a.e. x, For $\mathbb{P}_T^{(x)}$-a.e. orientation X, $\mathbb{P}(\{M_n = (0,0) \text{ i.o.}\}) = 0$.

The proof of Theorem 10.2 follows the same lines.

3.3 PROOFS OF THE LIMIT THEOREMS
3.3.1 STRONG LAW OF LARGE NUMBERS

We still use the decomposition of the simple random walk along the vertical and horizontal axis introduced in Section 3.1, and first prove the following

LEMMA 25 *(SLLN for the embedded random walk)*

$$\lim_{n \to +\infty} \frac{M_{T_n}}{n} = (0,0) \quad \tilde{\mathbb{P}}_\mu\text{-almost surely.} \quad (10.22)$$

Proof:
We have $M_{T_n} = (S_n, S'_n)$ and since $(S'_n)_{n \geq 0}$ is a simple random walk,

$$\lim_{n \to +\infty} \frac{S'_n}{n} = 0 \quad \tilde{\mathbb{P}}_\mu\text{-almost surely.} \tag{10.23}$$

So it is enough to prove that $(\frac{S_n}{n})$ converges almost surely to 0. Introduce the random variable

$$Z_n = \sum_{k=0}^{n-1} X_{S_k} = \sum_{y \in \mathbb{Z}} X_y N_{n-1}(y).$$

Under the probability measure $\tilde{\mathbb{P}}_\mu$, the stationary sequence $(X_{S'_k})_{k \geq 0}$ is ergodic ([91]), so from Birkhoff's theorem, as n tends to infinity,

$$\frac{Z_n}{n} \to \mathbb{E}(X_0) = 0 \quad \text{almost surely.} \tag{10.24}$$

Clearly,

$$S_n - mZ_n = \sum_{y \in \mathbb{Z}} X_y \sum_{i=1}^{N_{n-1}(y)} (\xi_i^{(y)} - m).$$

Let r be an even integer.

$$\mathbb{E}((S_n - mZ_n)^r)$$

$$= \sum_{y_1,\ldots,y_r \in \mathbb{Z}} \mathbb{E}\left(X_{y_1} \ldots X_{y_r} \sum_{i_1=1}^{N_{n-1}(y_1)} \cdots \sum_{i_r=1}^{N_{n-1}(y_r)} \mathbb{E}(\prod_{j=1}^{r} (\xi_{i_j}^{(y_j)} - m) | \mathcal{F} \vee \mathcal{G}) \right).$$

The $\xi_i^{(y)}$'s are independent of the vertical walk and the orientations; moreover, the random variables $\xi_i^{(y)} - m, i \geq 1, y \in \mathbb{Z}$ are i.i.d. and centered, so the summands are non zero if and only if $i_1 = \ldots = i_r$ and $y_1 = \ldots = y_r$. Then,

$$\mathbb{E}((S_n - mZ_n)^r) = n\mathbb{E}((\xi_1^{(0)} - m)^r) = nm_r \quad \text{(say)}.$$

Let $\delta > 0$. By Tchebychev's inequality,

$$\mathbb{P}\left(\left| \frac{S_n - mZ_n}{n} \right| \geq \delta \right) \leq \frac{1}{\delta^r n^r} \mathbb{E}((S_n - mZ_n)^r)$$

$$\leq \frac{m_r}{\delta^r n^{r-1}}.$$

We choose $r = 4$ and thus from Borel-Cantelli Lemma, we deduce that $\frac{S_n - mZ_n}{n}$ converges almost surely to 0 as n goes to infinity.

To get the result for the simple random walk $(M_n)_n$, we use the

LEMMA 26 *The sequence of random variables $(\frac{T_n}{n})_{n\geq 1}$ converges $\tilde{\mathbb{P}}_\mu$-a.s. to $(1+m)$ as $n \to +\infty$.*

Proof:
Let us remark that

$$T_n = n + \sum_{y \in \mathbb{Z}} \sum_{i=1}^{N_{n-1}(y)} (\xi_i^{(y)} - m) + m \sum_{y \in \mathbb{Z}} N_{n-1}(y).$$

Now, reasoning as in the proof of Lemma 25,

$$\mathbb{E}\left(\left(\sum_{y \in \mathbb{Z}} \sum_{i=1}^{N_{n-1}(y)} (\xi_i^{(y)} - m)\right)^3\right)$$

$$= \sum_{y_1 \in \mathbb{Z}, \ldots, y_3 \in \mathbb{Z}} \mathbb{E}\left(\sum_{i_1=1}^{N_{n-1}(y_1)} \cdots \sum_{i_3=1}^{N_{n-1}(y_3)} (\xi_{i_1}^{(y_1)} - m) \ldots (\xi_{i_3}^{(y_3)} - m)\right)$$

$$= \sum_{y_1 \in \mathbb{Z}, \ldots, y_3 \in \mathbb{Z}} \mathbb{E}\left(\sum_{i_1=1}^{N_{n-1}(y_1)} \cdots \sum_{i_3=1}^{N_{n-1}(y_3)} \mathbb{E}((\xi_{i_1}^{(y_1)} - m) \ldots (\xi_{i_3}^{(y_3)} - m)|\mathcal{F} \vee \mathcal{G})\right)$$

$$= m_3 \sum_{y_1 \in \mathbb{Z}, \ldots, y_3 \in \mathbb{Z}} \mathbb{E}\left(\sum_{i_1=1}^{N_{n-1}(y_1)} \cdots \sum_{i_3=1}^{N_{n-1}(y_3)} \mathbf{1}_{\{i_1=\ldots=i_3\}} \mathbf{1}_{\{y_1=\ldots=y_3\}}\right)$$

$$= m_3 \sum_{y \in \mathbb{Z}} \mathbb{E}(N_{n-1}(y)) = m_3 n \quad (\text{where } m_3 = \mathbb{E}((\xi_1^{(0)} - m)^3)).$$

From Tchebychev's inequality and Borel-Cantelli Lemma again, we deduce that, as $n \to +\infty$,

$$\frac{1}{n} \sum_{y \in \mathbb{Z}} \sum_{i=1}^{N_{n-1}(y)} (\xi_i^{(y)} - m) \to 0 \text{ a.s..}$$

Using the fact that $\sum_{y \in \mathbb{Z}} N_{n-1}(y) = n$, we deduce the lemma.

Let us prove now the almost sure convergence of the sequence $(\frac{M_n}{n})_{n\geq 1}$ to $(0,0)$. Since the sequence $(T_n)_{n\geq 1}$ is strictly increasing, for every $n \geq 1$, there exists a non-decreasing sequence of integers sequence $(U_n)_n$ such that

$$T_{U_n} \leq n < T_{U_n+1}.$$

From the definition of the embedding, and if we denote $M_n = (M_n^{(1)}, M_n^{(2)})$,

$$M_n^{(1)} \in [\min(M_{T_{U_n}}^{(1)}, M_{T_{U_n+1}}^{(1)}), \max(M_{T_{U_n}}^{(1)}, M_{T_{U_n+1}}^{(1)})]$$

and
$$M_n^{(2)} = M_{T_{U_n}}^{(2)}.$$

The (sub-)sequence $(U_n)_{n\geq 1}$ is nondecreasing and $\lim_{n\to+\infty} U_n = +\infty$, then by combining Lemmata 25 and 26, we get that as $n \to +\infty$,

$$\frac{M_{T_{U_n}}}{T_{U_n}} \to (0,0) \ \tilde{\mathbb{P}}_\mu \text{ a.s.} \qquad (10.25)$$

Now,

$$\left|\frac{M_n^{(1)}}{n}\right| \leq \max\left(\left|\frac{M_{T_{U_n}}^{(1)}}{n}\right|, \left|\frac{M_{T_{U_n+1}}^{(1)}}{n}\right|\right) \leq \max\left(\left|\frac{M_{T_{U_n}}^{(1)}}{T_{U_n}}\right|, \left|\frac{M_{T_{U_n+1}}^{(1)}}{T_{U_n}}\right|\right) \qquad (10.26)$$

and

$$\left|\frac{M_n^{(2)}}{n}\right| = \left|\frac{M_{T_{U_n}}^{(2)}}{T_{U_n}}\right| \frac{T_{U_n}}{n} \leq \left|\frac{M_{T_{U_n}}^{(2)}}{T_{U_n}}\right|.$$

From (10.25), we deduce the almost sure convergence of the coordinates to 0 and then this of the sequence $(\frac{M_n}{n})_{n\geq 1}$ to $(0,0)$ as $n \to \infty$.

3.3.2 PROOF OF THE FUNCTIONAL LIMIT THEOREM

PROPOSITION 13 *The sequence of random processes $n^{-3/4}(S_{[nt]})_{t\geq 0}$ weakly converges in the space $\mathcal{D}([0,\infty[,\mathbb{R})$ to the process $(m\Delta_t)_{t\geq 0}$.*

Proof:
Let us first prove that the finite dimensional distributions of $n^{-3/4}(S_{[nt]})_{t\geq 0}$ converge to those of $(m\Delta_t)_{t\geq 0}$ as $n \to \infty$. We can rewrite for every $n \in \mathbb{N}$,

$$S_n = S_n^{(1)} + S_n^{(2)}$$

where

$$S_n^{(1)} = \sum_{y\in\mathbb{Z}} X_y \left(\sum_{i=1}^{N_{n-1}(y)} \xi_i^{(y)} - m\right)$$

and

$$S_n^{(2)} = m \sum_{y\in\mathbb{Z}} X_y N_{n-1}(y).$$

Thanks to Theorem 4.3 the finite dimensional distributions of $n^{-3/4}(S_{[nt]}^{(2)})_{t\geq 0}$ converge to those of $(m\Delta_t)_{t\geq 0}$ as $n \to \infty$. To conclude we show that the sequence of random variables $n^{-3/4}(S_n^{(1)})_{n\in\mathbb{N}}$ converges for the L^2-norm to 0 as $n \to +\infty$. We have

$$\mathbb{E}\left((S_n^{(1)})^2\right) = \mathbb{E}\left(\sum_{x,y\in\mathbb{Z}} X_x X_y \sum_{i=1}^{N_{n-1}(x)} \sum_{j=1}^{N_{n-1}(y)} \mathbb{E}((\xi_i^{(x)}-m)(\xi_j^{(y)}-m)|\mathcal{F}\vee\mathcal{G})\right)$$

From the equality

$$\mathbb{E}((\xi_i^{(x)} - m)(\xi_j^{(y)} - m)|\mathcal{F}\vee\mathcal{G}) = m^2 \delta_{i,j}\delta_{x,y},$$

we obtain

$$n^{-3/2}\mathbb{E}\left((S_n^{(1)})^2\right) = m^2 n^{-3/2}\sum_{x\in\mathbb{Z}} N_{n-1}(x) = m^2 n^{-1/2} = o(1).$$

Let us recall that $M_{T_n} = (S_n, S'_n)$ for every $n \geq 1$. The sequence of random processes $n^{-3/4}(S'_{[nt]})_{t\geq 0}$ weakly converges in $\mathcal{D}([0,\infty[,\mathbb{R})$ to 0, thus the sequence of $\mathbb{R}^2$-valued random processes $n^{-3/4}(M_{T_{[nt]}})_{t\geq 0}$ weakly converges in $\mathcal{D}([0,\infty[,\mathbb{R}^2)$ to the process $(m\Delta_t, 0)_{t\geq 0}$. Theorem 10.4 follows from this remark and Lemma 26.

4. EXAMPLES

The main motivation of this work is the generalization of the transience of the i.i.d. case of [23] to dependent or inhomogeneous orientations. Depending on the original dynamical systems, we obtain various extensions corresponding to well known examples of dynamical systems such that Bernoulli and Markov shifts, Gibbs measures, SRB measures (Manneville-Pomeau maps), rotations on the torus, etc., our framework is very general from this point of view. Nevertheless, to get the transience of the walk, we need to generate the orientations by choosing a suitable function f satisfying (10.4). In some sense, this condition requires the model not to be too close to the deterministic case because to satisfy the condition, f should not be "μ-too often" 0 or 1. We describe now the examples providing extensions of the i.i.d. case to various disordered orientations.

1. Bernoulli shift

The first considered dynamical system S is the *Bernoulli shift* on the product space $E = [0,1]^{\mathbb{Z}}$ endowed with the Borel σ-algebra, the bilatere shift transformation T defined by

$$T : E \longrightarrow E$$
$$x = (x_y)_{y\in\mathbb{Z}} \longmapsto (Tx)_y = x_{y+1}, \forall y \in \mathbb{Z}.$$

The product Lebesgue measure $\mu = \lambda^{\otimes \mathbb{Z}}$ of the Lebesgue measure λ on $[0,1]$ is T-invariant and we choose as generating function f the projection on the zero coordinate:

$$f : E \longrightarrow [0,1]$$
$$x \longmapsto x_0.$$

For all $y \in \mathbb{Z}$, we then have

$$f \circ T^y(x) = x_y = \xi(y) \in [0,1].$$

We consider this ξ's as new random variables on E whose independence is inherited from the product structure of μ. The condition

$$\int_E \frac{d\mu}{\sqrt{f(1-f)}} < \infty$$

becomes

$$\int_0^1 \frac{d\lambda(x)}{\sqrt{x(1-x)}} < \infty$$

and the transience holds in this particular case. In fact, this product form of μ allows in this annealed case another description of the i.i.d. case of [23], for which we check $\xi(y) \equiv \frac{1}{2}$ for all $y \in \mathbb{Z}$ and

$$\operatorname{Cov}_\mu[X_0, X_y] = \mathbb{E}_\mu[X_0 X_y] = 4\mathbb{E}[\xi(0)\xi(y)] - 1 = 0. \qquad (10.27)$$

The result is also valid in the quenched case, for which the distribution of the orientation has an inhomogeneous product form.

2. **Markov shift**

If one considers a measure μ with correlations, then the same holds for $\mathbb{P}_\mu$. It is the case when one considers a Markovian measure instead of a product one on the space $[0,1]^{\mathbb{Z}}$ with stationary distribution π, whose correlations are given by (10.3).

The transience of the simple random walk on this particular dynamically oriented lattice holds for $\mathbb{P}_\mu$-a.e. environment as soon as the following condition is satisfied:

$$\int_0^1 \frac{d\pi(x)}{\sqrt{x(1-x)}} < \infty.$$

It is the case when the usual Lebesgue measure or Lebesgue measure of index p is the invariant measure.

In the quenched case, there are no correlations by construction and the law of the orientations depends on the measurable transformation only. This case is nevertheless different from this of the Bernoulli shift because the typical set of points x for which the transience holds depends on the measure μ.

3. Translation-invariant Gibbs measures

We consider now a measurable space of the form $E = \Sigma^{\mathbb{Z}}$ where Σ is a finite alphabet and T is again the bilatere shift defined above. We focus on the Ising model for which $\Sigma = \{-1, +1\}$, and the function f used to generate the transition probabilities and to come back in $[0, 1]$ is a dyadic transformation. In ergodic theory, Gibbs measures can be defined as equilibrium states or directly in term of an *energy function* $\Psi : E \longrightarrow \mathbb{R}$, regular enough and chosen here to be Hölder continuous (more details can be found in [93]). A Borel probability measure on E is a Gibbs measure for Ψ if for every homeomorphism τ that affects only finitely many coordinates,

$$\tau \Psi = \mu e^{\Psi_\tau}$$

where

$$\Psi_\tau = \lim_n \Psi_n \circ \tau^{-1} - \Psi_n$$

and Ψ_n is the restriction of Ψ on $\Sigma^{\{-n,\ldots,n\}}$. There exist many equivalent definitions of Gibbs measures in ergodic theory, see [93]. We focus here on the example of the Ising model where the energy function is

$$\Psi(\omega) = -J\omega_0\omega_1$$

with a coupling $J \in \mathbb{R}$. The Gibbs measures are very different depending on the sign of the coupling; if $J > 0$, the model is said to be ferromagnetic and has positive correlations (one orientation is likely to agree with its neighbors), while in the antiferromagnetic case ($J < 0$) the sign of the correlations can differ. The case $J = 0$ correspond to the i.i.d. case, already known to be a transient case.

To go back in $[0, 1]$, we introduce $f = d \circ \pi$ with $\pi : \{-1, +1\}^{\mathbb{Z}} \longrightarrow \{0, 1\}^{\mathbb{Z}}$; $\omega \longmapsto \sigma$ with $\sigma_y = \frac{1+\omega_y}{2}$ for all $y \in \mathbb{Z}$, and

$$d : \{0, +1\}^{\mathbb{Z}} \longrightarrow [0, 1]$$
$$\omega \longmapsto \sum_{y \in \mathbb{Z}} \frac{\sigma_y}{2^y}.$$

Due to the absence of phase transition in this one dimensional model, the average of f under μ is $\frac{1}{2}$. Condition (10.4) is believed to be true as soon as the energy function is finite range. This would extend in particular the transience of the i.i.d. case to more general models with exponential decays of (positive or negative) correlations.

4. SRB measures, Manneville-Pomeau maps

SRB measures provide another source of examples for dependent orientations. When E is the interval $[0,1]$, a measure μ of the dynamical system S is said to be an *SRB* measure if the empirical measure $\frac{1}{n}\sum_{i=1}^{n}\delta_{T^i(x)}$ converge weakly to μ for Lebesgue a.e. x. There exist many other definitions of SRB measures, see e.g. [172, 89]. In particular, it has the Bowen boundedness property in the sense that it is close to a Gibbs measure on some increasing cylinder, i.e. there exists a constant $C > 0$ such that for all $x \in [0,1]$ and every $n \geq 1$

$$\frac{1}{C} \leq \frac{\mu(I_{i_1,\ldots,i_n}(x))}{\exp\left(\sum_{k=0}^{n-1} \Phi(T^k(x))\right)} \leq C$$

where $\Phi = -\log|T'|$ and $I_{i_1,\ldots,i_n}$ is the interval of monotonicity for T^n which contains x.

In some cases, it is possible to control the correlations for SRB measures and we detail now an example where our transience result holds, the *Manneville-Pomeau maps*. These maps have been introduced in the 1980's to study intermittency phenomenon in the study of turbulence in chaotic systems ([7] and references therein) and has been recently identified as weakly Gibbsian measures, see [126]. They are expanding interval maps and in this example we describe the original MP map. The measurable space E is the unit interval $[0,1]$ and for $\alpha \in]0,1[$ the map is given by

$$\begin{aligned} T : [0,1] &\longrightarrow [0,1] \\ x &\longmapsto T(x) = x + x^{1+\alpha} \bmod 1. \end{aligned}$$

The existence of an absolutely continuous (w.r.t. the Lebesgue measure on $[0,1]$) SRB invariant measure μ has been established by [150] and the following bounds of Radon-Nikodym derivative $h = \frac{d\mu}{d\lambda}$ has been proved (see [126, 185]):

$$\exists C_\star, C^\star > 0 \text{ s.t. } \frac{C_\star}{x^\alpha} < h(x) < \frac{C^\star}{x^\alpha}. \tag{10.28}$$

This measure is known to be mixing, and a polynomial decay of correlation, with a power $\beta > 0$, has even been proved for g regular enough ([85, 118, 126, 189]):

$$|C_\mu^g(y)| = \mathcal{O}(|y|^{-\beta}). \tag{10.29}$$

The map T is not invertible but we use Theorem 10.2. It remains to find suitable function f who generates orientations for which the simple random walk is transient. By (10.28), a sufficient condition for the condition (10.4) to hold is

$$\int_0^1 \frac{dx}{x^\alpha \sqrt{f(x)(1-f(x))}} < \infty$$

and this is for example true for the function $f(x) = \frac{1}{2}(1 + x - T(x))$ and the choice of an $\alpha < \frac{1}{3}$.

5. The rotation on the torus

We consider the dynamical system $S = ([0,1], \mathcal{B}([0,1]), \lambda, T_\alpha)$ where T_α is the rotation on the torus $[0,1]$ with angle $\alpha \in \mathbb{R}$ defined by

$$x \longmapsto x + \alpha \mod 1$$

and λ is the Lebesgue measure on $[0,1]$. For every function $f : [0,1] \mapsto [0,1]$ such that $\int_0^1 f(x)\,dx = \frac{1}{2}$ and

$$\int_0^1 \frac{dx}{\sqrt{f(x)(1-f(x))}} < \infty,$$

conclusions of Theorem 10.1 hold. Such functions are called *admissible*. Every function uniformly bounded from 0 and 1, with integral $\frac{1}{2}$ is admissible. We also allow functions f to take values 0 and 1: for instance, $f_1(x) = x$ is admissible although $f_2(x) = \cos^2(2\pi x)$ is not. We actually have no explanations about this phenomenon, moreover we do not know the behavior (recurrence or transience) of the simple random walk on the dynamically oriented lattice generated by f_2. Nevertheless, we can construct particular angles and functions for which the random walk on the corresponding lattice is recurrent. Take $\alpha = \frac{1}{2q}$ for q an integer larger or equal to 1 and $f = \mathbf{1}_{[0,1/2[}$, then the lattice we obtain is $\mathbb{Z}^2$ with undirected vertical lines and horizontal strips of height q, alternatively oriented to the left then to the right. The simple random walk on this deterministic and periodic lattice is known to be recurrent, see [23]. A deeper study is needed for this particular choice of dynamical system.

Reference.
This chapter is based on [77].

Chapter 11

ASSET PRICING IN DYNAMIC (B, S)-MARKETS

1. INTRODUCTION

Many authors have investigated the discrete (B, S)-model for the dynamic pricing of derivative securities in financial markets, e.g. see Föllmer and Schied [57], Melnikov [134] and Shiryaev [171], originating from the analysis of Cox, Ross and Rubinstein [27]. Usually, the dynamics of the current amount hold in the fairly risky asset B is modelled by the stochastic process $B = (B_n)_{n \in \mathbb{N}}$ and the current amount of the risky asset S by the stochastic process $S = (S_n)_{n \in \mathbb{N}}$. Under some fairly mild assumption of having nonnegative semimartingale structure for B and S (cf. multiplicative decompositions of nonnegative semimartingales, e.g. in Melnikov [134]), one can show that both B and S must be governed by the linear stochastic difference equations

$$\Delta B_n = r_n B_{n-1}, \qquad (11.1)$$
$$\Delta S_n = \rho_n S_{n-1}, \qquad (11.2)$$

where $\Delta B_n = B_n - B_{n-1}$ and $\Delta S_n = S_n - S_{n-1}$ for $n \geq 1$, started at the nonrandom initial vector $(B_0, S_0) \in \mathbb{R}_+^2$ at $n = 0$ and driven by the stochastic processes $r = (r_n)_{n \in \mathbb{N}}$ and $\rho = (\rho_n)_{n \in \mathbb{N}}$. Throughout this exposition, we suppose that B, S, r, ρ are defined over the complete, filtered probability space $(\Omega, \mathcal{F}, (\mathcal{F}_n)_{n \in \mathbb{N}}, \mathbb{P})$. The stochastic process r is interpreted as the interest rate of dynamics of the fairly riskless asset B and ρ as the interest rate belonging to the risky asset S. For the sake of simplicity, we confine us to the more practice-relevant case of finite atomic sample spaces Ω (i.e. $\#(\Omega) < +\infty$) with $\mathbb{P}(\{\omega_i\}) > 0$ for all $\omega_i \in \Omega$.

We are aiming at deriving pricing and hedging formulas for the (B, S)-market governed by (11.1) and (11.2) with dynamic probabilities for the interest rates ρ instead of the commonly studied binomial model with static probabilities for the distribution of ρ. For this purpose, we shall study existence and equivalence of martingale measures, the fundamental theorem of dynamic asset pricing with dynamic multinomial distributions, the completeness of the dynamic (B, S)-market, the fair price and optimal hedging strategies. In order to compute a fair price and hedging strategies, conditions on the parameters of ρ and r need to be found to guarantee the (unique) existence of equivalent martingale-measure $\mathbb{P}^*$ to the original measure $\mathbb{P}$. The approach we pursue should be computationally feasible, efficient and practice-relevant. That is why we prefer the discrete model of (B, S)-market instead of continuous models. The dynamic aspect is obvious since the real data in the market undergo time-varying fluctuations and gives us more flexibility. Most of the commonly used models in mathematical finance suppose that the sums of the interest rates

$$V = \{V_n : V_n = \sum_{k=0}^{n} \rho_k\} \tag{11.3}$$

follow the classical symmetric random walk with constant probability transitions such that $(\rho_n)_{n \in \mathbb{N}}$ are independently identically distributed (i.i.d.) random variables. Therefore, an i.i.d. sequence of random variables ρ_n is generated to model the dynamics of $(S_n)_{n \in \mathbb{N}}$ governed by (11.2). We rather want to consider the more general case when ρ_n are just independent, but nonidentically distributed. As the consequence, the sequence of $(r_n)_{n \in \mathbb{N}}$ cannot be modelled by i.i.d. random variables r_n either. Thus, the process

$$U = \{U_n : U_n = \sum_{k=0}^{n} r_k\} \tag{11.4}$$

must be modelled by nonidentically distributed variables r_n too.

In particular we are interested in computations when V is modelled by the dynamic random walk and its consequences for the pricing theory.

DEFINITION 11.1 *The random sequence $V = (V_n)_{n \in \mathbb{N}}$ on $(\Omega, \mathcal{F}, (\mathcal{F}_n)_{n \in \mathbb{N}}, \mathbb{P})$ is called dynamic random walk over the dynamical system (E, T) where E is a metric space and $T : E \to E$ is a transformation iff*

(i) $V_n = \sum_{k=1}^{n} Z_k$ *for* $n \in \mathbb{N}$*, where* $Z_k : (\Omega, \mathcal{F}_k, \mathbb{P}) \to (\mathbb{R}^d, \mathcal{B}(\mathbb{R}^d))$

(ii) $V_0 = Z_0 = 0$ *(a.s.)*

(iii) $\exists m \in \mathbb{N} : m \geq 2 \, \exists a_k^{(j)} \in \mathbb{R}^d : a_k^{(i)} \neq a_k^{(j)}$ (nonrandom) for $i \neq j$ and $i, j \in \{1, 2, ..., m\}$ such that

$$p_k = \mathbb{P}(Z_k = z) = \begin{cases} \frac{1}{m}(1 + g_1(T^k e)) & \text{if } z = a_k^{(1)} \\ \frac{1}{m}(1 + g_2(T^k e)) & \text{if } z = a_k^{(2)} \\ \cdots\cdots & \cdots\cdots \\ \frac{1}{m}(1 + g_m(T^k e)) & \text{if } z = a_k^{(m)} \end{cases}$$

where $e \in E$ and $g_j : E \to [-1, +1]$ as Borel-measurable functions with identity $\sum_{j=1}^{m} g_j(x) = 0$ for all $x \in E$ are fixed,

(iv) $Z = (Z_k)_{k \in \mathbb{N}}$ is a sequence of independent random variables.

This allows us to model the probability distributions of V and ρ_n in a time-dependent manner. If the (B, S)-market is modelled by $V_n = \sum_{k=0}^{n} \rho_k$ as a dynamic random walk then we call it the model of dynamic (B, S)-market. We concentrate us on the univariate dynamic random walk with two- or three-point distributed increments at first, i.e. $d = 1$ and $m = 2, 3$. In particular, the specific choices of

$$p_k = \mathbb{P}(Z_k = z) = \begin{cases} \frac{1}{2}(1 + g(T^k e)) & \text{if } z = a_k \\ \frac{1}{2}(1 - g(T^k e)) & \text{if } z = b_k \end{cases}$$

with $a_k < b_k$ and

$$p_k = \mathbb{P}(Z_k = z) = \begin{cases} \frac{1}{3}(1 + g(T^k e)) & \text{if } z = a_k \\ \frac{1}{3}(1 - g(T^k e)) & \text{if } z = b_k \\ \frac{1}{3} & \text{if } z = c_k \end{cases}$$

with $a_k < c_k < b_k$ are in the center of our interest. This can replicate the practically observed scenario of values of assets B and S to go up, down or stay at the same level when compared at consecutive time-steps (effected by their interest rates r and ρ, respectively). However, the related (B, S)-market model is considered to be a small step towards the analysis of financial markets with time-dependent parameters. The more general model using dynamic random walks with multipoint-distributions for Z_k is conceivable, but left to future research due to its complexity.

The following aims of mathematical analysis of (B, S)-market are pursued during this paper.

1. Thorough computation for fair price and hedging strategies for one example of (B, S)-market which allows dynamic flexibility in the

interest rates r, ρ and which is more general than the Cox-Ross-Rubinstein market model based on static binomial probabilities for the random variables $(\rho_n)_{n\in\mathbb{N}}$.

2. Construction of equivalent martingale-measure $\mathbb{P}^*$ to the original probability measure $\mathbb{P}$ which renders $(S_n/B_n)_{n\in\mathbb{N}}$ to be a $(\mathcal{F}_n)_n$-martingale under $\mathbb{P}^*$ and Proof of absence of arbitrage by the fundamental theorem of asset pricing for dynamic (B,S)-markets.

3. Completeness of the dynamic (B,S)-market.

4. Calculation of fair call prices $C = C(N)$ with time $N > 0$ of maturity (i.e. the exercising time).

5. Construction of a.s. minimal hedging strategies π^* and its value X^{π^*} with probability $\mathbb{P}(X_N^{\pi^*} = f(S_N)) = 1$ for given pay-off function f.

6. γ-Hedging and pricing with positive probability, i.e. when $\mathbb{P}(X_N^{\pi^*} \geq f(S_N)) \geq 1-\gamma$ and $X_0^{\pi^*} = (1-\gamma)C(N)$.

7. Asymptotic behavior of the results as $N \to +\infty$ or parameters in dynamic random walk converge to certain values.

As a simple example of pay-off functions we are motivated by the European call option with $f(S_N) = (S_N - K(N))_+$ where $K(N)$ is the striking price to be agreed with by market participants and N is supposed to be a nonrandom finite exercising time. Due to the well-known Call-Put parity relation, the presented analysis shall be relevant for both call and put options.

2. ABSENCE OF ARBITRAGE OF DYNAMIC (B,S)-MARKETS

Let $\Delta Z_n = Z_n - Z_{n-1}$ denote the n-th increment of random sequence $Z = (Z_n)_{n\in\mathbb{N}}$. For the sake of simplicity, we assume that the underlying σ-algebra $\mathcal{F} = \sigma(\mathcal{F}^S, \mathcal{F}^B)$ is determined by

$$\mathcal{F}_n^S = \sigma(S_0, S_1, ..., S_n), \quad \mathcal{F}_n^B = \sigma(B_0, B_1, ..., B_n)$$

as the smallest σ-algebras generated by the sequences $S = (S_k)_{0 \leq k \leq n}$ and $B = (B_k)_{0 \leq k \leq n}$ governed by (11.2) and (11.1), respectively. We shall follow a notation which is similar to that of Melnikov [134].

2.1 KEY DEFINITIONS AND AUXILIARY RESULTS

Fix N as the time of maturity and let $d = 1$. We recall some standard definitions at first.

DEFINITION 11.2 *An investment strategy or portfolio $\pi = (\pi_n)_{0 \le n \le N}$ of (B, S)-market is defined by $\pi_n = (\beta_n, \alpha_n)$ where the random variables $\alpha_n, \beta_n : (\Omega, \mathcal{F}_n, \mathbb{P}) \to (\mathbb{R}, \mathcal{B}(\mathbb{R}))$ are interpreted as the amount β_n of riskless asset B_n hold in the portfolio π at the time n and α_n of risky asset S_n hold in the portfolio π at the time n, respectively. An investment strategy π is called self-financing for the (B, S)-market iff*

$$\forall n : 1 \le n \le N \quad \Delta\beta_n B_{n-1} + \Delta\alpha_n S_{n-1} = 0. \quad (11.5)$$

The set of all self-financing strategies is defined by

$$SF = \Big\{\pi : \pi \text{ portfolio satisfying } (11.5)\Big\}.$$

The quantity $X^\pi = \{X_n^\pi : 0 \le n \le N\}$ is called the value of the investment strategy $\pi = (\beta_n, \alpha_n)_{0 \le n \le N}$ iff

$$\forall n : 0 \le n \le N \quad X_n^\pi = \beta_n B_n + \alpha_n S_n. \quad (11.6)$$

LEMMA 27 *Any portfolio $\pi = (\beta_n, \alpha_n)_{0 \le n \le N} \in SF$ admits the a.s. representation*

$$X_n^\pi = X_0^\pi + \sum_{k=1}^n \Big(\beta_k \Delta B_k + \alpha_k \Delta S_k\Big) \quad (11.7)$$

for all $1 \le n \le N$. If additionally $\Delta B_0 = \Delta S_0 = 0$ then we have the equivalence

$$\forall n \in \{0, 1, ..., N\}, \ X_n^\pi = X_0^\pi + \sum_{k=1}^n \Big(\beta_k \Delta B_k + \alpha_k \Delta S_k\Big)$$
$$\iff \pi \in SF \ (i.e. \ (11.5)).$$

Proof: Observe that

$$X_n^\pi = \alpha_n S_n + \beta_n B_n = \alpha_n(S_n - S_{n-1}) + \beta_n(B_n - B_{n-1}) + \alpha_n S_{n-1} + \beta_n B_{n-1}$$
$$= \alpha_n \Delta S_n + \beta_n \Delta B_n + \alpha_{n-1} S_{n-1} + \beta_{n-1} B_{n-1} + (\alpha_n - \alpha_{n-1}) S_{n-1}$$
$$+ (\beta_n - \beta_{n-1}) B_{n-1}$$
$$= \alpha_n \Delta S_n + \beta_n \Delta B_n + X_{n-1}^\pi + \Delta\alpha_n S_{n-1} + \Delta\beta_n B_{n-1}.$$

This is equivalent to

$$\Delta X_n^\pi - \alpha_n \Delta S_n - \beta_n \Delta B_n = \Delta \alpha_n S_{n-1} + \Delta \beta_n B_{n-1}.$$

Consequently, we have the equivalence

$$\Delta X_n^\pi - \alpha_n \Delta S_n + \beta_n \Delta B_n = 0 \iff \Delta \alpha_n S_{n-1} + \Delta \beta_n B_{n-1} = 0$$

for all $1 \leq n \leq N$. However, by telescoping, the case $\Delta X_n^\pi - \alpha_n \Delta S_n + \beta_n \Delta B_n = 0$ for all $1 \leq n \leq N$ is equivalent to $X_n^\pi = X_0^\pi + \sum_{k=1}^n \left(\alpha_n \Delta S_n + \beta_n \Delta B_n \right)$ for all $1 \leq n \leq N$. Therefore, the representation (11.7) is verified. Now, suppose that $\Delta B_0 = \Delta S_0 = 0$ and $S_{-1} = 0$, $B_{-1} = 0$. Then the above equivalence is even true for all $0 \leq n \leq N$, and the proof of Lemma 27 is complete.

DEFINITION 11.3 *An investment strategy $\pi \in SF$ is said to realize arbitrage iff there exists a number $N > 0$ such that $\mathbb{P}(X_N^\pi > 0) > 0$ with $X_0^\pi = 0$ and $X_n^\pi \geq 0$ for all $1 \leq n \leq N$. The set SF_{arb} of all arbitrage-realizing strategies $\pi \in SF$ is defined by*

$$SF_{arb} = \left\{ \pi \in SF : \pi \text{ realizes arbitrage} \right\}.$$

A (B, S)-market is called arbitrage-free or no-arbitrage market iff $SF_{arb} = \emptyset$ (the empty set).

Solutions to linear systems of homogeneous stochastic difference equations such as (11.1) and (11.2) can be expressed explicitly in terms of stochastic exponentials under some mild regularity conditions. For this purpose, consider the following definition.

DEFINITION 11.4 *Given a sequence $Z = (Z_n)_{n \in \mathbb{N}}$ of random variables $Z_n : (\Omega, \mathcal{F}_n, \mathbb{P}) \to (\mathbb{R}, \mathcal{B}(\mathbb{R}))$. Then, the random sequence $\mathcal{E}(Z) = (\mathcal{E}_n(Z))_{n \in \mathbb{N}}$ defined by*

$$\mathcal{E}_n(Z) = \prod_{k=0}^n (1 + \Delta Z_k)$$

for $n \in \mathbb{N}$, where $Z_n = Z_0 + \sum_{k=1}^n \Delta Z_k$ and $\Delta Z_0 = 0$, is called discrete stochastic exponential of Z.

LEMMA 28 *Assume that*

(i) $U = (U_n)_{n \in \mathbb{N}}$ *is a random sequence with*

$$U_n : (\Omega, \mathcal{F}_n, \mathbb{P}) \to (\mathbb{R}, \mathcal{B}(\mathbb{R}))$$

(ii) $N = (N_n)_{n \in \mathbb{N}}$ *is a random sequence with*

$$N_n : (\Omega, \mathcal{F}_n, \mathbb{P}) \to (\mathbb{R}^d, \mathcal{B}(\mathbb{R}^d))$$

(iii) $Z = (Z_n)_{n \in \mathbb{N}}$ *as a random sequence with*

$$Z_n : (\Omega, \mathcal{F}_n, \mathbb{P}) \to (\mathbb{R}^d, \mathcal{B}(\mathbb{R}^d))$$

satisfies the initial value problem (IVP)

$$\begin{aligned} \Delta Z_n &= \Delta N_n + Z_{n-1} \Delta U_n, n \geq 1, \\ Z_0 &= N_0, \end{aligned}$$

(iv) $\forall k \geq 1$, $\mathbb{P}(\Delta U_k = -1) = 0$.

Then, the sequence $Z = (Z_n)_{n \in \mathbb{N}}$ possesses the a.s. $(\mathcal{F}_n, \mathcal{B}(\mathbb{R}^d))$-measurable representation

$$Z_n = \mathcal{E}_n(U) \cdot \left\{ N_0 + \sum_{k=1}^n \mathcal{E}_k^{-1}(U) \Delta N_k \right\} \quad (11.8)$$

for all $n \geq 1$. In particular, for the case $N_n = X_0 = N_0$ $(n \geq 1)$, we have

$$Z_n = \mathcal{E}_n(U) \cdot Z_0, \ n \geq 0.$$

Proof: (Borrowed from Melnikov [134], p. 14-15). By complete induction on $n \in \mathbb{N}$. Suppose that $n = 1$ (i.e. induction beginning). Then, one easily recognizes that

$$\begin{aligned} \mathcal{E}_1(U) N_0 + \mathcal{E}_1(U) \mathcal{E}_1^{-1}(U) \Delta N_1 &= (1 + \Delta U_1) N_0 + \Delta N_1 \\ &= N_1 + N_0 \Delta U_1 \\ &= N_0 + N_1 - N_0 + N_0 \Delta U_1 \\ &= Z_0 + \Delta N_1 + Z_0 \Delta U_1 \\ &= Z_1 \end{aligned}$$

since $Z_0 = N_0$. Next, suppose that the claim is true for any $n-1$ with $n \geq 1$ (induction assumption). Now, consider the case n (induction step). One arrives at

$$\begin{aligned} Z_n &= Z_{n-1} + \Delta N_n + Z_{n-1}\Delta U_n = Z_{n-1}(1+\Delta U_n) + \Delta N_n \\ &= (1+\Delta U_n) \cdot \mathcal{E}_{n-1}(U) \cdot \left\{ N_0 + \sum_{k=1}^{n-1} \mathcal{E}_k^{-1}(U) \Delta N_k \right\} + \Delta N_n \\ &= \mathcal{E}_n(U) \cdot \left\{ N_0 + \sum_{k=1}^{n-1} \mathcal{E}_k^{-1}(U) \Delta N_k \right\} + \mathcal{E}_n(U) \cdot \mathcal{E}_n^{-1}(U) \cdot \Delta N_n \\ &= \mathcal{E}_n(U) \cdot \left\{ N_0 + \sum_{k=1}^{n} \mathcal{E}_k^{-1}(U) \Delta N_k \right\}. \end{aligned}$$

This proves Lemma 28.

Remark: It is rather obvious that

(i) $\mathcal{E}(U) = (\mathcal{E}_n(U))_{n \in \mathbb{N}}$ is a $(\mathcal{F}_n)_{n \in \mathbb{N}}$-martingale $\iff U = (U_n)_{n \in \mathbb{N}}$ is a $(\mathcal{F}_n)_{n \in \mathbb{N}}$-martingale,

(ii) $\forall n \in \mathbb{N}: \mathcal{E}_n(U) \cdot \mathcal{E}_n(V) = \mathcal{E}_n(U+V+ <U,V>)$ where $<U,V>$ is the quadratic covariation defined by $<U,V>_n = \sum_{k=0}^{n} \Delta U_k \cdot \Delta V_k$ where $\Delta U_0 = \Delta V_0 = 0$, and

(iii) regularity condition (iv) in Lemma 28 guarantees the invertibility of discrete exponentials $\mathcal{E}$ at all times n.

For proving a version of the fundamental theorem of dynamic asset pricing within the framework of dynamic (B,S)-markets, we recall some results on the relation of $(\mathcal{F}_n)_{n \in \mathbb{N}}$-martingales and risk-neutral (martingale) probability measures.

DEFINITION 11.5 *A probability measure* $\mathbb{P}^* \sim \mathbb{P}$ *over* $(\Omega, \mathcal{F}, (\mathcal{F}_n)_{n \in \mathbb{N}})$ *is said to be a martingale-measure or risk-neutral for the* (B,S)-*market iff the random sequence* $R = \{R_n : 0 \leq n \leq N\}$ *defined by* $R_n = S_n/B_n$ *is a* $(\mathcal{F}_n)_{n \in \mathbb{N}}$-*martingale with respect to* $\mathbb{P}^*$. *The set of all equivalent martingale-measures for the* (B,S)-*market is defined by* $\mathcal{P}^*$.

LEMMA 29 *Assume that* (B,S)-*market over* $(\Omega, \mathcal{F}, (\mathcal{F}_n)_{n \in \mathbb{N}}, \mathbb{P})$ *satisfies that*

(i) $r = (r_n)_{n \in \mathbb{N}}$ is predictable,

(ii) $r_n > -1$ for all $0 \leq n \leq N$ (a.s.), and

(iii) $\mathbb{P}^*$ is a probability measure on $(\Omega, (\mathcal{F}_n)_{n \in \mathbb{N}})$.

Then, we have that

$$R = \{R_n : 0 \leq n \leq N \text{ with } R_n = \frac{S_n}{B_n} \text{ is a } (\mathcal{F}_n)_{0 \leq n \leq N} - \text{martingale}$$
$$\text{w.r.t. } \mathbb{P}^*\}$$
$$\iff \left(\sum_{k=0}^{n}(\rho_k - r_k)\right)_{0 \leq n \leq N} \text{ is a } (\mathcal{F}_n)_{0 \leq n \leq N} - \text{martingale w.r.t. to } \mathbb{P}^*.$$

Proof: (Borrowed in parts from Melnikov [134], Theorem 3.1, p. 22-23). Define $U_0 = V_0 = 0$ and $\mathcal{E}_0(U) = \mathcal{E}_0(V) = 1$. Let $U_n = \sum_{k=1}^{n} r_k$, $V_n = \sum_{k=1}^{n} \rho_k$ and

$$\mathcal{E}_n(U) = \prod_{k=1}^{N}(1 + r_k), \quad \mathcal{E}_n(V) = \prod_{k=1}^{N}(1 + \rho_k)$$

for all $1 \leq n \leq N$. Set $U_0^* = 0$ and $U_n^* = U_0^* + \sum_{k=1}^{n} \Delta U_k^*$ where

$$\Delta U_k^* = \Delta U_k - \frac{(\Delta U_k)^2}{1 + \Delta U_k}$$

for $1 \leq k \leq N$. Then, using properties of stochastic exponentials as in Melnikov [134] and remark (ii) from above, we find that

$$\begin{aligned}
R_n &= \frac{S_n}{B_n} = \frac{S_0 \cdot \mathcal{E}_n(V)}{B_0 \cdot \mathcal{E}_n(U)} = R_0 \cdot \mathcal{E}_n(V) \cdot \mathcal{E}_n^{-1}(U) \\
&= R_0 \cdot \mathcal{E}_n(V - U^* - <V, U^*>) \\
&= R_0 \cdot \mathcal{E}_n\left(V - U + \sum_{k=1}^{\cdot}\frac{(\Delta U_k)^2}{1 + \Delta U_k} - <V, U> + \sum_{k=1}^{\cdot}\frac{\Delta V_k(\Delta U_k)^2}{1 + \Delta U_k}\right) \\
&= R_0 \cdot \mathcal{E}_n\left(\sum_{k=1}^{\cdot}\Delta V_k - \sum_{k=1}^{\cdot}\Delta U_k - \sum_{k=1}^{\cdot}\Delta V_k \cdot \Delta U_k \right. \\
&\quad \left. + \sum_{k=1}^{\cdot}\frac{(1 + \Delta V_k)(\Delta U_k)^2}{1 + \Delta U_k}\right) \\
&= R_0 \cdot \mathcal{E}_n\left(\sum_{k=1}^{\cdot}\Delta V_k - \sum_{k=1}^{\cdot}\Delta U_k - \sum_{k=1}^{\cdot}\frac{\Delta V_k \cdot \Delta U_k}{1 + \Delta U_k} + \sum_{k=1}^{\cdot}\frac{(\Delta U_k)^2}{1 + \Delta U_k}\right) \\
&= R_0 \cdot \mathcal{E}_n\left(\sum_{k=1}^{\cdot}\frac{\Delta V_k - \Delta U_k}{1 + \Delta U_k}\right)
\end{aligned}$$

for $1 \leq n \leq N$. Now, calculate

$$\mathbb{E}(R_n - R_{n-1}|\mathcal{F}_n)$$
$$= \mathbb{E}\left(R_0 \cdot \mathcal{E}_n\left(\sum_{k=1}^{\cdot} \frac{\Delta V_k - \Delta U_k}{1 + \Delta U_k}\right) - R_0 \cdot \mathcal{E}_{n-1}\left(\sum_{k=1}^{\cdot} \frac{\Delta V_k - \Delta U_k}{1 + \Delta U_k}\right)\bigg|\mathcal{F}_{n-1}\right)$$
$$= R_0 \cdot \mathcal{E}_{n-1}\left(\sum_{k=1}^{\cdot} \frac{\Delta V_k - \Delta U_k}{1 + \Delta U_k}\right) \mathbb{E}\left(1 + \frac{\Delta V_n - \Delta U_n}{1 + \Delta U_n} - 1\bigg|\mathcal{F}_{n-1}\right)$$
$$= R_0 \cdot \mathcal{E}_{n-1}\left(\sum_{k=1}^{\cdot} \frac{\Delta V_k - \Delta U_k}{1 + \Delta U_k}\right) \cdot (1 + r_n)^{-1} \cdot \mathbb{E}(\Delta V_n - \Delta U_n|\mathcal{F}_{n-1})$$
$$= R_0 \cdot \mathcal{E}_{n-1}\left(\sum_{k=1}^{\cdot} \frac{\Delta V_k - \Delta U_k}{1 + \Delta U_k}\right) \cdot (1 + r_n)^{-1} \cdot \mathbb{E}(\rho_n - r_n|\mathcal{F}_{n-1}).$$

Thus, $\mathbb{E}(R_n - R_{n-1}|\mathcal{F}_n) = 0$ iff $\mathbb{E}(\rho_n - r_n|\mathcal{F}_n) = 0$ for all $1 \leq n \leq N$. This fact confirms the claim of Lemma 29, hence the proof is complete.

Remark: Lemma 29 provides us an important criteria for martingale-measures $\mathbb{P}^*$. Suppose that $r = (r_n)_{n \in \mathbb{N}}$ is predictable with respect to the filtration $(\mathcal{F}_n)_{n \in \mathbb{N}}$. Then, R is a martingale iff

$$\mathbb{E}(\rho_n|\mathcal{F}_{n-1}) = r_{n-1}, \quad n \geq 1$$

Thus, for martingale-measures, there is a well-determined relation between the dynamics of interest rates r and ρ, and hence they cannot be chosen arbitrarily independent from each other within linear (B, S)-market models.

2.2 FUNDAMENTAL THEOREM OF ASSET PRICING IN (B, S)-MARKETS

A theorem which connects the important concepts of arbitrage (economics) and martingale measures (probability) plays a fundamental rôle in the analysis of financial markets. This is particularly true for our (B, S)-market model with dynamic probabilities. For a more general context, see Schachermeyer [158], [159], and Delbaen and Schachermeyer [35] who build up their ideas based on the fundamental works of Merton [136], Kreps [106], Stricker [180] and Dalang, Morton and Willinger [33] who proved the equivalence between the existence of martingale-measures and the absence of arbitrage in discrete time with finite time-horizons. Note that there are counterexamples when the following equivalence break down in the case of infinite time-horizons, e.g. see Back and Pliska [3].

THEOREM 11.1 *Let* (B, S, π, X^π) *be a financial market over the filtered probability space* $(\Omega, \mathcal{F}, (\mathcal{F}_n)_{n \in \mathbb{N}}, \mathbb{P})$. *Assume that*

(i) $r = (r_n)_{n \in \mathbb{N}}$ *is predictable and* $r_n > -1$ *for all* $0 \leq n \leq N$ *(a.s.)*,

(ii) $\sigma(r_n)$ *is independent of* $\mathcal{F}_k$ *for all* $n > k$,

(iii) $\alpha = (\alpha_n)_{n \in \mathbb{N}}$ *is predictable*,

(iv) $B_0 > 0$ *and* B_0 *is* $(\mathcal{F}_0, \mathcal{B}(\mathbb{R}))$-*measurable, and*

(v) *cardinality* $\#(\Omega) < +\infty$ *with* $\mathbb{P}(\{\omega_i\}) > 0$ *for all* $\omega_i \in \Omega$.

Then, we have that

$$\mathbb{P}^* \neq \emptyset \quad \Longleftrightarrow \quad SF_{arb} = \emptyset.$$

Proof: (Borrowed in parts from Melnikov [134], Theorem 3.2, p. 24-27). First, suppose that $\mathbb{P}^* \neq \emptyset$. Let $\mathbb{P}^* \in \mathbb{P}^*$. Then, for all $\pi \in SF$, we have

$$\begin{aligned}
\Delta X_n^\pi &= \beta_n \Delta B_n + \alpha_n \Delta S_n = \beta_n r_n B_{n-1} + \alpha_n \rho_n S_{n-1} \quad (11.9) \\
&= \beta_n r_n B_{n-1} + \alpha_n r_n S_{n-1} + \alpha_n \rho_n S_{n-1} - \alpha_n r_n S_{n-1} \\
&= r_n (\beta_n B_{n-1} + \alpha_n S_{n-1}) + \alpha_n S_{n-1} (\rho_n - r_n) \\
&= r_n (\beta_{n-1} B_{n-1} + \alpha_{n-1} S_{n-1}) + \alpha_n S_{n-1} (\rho_n - r_n) \\
&\quad + r_n (\Delta \beta_n B_{n-1} + \Delta \alpha_n S_{n-1}) \\
&= r_n X_{n-1}^\pi + \Delta N_n
\end{aligned}$$

where $\Delta N_n = \alpha_n S_{n-1} (\rho_n - r_n)$. That is that $X^\pi = (X_n^\pi)_{n \in \mathbb{N}}$ satisfies the initial value problem for nonhomogeneous linear stochastic difference equation

$$\Delta X_n^\pi = r_n X_{n-1}^\pi + \Delta N_n, \quad X_0^\pi = X_0 \quad (11.10)$$

with nonhomogeneous terms $\Delta N_n = \alpha_n S_{n-1} (\Delta V_n - \Delta U_n)$ and initial value $N_0 = X_0$. Therefore, by applying Lemma 28 under $\Delta U_n = r_n > -1$ with $U_n = \sum_{k=1}^n \Delta U_k = \sum_{k=1}^n r_k$, we find that

$$\begin{aligned}
X_n^\pi &= \mathcal{E}_n(U) \left\{ N_0 + \sum_{k=1}^n [\mathcal{E}_k(U)]^{-1} \Delta N_k \right\} \\
&= \mathcal{E}_n(\sum_{l=1}^\cdot r_l) \left\{ X_0 + \sum_{k=1}^n [\mathcal{E}_k(\sum_{l=1}^\cdot r_l)]^{-1} \alpha_k S_{k-1} (\rho_k - r_k) \right\}
\end{aligned}$$

for $0 \leq n \leq N$. Recall that $U_0 = 0$, $U_n = \sum_{k=1}^{n} r_k$, $V_0 = 0$ and $V_n = \sum_{k=1}^{n} \rho_k$. Note also that the inverses $\mathcal{E}_k^{-1}(U)$ exist for all k since condition all $r_l > -1$. Now, choose $\mathbb{P}^* \in \mathcal{P}^*$ such that

$$r_n = \mathbb{E}\left(Z_{n-1}^{-1} Z_n \rho_n | \mathcal{F}_{n-1}\right) \quad \text{where} \quad Z_n = \left.\frac{d\mathbb{P}^*}{d\mathbb{P}}\right|_{\mathcal{F}_n}. \quad (11.11)$$

Hence, $\mathbb{P}^* \sim \mathbb{P}$ and $(\sum_{k=1}^{n}(\rho_k - r_k))_{n \in \mathbb{N}}$ forms a $(\mathcal{F}_n)_{n \in \mathbb{N}}$-martingale under $\mathbb{P}^*$ by Lemma 29. Now, a contradiction is verified as follows. Set $X_0^\pi = 0$ on Ω. Then, for all $0 \leq n \leq N$, compute

$$\mathbb{E}_{\mathbb{P}^*}(X_n^\pi)$$

$$= \mathbb{E}_{\mathbb{P}^*}\left(\mathcal{E}_n(\sum_{k=1}^{n} r_k)\left\{X_0^\pi + \sum_{k=1}^{n} \mathcal{E}_k^{-1}(\sum_{l=1}^{\cdot} r_l)\alpha_k S_{k-1}(\rho_k - r_k)\right\}\right)$$

$$= \sum_{k=1}^{n} \mathbb{E}_{\mathbb{P}^*}\left(\mathcal{E}_n(\sum_{k=1}^{n} r_k)\mathcal{E}_k^{-1}(\sum_{l=1}^{\cdot} r_l)\alpha_k S_{k-1}(\rho_k - r_k)\right)$$

$$= \sum_{k=1}^{n} \mathbb{E}_{\mathbb{P}^*}\left(\mathbb{E}_{\mathbb{P}^*}\left(\prod_{l=k+1}^{n}(1+r_l)\alpha_k S_{k-1}(\rho_k - r_k) \Big| \mathcal{F}_{k-1}\right)\right)$$

$$= \sum_{k=1}^{n} \mathbb{E}_{\mathbb{P}^*}\left(\prod_{l=k+1}^{n}(1+r_l)\right) \mathbb{E}_{\mathbb{P}^*}\left(\alpha_k S_{k-1} \mathbb{E}_{\mathbb{P}^*}\left(\rho_k - r_k | \mathcal{F}_{k-1}\right)\right)$$

$$= \sum_{k=1}^{n} \mathbb{E}_{\mathbb{P}^*}\left(\prod_{l=k+1}^{n}(1+r_l)\right) \mathbb{E}_{\mathbb{P}^*}\left(\alpha_k S_{k-1} \mathbb{E}_{\mathbb{P}}\left(\rho_k Z_{k-1}^{-1} Z_k - r_k \Big| \mathcal{F}_{k-1}\right)\right)$$

$$= 0$$

since (11.11) and $(\alpha_k)_{k \in \mathbb{N}}$ is predictable. Indirectly, suppose that there is $SF_{arb} \neq \emptyset$ and $\pi \in SF_{arb}$. Then, by definition of set SF_{arb}, we must find a finite number N such that $\mathbb{E}_{\mathbb{P}}(X_N^\pi) > 0$ for $X_0^\pi = 0$. Recall that $\mathbb{P}^* \sim \mathbb{P}$. Therefore, $\mathbb{E}_{\mathbb{P}^*}(X_N^\pi) > 0$ with $X_0^\pi = 0$. This fact contradicts to our prior observation that $\mathbb{E}_{\mathbb{P}^*}(X_n^\pi) = 0$ for all $0 < n \leq N$, provided that $X_0^\pi = 0$ (a.s.).

Now, the backward direction. The key idea is to apply the separation theorem for convex sets on Euclidean linear vector spaces. Suppose that $SF_{arb} = \emptyset$. Introduce the following two nonempty sets of random variables $\xi : (\Omega, \mathcal{F}_N, \mathbb{P}) \to (\mathbb{R}, \mathcal{B}(\mathbb{R}))$ by

$$\mathcal{H}_0 = \{\xi : \exists \pi \in SF \, \exists N > 0 \text{ such that } X_0^\pi = 0, X_N^\pi = \xi\},$$
$$\mathcal{H}_1 = \{\xi : \xi \geq 0 \text{ and } \mathbb{E}(\xi) \geq 1\}.$$

First, we show that $\mathcal{H}_0 \cap \mathcal{H}_1 = \emptyset$ if $SF_{arb} = \emptyset$. Indirectly, suppose that $\mathcal{H}_0 \cap \mathcal{H}_1 \neq \emptyset$. Hence, $\exists \pi \in SF$ such that $X_0^\pi = 0$, $X_N^\pi \geq 0$, and

$\mathbb{P}(\{X_N^\pi > 0\}) > 0$. We are going to construct a $\bar{\pi} \in SF_{arb}$ based on that $\pi \in SF$ and contradicting to $SF_{arb} = \emptyset$. If $X_n^\pi \geq 0$ for all $0 \leq n \leq N$ then $\bar{\pi} = \pi$. If there is an integer $n < N$ with $\mathbb{P}(\{X_n^\pi < 0\}) > 0$ then we must have a $\omega_0 \in \Omega$ with $\mathbb{P}(\{\omega_0\}) > 0$ and $X_n^\pi(\omega_0) < 0$. That means that $\exists m < N$ such that

$$X_m^\pi(\omega_0) = \beta_m(\omega_0)B_m(\omega_0) + \alpha_m(\omega_0)S_m(\omega_0) < 0,$$
$$X_n^\pi(\omega) \geq 0 \text{ for all } n \text{ with } m < n \leq N, \omega \in \Omega.$$

Define $\bar{\pi} = (\bar{\pi}_n)_{0 \leq n \leq N}$ with $\bar{\pi}_n = (\bar{\beta}_n, \bar{\alpha}_n)$ as follows. Set $a = X_m^\pi(\omega_0) < 0$, $A = \{\omega \in \Omega : X_m^\pi(\omega) = a\}$, and

$$\bar{\beta}_n(\omega) = \mathbf{1}_A(\omega)\left[\beta_m(\omega) - \frac{a}{B_m}\right]\mathbf{1}_{\{n > m\}} \quad (11.12)$$
$$\bar{\alpha}_n(\omega) = \mathbf{1}_A(\omega)\alpha_n(\omega)\mathbf{1}_{\{n > m\}} \quad (11.13)$$

where $\mathbf{1}_S$ denotes the indicator function of subscribed set S. Obviously, the pair $(\bar{\beta}_n, \bar{\alpha}_n)$ is $(\mathcal{F}_{n-1}, \mathcal{B}(\mathbb{R}))$-measurable. It remains to show $\bar{\pi} \in SF$ for that $\pi \in SF$, i.e.

$$B_m \Delta\bar{\beta}_{m+1} + S_m \Delta\bar{\alpha}_{m+1} = 0. \quad (11.14)$$

Clearly, (11.14) holds on the complement $A^c (= \bar{A})$. On A, we have

$$\Delta\bar{\beta}_{m+1} = \bar{\beta}_{m+1} = \beta_{m+1} - \frac{a}{B_{m+1}},$$
$$B_m \Delta\bar{\beta}_{m+1} = B_m \beta_{m+1} - a, \quad \Delta\bar{\alpha}_{m+1} = \alpha_{m+1}.$$

Therefore, since $\pi \in SF$, we have

$$B_m \Delta\bar{\beta}_{m+1} + S_m \Delta\bar{\alpha}_{m+1} = B_m \beta_{m+1} - a + S_m \alpha_{m+1}$$
$$= B_m \Delta\beta_{m+1} + S_m \Delta\alpha_{m+1} + B_m \beta_m - a + S_m \alpha_m$$
$$= B_m \beta_m + S_m \alpha_m - a = X_m^\pi - a$$
$$= 0$$

on A. Consequently, (11.14) holds and $\bar{\pi} \in SF$. Next, we show that $X_n^{\bar{\pi}} \geq 0$ (a.s.) for all integer n with $0 \leq n \leq N$. Note, if $\omega \in A^c$ then $X_n^{\bar{\pi}}(\omega) = 0$ for all $0 \leq n \leq N$. Furthermore, if $\omega \in A$ then, for all $n \leq m$, we have $X_n^{\bar{\pi}}(\omega) = 0$ due to the construction of $\bar{\pi}$, and, for all $n > m$, we have

$$X_n^{\bar{\pi}}(\omega) = \bar{\beta}_n(\omega)B_n + \bar{\alpha}_n(\omega)S_n = \beta_n(\omega)B_n + \alpha_n(\omega)S_n - a\frac{B_n}{B_m}$$
$$= X_n^\pi(\omega) - a\frac{B_n}{B_m} \geq 0$$

because of $X_n^\pi \geq 0$ on A, $a < 0$ and all $B_k > 0$ started at $B_0 > 0$. Eventually, for $\omega \in A$, we obtain

$$\begin{aligned} X_N^{\bar{\pi}}(\omega) &= \bar{\beta}_N(\omega)B_N + \bar{\alpha}_N(\omega)S_N \\ &= \beta_N(\omega)B_N + \alpha_N(\omega)S_N - a\frac{B_N}{B_m} = X_N^\pi(\omega) - a\frac{B_N}{B_m} > 0, \end{aligned}$$

thanks to $B_0 > 0$ and all $r_k > -1$. Thus, with measures $\mathbb{P}^* \sim \mathbb{P}$ and strategy $\bar{\pi}$ we realize arbitrage, $\bar{\pi} \in SF_{arb}$, i.e. $SF_{arb} \neq \emptyset$. Consequently, $SF_{arb} = \emptyset$ implies that $\mathcal{H}_0 \cap \mathcal{H}_1 = \emptyset$.

Second, we apply the separation theorem to the disjoint sets $\mathcal{H}_0$ and $\mathcal{H}_1$ in order to construct $\mathbb{P}^* \sim \mathbb{P}$. Recall that the sample space Ω is finite, $\mathbb{P}(\{\omega_i\}) > 0$ for all $\omega_i \in \Omega$ and $\xi : (\Omega, \mathcal{F}_N, \mathbb{P}) \to \mathbb{R}$. Then, each random variable ξ can be identified with a vector $x = (\xi(\omega_1), \xi(\omega_2), ..., \xi(\omega_k))^T \in \mathbb{R}^k$ where $k = \#(\Omega)$. Hence, $\mathcal{H}_0$ and $\mathcal{H}_1$ can be regarded as disjoint subsets of the finite-dimensional vector space $\mathbb{R}^k$. Moreover, note that $\mathcal{H}_1$ is a convex subset and $\mathcal{H}_0$ is a linear subspace of $\mathbb{R}^k$. Thanks to the separation theorem on Euclidean spaces, there must exist a linear functional $l = l(x)$ such that, for all $x \in \mathbb{R}^k$, we have the separation property

$$l(x) \begin{cases} = 0 \text{ if } x \in \mathcal{H}_0 \\ > 0 \text{ if } x \in \mathcal{H}_1 \end{cases}$$

and this linear functional l can be written as an inner product

$$l(x) = <x, q> = \sum_{i=1}^k x_i q_i$$

with some constant vector $q = (q_1, q_2, ..., q_k)^T \in \mathbb{R}^k$ for all $x \in \mathbb{R}^k$. Define $\vec{p}_1 = (p_1^{-1}, 0, ..., 0)^T$, $\vec{p}_2 = (0, p_2^{-1}, 0, ..., 0)^T$, $\cdot$, $\vec{p}_k = (0, 0, ..., p_k^{-1})^T$ where $k = \#(\Omega)$. Note that $\vec{p}_i \in \mathcal{H}_1$ for all $i = 1, 2, ..., k$. Compute $l(\vec{p}_m) = \sum_{i=1}^k p_i^{-1} q_i = p_m^{-1} q_m > 0$ for all $1 \leq m \leq k$ since all q_i and $p_i > 0$. Let $\tilde{q} = (\tilde{q}_1, \tilde{q}_2, ..., \tilde{q}_k)^T$ be the normalized vector to $q \in \mathbb{R}^k$ originating from the separation theorem such that $\sum_{i=1}^k \tilde{q}_i = 1$. In what follows, we show that $M = (M_n)_{0 \leq n \leq N}$ with $M_n = S_n/B_n$ forms a $(\mathcal{F}_n)_{0 \leq n \leq N}$-martingale over $(\Omega, \mathcal{F}, (\mathcal{F}_n)_{0 \leq n \leq N}, \mathbb{P}^*)$ with martingale-measure $\mathbb{P}^*(\{\omega_i\}) = \tilde{q}_i$ which satisfies $\mathbb{P}^* \sim \mathbb{P}$. Due to the separation property of the linear functional l, we have $<x, \tilde{q}>$ for all $x \in \mathcal{H}_0$. Moreover, for all random variables $\xi \in \mathcal{H}_0$, $\mathbb{E}_{\mathbb{P}^*}(\xi) = <x, \tilde{q}> = 0$ holds. That is that, thanks to the definition of $\mathcal{H}_0$, we have $\mathbb{E}_{\mathbb{P}^*}(X_n^\pi) = \mathbb{E}((X_n^\pi)^*) = 0$ for all $\pi \in SF$ with $X_0^\pi = 0$. For the martingale-property of M, it suffices to show that $\mathbb{E}_{\mathbb{P}^*}(S_\tau/B_\tau) = \mathbb{E}_{\mathbb{P}^*}(S_0/B_0)$ or equivalently

$$\mathbb{E}_{\mathbb{P}^*}\left(\frac{S_\tau}{B_\tau} - \frac{S_0}{B_0}\right) = 0 \qquad (11.15)$$

for all $(\mathcal{F}_n)_{0 \leq n \leq N}$-stopping-times τ with $0 \leq \tau \leq N$. Suppose that τ is any $(\mathcal{F}_n)_{0 \leq n \leq N}$-stopping-time with $0 \leq \tau \leq N$ (a.s.). Now, we construct a $(\mathcal{F}_n)_{0 \leq n \leq N}$-stopping-time $\tilde{\tau}$ with $0 \leq \tilde{\tau} \leq N$ and $\tau = \tilde{\tau}$ (a.s.) such that $\tilde{\pi} \in SF$ and $\mathbb{E}_{\mathbb{P}^*}(X_N^{\tilde{\pi}}) = 0$ iff (11.15) holds. For this purpose, $\tilde{\pi} = (\tilde{\beta}_n, \tilde{\alpha}_n)_{0 \leq n \leq N}$ is defined by

$$\tilde{\beta}_n = \frac{S_{\tilde{\tau}}}{B_{\tilde{\tau}}}\mathbf{1}_{\{n > \tilde{\tau}\}} - \frac{S_0}{B_0}, \qquad \tilde{\alpha}_n = \mathbf{1}_{\{n \leq \tilde{\tau}\}}. \qquad (11.16)$$

Note that $\tilde{\beta}_n$ and $\tilde{\alpha}_n$ are $(\mathcal{F}_{n-1}, \mathcal{B}(\mathbb{R}))$-measurable since $\tilde{\tau}$ is a $(\mathcal{F}_n)_{0 \leq n \leq N}$-stopping-time. Moreover, $\tilde{\pi} = (\tilde{\beta}_n, \tilde{\alpha}_n)_{0 \leq n \leq N} \in SF$ because using the definition of $\tilde{\pi}$ in the simple substraction of

$$\tilde{\beta}_n B_n + \tilde{\alpha}_n S_n = \frac{S_{\tilde{\tau}}}{B_{\tilde{\tau}}}\mathbf{1}_{\{n > \tilde{\tau}\}} B_n + S_n \mathbf{1}_{\{n \leq \tilde{\tau}\}} - \frac{S_0}{B_0} B_n \quad \text{and}$$

$$\tilde{\beta}_{n+1} B_n + \tilde{\alpha}_{n+1} S_n = \frac{S_{\tilde{\tau}}}{B_{\tilde{\tau}}}\mathbf{1}_{\{n+1 > \tilde{\tau}\}} B_n + S_n \mathbf{1}_{\{n+1 \leq \tilde{\tau}\}} - \frac{S_0}{B_0} B_n$$

leads to

$$B_n \Delta\tilde{\beta}_{n+1} + S_n \Delta\tilde{\alpha}_{n+1} = \frac{S_{\tilde{\tau}}}{B_{\tilde{\tau}}}\mathbf{1}_{\{\tilde{\tau}=n\}} B_n + S_n \mathbf{1}_{\{\tilde{\tau}=n\}},$$

which confirms the self-financing character of $\tilde{\pi} \in SF$. Finally, we establish the equivalence $\mathbb{E}_{\mathbb{P}^*}(X_N^{\tilde{\pi}}) = 0$ iff (11.15). For this purpose, calculate

$$\begin{aligned}
0 &= \mathbb{E}_{\mathbb{P}^*}(X_N^{\tilde{\pi}}) \\
&= \mathbb{E}_{\mathbb{P}^*}(\tilde{\beta}_N B_N + \tilde{\alpha}_N S_N) \\
&= \mathbb{E}_{\mathbb{P}^*}\left(\left(\frac{S_{\tilde{\tau}}}{B_{\tilde{\tau}}}\mathbf{1}_{\{\tilde{\tau} < N\}} - \frac{S_0}{B_0}\right) B_N + \mathbf{1}_{\{\tilde{\tau}=N\}} S_N\right) \\
&= \mathbb{E}_{\mathbb{P}^*}\left(\left(\frac{S_{\tilde{\tau}}}{B_{\tilde{\tau}}} - \frac{S_0}{B_0}\right) B_N\right) + \mathbb{E}_{\mathbb{P}^*}\left(S_N \mathbf{1}_{\{\tilde{\tau}=N\}} - \frac{S_{\tilde{\tau}}}{B_{\tilde{\tau}}}\mathbf{1}_{\{\tilde{\tau}=N\}} B_N\right) \\
&= \mathbb{E}_{\mathbb{P}^*}\left(\left(\frac{S_{\tilde{\tau}}}{B_{\tilde{\tau}}} - \frac{S_0}{B_0}\right) B_N\right) = B_N \mathbb{E}_{\mathbb{P}^*}\left(\frac{S_{\tilde{\tau}}}{B_{\tilde{\tau}}} - \frac{S_0}{B_0}\right)
\end{aligned}$$

since $B_N = B_0 \mathcal{E}_N(U)$ with $U_n = \sum_{k=1}^n r_k$. This identity verifies that

$$\mathbb{E}_{\mathbb{P}^*}(X_N^{\tilde{\pi}}) = 0 \iff M = (M_n)_{0 \leq n \leq N} \text{ is a } (\mathcal{F}_n)_{0 \leq n \leq N} - \text{martingale},$$

hence the proof of Theorem 11.1 is complete.

3. COMPLETENESS OF DYNAMIC (B, S)-MARKETS

A very natural requirement of "fair" financial markets is that every yield (value) can be reproduced. This property relates to the concept of completeness.

DEFINITION 11.6 *A (B, S, π, X^π)-market defined on the filtered, discrete time probability space $(\Omega, \mathcal{F}, (\mathcal{F}_n)_{0 \leq n \leq N}, \mathbb{P})$ is said to be a (pathwise) complete iff*

$$\forall (\mathcal{F}_N, \mathcal{B}(\mathbb{R})) - \text{measurable } f : \Omega \to \mathbb{R} \; \exists \pi \in SF \; \forall \omega \in \Omega : X_N^\pi(\omega) = f(\omega).$$

Remark: Completeness also means the absence of any constraints for investing in related assets and assures the accessibility of all assets involved.

THEOREM 11.2 *Fix $N \in \mathbb{N} \setminus \{0\}$ as the time of maturity. Assume that $\mathcal{P}^* \neq \emptyset$ and $\mathbb{P}^* \in \mathcal{P}^*$, and $r = (r_n)_{0 \leq n \leq N}$ is predictable. Then the following assertions are equivalent:*

(1) *Completeness of the (B, S, π, X^π)-market,*

(2) *Uniqueness of the equivalent martingale-measure, i.e. $\mathcal{P}^* = \{\mathbb{P}^*\}$,*

(3) *All $(\mathcal{F}_n)_{0 \leq n \leq N}$-martingales $M = (M_n)_{0 \leq n \leq N}$ possess the representation*

$$M_n = M_0 + \sum_{k=1}^{n} \alpha_k \Delta m_k \qquad (11.17)$$

for $1 \leq n \leq N$, where α_k is predictable and $\Delta m_k = S_k/B_k - S_{k-1}/B_{k-1}$.

Remark: Note that $\Delta m_k = S_{k-1} B_k^{-1}(\rho_k - r_k)$. Also, in view of Lemma 29, the martingale representation (11.17) is equivalent to

$$M_n = M_0 + \sum_{k=1}^{n} \tilde{\alpha}_k (\rho_k - r_k)$$

where $\tilde{\alpha}_k = \alpha_k S_{k-1} B_k^{-1}$ which is predictable under the above assumptions.

Proof: (Adapted in parts from Melnikov [134], Theorem 3.3, p. 27-29). First, proof of (1) $\Longrightarrow$ (2). Indirectly, suppose that $\exists \mathbb{P}^{**} \neq \mathbb{P}^*$ and

$\mathbb{P}^{**} \in \mathcal{P}^*$. This implies that there is an event $A \in \mathcal{F}$ such that $\mathbb{P}^{**}(A) \neq \mathbb{P}^*(A)$. Now, set $f(\omega) = \mathbf{1}_A(\omega)B_N(\omega)$ for all $\omega \in \Omega$. Completeness of the (B, S, π, X^π)-market implies that there is a $\pi \in SF$ such that

$$\mathbb{P}(\{\omega \in \Omega : X_N^\pi(\omega) = \mathbf{1}_A(\omega)B_N(\omega)\}) = 1.$$

Recall that $\mathbb{P}^{**} \sim \mathbb{P}$ and $\mathbb{P}^* \sim \mathbb{P}$. Thus, we must also have

$$\mathbb{P}^*(\{\omega \in \Omega : X_N^\pi(\omega) = \mathbf{1}_A(\omega)B_N(\omega)\}) = 1$$
$$= \mathbb{P}^{**}(\{\omega \in \Omega : X_N^\pi(\omega) = \mathbf{1}_A(\omega)B_N(\omega)\}).$$

Note that both $\mathbb{P}^*$ and $\mathbb{P}^{**}$ guarantee the martingale-property of discounted sequence X/B in the (B, S, π, X^π)-market. Hence

$$\mathbb{P}^*(A) = \mathbb{E}_{\mathbb{P}^*}(\mathbf{1}_A) = \mathbb{E}_{\mathbb{P}^*}\left(\frac{X_N^\pi}{B_N}\right) = \mathbb{E}_{\mathbb{P}^*}\left(\frac{X_0^\pi}{B_0}\right) = \mathbb{E}_{\mathbb{P}}\left(\frac{X_0^\pi}{B_0}\right)$$
$$= \mathbb{E}_{\mathbb{P}^{**}}\left(\frac{X_0^\pi}{B_0}\right) = \mathbb{E}_{\mathbb{P}^{**}}\left(\frac{X_N^\pi}{B_N}\right) = \mathbb{E}_{\mathbb{P}^{**}}(\mathbf{1}_A) = \mathbb{P}^{**}(A)$$

for all events $A \in \mathcal{F}_N$. This obviously contradicts to $\mathbb{P}^{**} \neq \mathbb{P}^*$ on $\mathcal{F}_N$. Consequently, (1) $\Longrightarrow$ (2).

Second, proof of (2) $\Longrightarrow$ (1). Suppose that $\mathbb{P}^* \in \mathcal{P}^*$ is the unique martingale-measure. Now, define the following sets of random variables $\xi : (\Omega, \mathcal{F}_N, \mathbb{P}) \to (\mathbb{R}, \mathcal{B}(\mathbb{R}))$ by

$$\mathcal{H}_0 = \{\xi : (\Omega, \mathcal{F}_N, \mathbb{P}) \to (\mathbb{R}, \mathcal{B}(\mathbb{R})) : \exists \pi \in SF \, \exists N > 0 \text{ such that}$$
$$X_0^\pi = 0, X_N^\pi = \xi\},$$
$$\mathcal{H}_2 = \{\xi : (\Omega, \mathcal{F}_N, \mathbb{P}) \to (\mathbb{R}, \mathcal{B}(\mathbb{R})) : \mathbb{E}_{\mathbb{P}^*}(\xi) = 0\}.$$

Note that $\mathcal{H}_0 \subseteq \mathcal{H}_2$. It suffices to show that $\mathcal{H}_0 = \mathcal{H}_2$ since then (1) follows as follows: Suppose that we have shown that $\mathcal{H}_0 = \mathcal{H}_2$. Fix a random variable $f : (\Omega, \mathcal{F}_N, \mathbb{P}) \to (\mathbb{R}, \mathcal{B}(\mathbb{R}))$ as the payoff at time N of maturity. Define $\xi = f - \mathbb{E}_{\mathbb{P}^*}(f)$. It is clear that $\xi \in \mathcal{H}_2 = \mathcal{H}_0$. That implies that there is a $\pi = (\beta_n, \alpha_n)_{0 \leq n \leq N} \in SF$ such that $X_N^\pi = \xi$. Define $\tilde{\pi} = (\tilde{\beta}_n, \tilde{\alpha}_n)_{0 \leq n \leq N}$ by

$$\tilde{\alpha}_n = \alpha_n, \quad \tilde{\beta}_n = \mathbb{E}_{\mathbb{P}^*}\left(\frac{f}{B_N}\right) + \beta_N.$$

Note that $\tilde{\pi} \in SF$ and $X_N^{\tilde{\pi}} = f$ by construction. Thus, it remains to show that the uniqueness of $\mathbb{P}^* \in \mathcal{P}^*$ implies that $\mathcal{H}_0 = \mathcal{H}_2$. As in the proof of Theorem 11.1, identify each random variable ξ defined on $(\Omega, \mathcal{F}_N)$ with the vector $x = (\xi(\omega_1), \xi(\omega_2), ..., \xi(\omega_k))^T \in \mathbb{R}^k$ where $k = \#(\Omega)$. Also $\mathbb{P}^*$ is identified on $(\Omega, \mathcal{F}_N)$ by the vector $q^* = (q_1^*, q_2^*, ..., q_k^*)^T \in$

$\mathbb{R}_+^k$ where $q_i^* = \mathbb{P}^*(\{\omega_i\})$ for $1 \leq i \leq k$. Trivially, $\mathcal{H}_0$ and $\mathcal{H}_2$ are linear subspaces of $\mathbb{R}^k$. Indirectly, suppose that $\mathcal{H}_0 \subset \mathcal{H}_2$ and $\exists \tilde{x} \in \mathcal{H}_2 \setminus \mathcal{H}_0$ with $\tilde{x} \perp \mathcal{H}_0$, i.e.

$$\forall x \in \mathcal{H}_0 \;:\; <\tilde{x}, x> = \sum_{i=1}^k \tilde{x}_i x_i = 0.$$

Define $\tilde{q}_i = q_i^* - \varepsilon \tilde{x}_i$ for sufficiently small $\varepsilon > 0$ such that $\tilde{q}_i > 0$ for all $1 \leq i \leq k$. Form $\tilde{q} = (\tilde{q}_1, \tilde{q}_2, ..., \tilde{q}_k)^T \in \mathbb{R}^k$. Then, it is possible to choose $\varepsilon > 0$ so small such that also

$$\forall x \in \mathcal{H}_0 \;:\; <\tilde{q}, x> = \sum_{i=1}^k \tilde{q}_i x_i = <q^*, x> = 0$$

holds, thanks to the separation property. As in proof of Theorem 11.1, we can show that the normalized measure $\tilde{\mathbb{P}}$ on $(\Omega, \mathcal{F}_N)$ defined by $\tilde{\mathbb{P}}(\{\omega_i\}) = \delta \tilde{q}_i$ with $\delta = 1/(\tilde{q}_1 + \tilde{q}_2 + ... + \tilde{q}_k)$ exhibits a martingale-measure. Consequently, $M = (M_n)_{0 \leq n \leq N}$ defined by $M_n = S_n/B_n$ defines a $(\mathcal{F}_n)_{0 \leq n \leq N}$-martingale with respect to $\tilde{\mathbb{P}}$. Thanks to uniqueness of the existing martingale-measure, we have $\tilde{\mathbb{P}} = \mathbb{P}^*$. This implies that the elementary probabilities

$$q^* \;=\; \delta \tilde{q} = \delta(q^* - \varepsilon \tilde{x})$$

coincide, hence

$$(1 - \delta) q^* \;=\; \delta \varepsilon \tilde{x}. \tag{11.18}$$

Recall that $\tilde{x} \in \mathcal{H}_2 \setminus \mathcal{H}_0$ and $q^* \in \mathcal{H}_0$. Note that $\tilde{x} \perp q^*$, hence $\delta = 1$ and $\tilde{x} = 0$ is the only possible choice to satisfy (11.18). This contradicts to $\tilde{x} \neq 0$.

Third, proof of (1) $\Longrightarrow$ (3). Suppose that the (B, S, π, X^π)-market is complete and $M = (M_n)_{0 \leq n \leq N}$ is a $(\mathcal{F}_n)_{0 \leq n \leq N}$-martingale on $(\Omega, \mathcal{F}, (\mathcal{F}_n)_{0 \leq n \leq N}, \mathbb{P}^*)$.
Define $f : (\Omega, \mathcal{F}_N) \to (\mathbb{R}, \mathcal{B}(\mathbb{R}))$ by $f(\omega) = M_N(\omega) B_N(\omega)$ on Ω. Due to the completeness, $\exists \pi \in SF$ such that $X_N^\pi(\omega) = f(\omega)$ for all $\omega \in \Omega$ (a.s.). The martingale-property of sequence $(X_n^\pi/B_n)_{0 \leq n \leq N}$ implies that $M_n = X_n^\pi/B_n$ under the measure $\mathbb{P}^*$ acting on $(\Omega, \mathcal{F}_n)$. Now, using the definition of X^π, calculate

$$M_{n+1} - M_n$$
$$= \frac{X_{n+1}^\pi}{B_{n+1}} - \frac{X_n^\pi}{B_n}$$

$$= \frac{\beta_{n+1}B_{n+1} + \alpha_{n+1}S_{n+1}}{B_{n+1}} - \frac{\beta_n B_n + \alpha_n S_n}{B_n}$$

$$= \frac{\beta_{n+1}B_{n+1} + \alpha_{n+1}S_{n+1}}{B_{n+1}}$$

$$- \frac{(\beta_n - \beta_{n+1})B_n + (\alpha_n - \alpha_{n+1})S_n + \beta_{n+1}B_n + \alpha_{n+1}S_n}{B_n}$$

$$= \alpha_{n+1}\left(\frac{S_{n+1}}{B_{n+1}} - \frac{S_n}{B_n}\right)$$

$$= \alpha_{n+1}\Delta m_{n+1}$$

where $\Delta m_{n+1} = S_{n+1}/B_{n+1} - S_n/B_n$. Consequently, we have found that

$$M_{n+1} = M_n + \alpha_{n+1}\Delta m_{n+1} = M_0 + \sum_{i=1}^{n+1} \alpha_i \Delta m_i,$$

hence, representation (11.17) is verified.

Fourth, proof of (3) $\Longrightarrow$ (1). Suppose that every martingale M can be represented by (11.17). We need to show the completeness of the (B, S, π, X^π)-market. For this purpose, suppose that $f : (\Omega, \mathcal{F}_N, \mathbb{P}) \to (\mathbb{R}, \mathcal{B}(\mathbb{R}))$. Define

$$M_n = \mathbb{E}_{\mathbb{P}^*}\left(\frac{f}{B_N}\Big|\mathcal{F}_n\right).$$

Thus, $M = (M_n)_{0 \leq n \leq N}$ forms a $(\mathcal{F}_n)_{0 \leq n \leq N}$-martingale. Therefore, M possesses the representation

$$M_n = M_0 + \sum_{k=1}^{n} \alpha_k \left(\frac{S_k}{B_k} - \frac{S_{k-1}}{B_{k-1}}\right).$$

Introduce a new portfolio $\pi^* = (\beta_n^*, \alpha_n^*)_{0 \leq n \leq N}$ satisfying

$$\alpha_n^* = \alpha_n, \quad \beta_n^* = M_n - \alpha_n \frac{S_n}{B_n}.$$

Then, for all $0 \leq n \leq N$, we have

$$S_{n-1}\Delta\alpha_n^* + B_{n-1}\Delta\beta_n^*$$

$$= S_{n-1}\Delta\alpha_n + B_{n-1}\left(\Delta M_n - \Delta(\alpha\frac{S}{B})_n\right)$$

$$= S_{n-1}\Delta\alpha_n + B_{n-1}\left(\alpha_n\Delta(\frac{S}{B})_n - \Delta(\alpha\frac{S}{B})_n\right)$$

$$= S_{n-1}(\alpha_n - \alpha_{n-1}) + \alpha_n\left(B_{n-1}\frac{S_n}{B_n} - B_{n-1}\frac{S_{n-1}}{B_{n-1}}\right)$$

$$-\alpha_n B_{n-1}\frac{S_n}{B_n} + \alpha_{n-1}B_{n-1}\frac{S_{n-1}}{B_{n-1}}$$
$$= 0,$$

hence $\pi^* \in SF$. Moreover, we can compute

$$\begin{aligned}X_n^{\pi^*} &= \beta_n^* B_n + \alpha_n^* S_n \\ &= \left(M_n - \alpha_n \frac{S_n}{B_n}\right) B_n + \alpha_n S_n \\ &= M_n B_n \\ &\geq 0\end{aligned}$$

for all $0 \leq n \leq N$. In particular, we have $X_N^{\pi^*} = M_N B_N = f$, hence the claim f is replicated at time N. This completes the proof of Theorem 11.2.

4. FAIR PRICING AND HEDGING STRATEGIES IN COMPLETE DYNAMIC MARKETS

This section aims at the calculation of fair prices and hedging strategies for European options in complete dynamic (B,S)-markets.

4.1 DEFINITION OF CONTINGENT CLAIMS, EUROPEAN OPTIONS, AND RELATED NOTIONS

Consider our dynamic (B, S, π, X^π)-market model over the complete probability space $(\Omega, \mathcal{F}, (\mathcal{F}_n)_{0 \leq n \leq N}, \mathbb{P})$. Recall the following notions from the literature, e.g. Shiryaev [171].

DEFINITION 11.7 *The pair (f, N) is called contingent claim with repayment date N iff $f : (\Omega, \mathcal{F}_N, \mathbb{P}) \to (\mathbb{R}_+, \mathcal{B}(\mathbb{R}_+))$, i.e. f is a $(\mathcal{F}_N, \mathcal{B}(\mathbb{R}_+))$-measurable random variable. The contingent claim (f, N) is said to be (pathwise) attainable for all market participants iff $\exists$ portfolio π such that $X_N^\pi \geq f$ on Ω (a.s.). The procedure of building up attainable contingent claims with appropriate portfolios π is called hedging and its result π is called hedging portfolio.*

Remark: One of the most important problems in financial markets is to calculate the fair premium c at time 0 which guarantees to purchase an asset with a derivative security (for selling, buying, etc.) at later time $N > 0$. Such an example is provided with one of the most studied options, the European option. This exhibits the right to exercise a given security at deterministic time $N > 0$ and the guarantee to receive a payment in the amount of f which is fixed by contracts with options.

DEFINITION 11.8 *Suppose that a* (B, S, π, X^π)-*market is given with initial capital* $x > 0$ *and contingent claim* (f, N). *Then, a portfolio* $\pi \in SF$ *is called* (x, f, N)-*hedge iff*

$$\forall \omega \in \Omega \ : \ X_0^\pi(\omega) = x \quad \text{and} \quad X_N^\pi(\omega) \geq f(\omega). \quad (11.19)$$

The related (x, f, N)-*hedge* $\pi \in SF$ *is said to be minimal iff* $X_N^\pi(\omega) = f(\omega)$ *for all* $\omega \in \Omega$. $H(x, f, N)$ *is defined to be the set of all* (x, f, N)-*hedges. The minimal investment cost of* (f, N) *is identified by*

$$c(N) \ = \ \inf\{x > 0 : H(x, f, N) \neq \emptyset\} \quad (11.20)$$

and is also called the fair price of the option in case of options.

Remark: Note that $c(N)$ is bounded since $\#(\Omega) < +\infty$ in this paper. Suppose that (f, N) is a contingent claim on an option. Then, (11.20) corresponds to a "fair contract" that both the seller can attain the claim (f, N) acting on the (B, S)-market and the buyer pays a minimal premium to the seller.

DEFINITION 11.9 *Given a dynamic* (B, S, π, X^π)-*market. Then, an European call (put) option is an option to buy an asset (shares) with the contingent claim* $f = (S_N - K)_+$ $(f = (K - S_N)_+$ *in case of put) at the strike price* K *at the exercise time* N.

Remark: It is clear that the option is exercised by the buyer if indeed $S_N > K$, but certainly not if $S_N < K$.

4.2 COMPUTING PRICES AND HEDGING STRATEGIES FOR EUROPEAN OPTIONS

There is an obvious need for general formulas for the fair price, minimal hedge and the related value for European options on nonarbitrage dynamic (B, S)-markets with interest rates $r = (r_n)_{0 \leq n \leq N}$.

THEOREM 11.3 *Let* (B, S, π, X^π) *be a financial market over the filtered probability space* $(\Omega, \mathcal{F}, \mathcal{F}_{n\in\mathbb{N}}, \mathbb{P})$. *Assume that*

(i) $r = (r_n)_{n\in\mathbb{N}}$ *is predictable and* $r_n > -1$ *for all* $0 \leq n \leq N$ *(a.s.),*

(ii) σ-*algebra* $\sigma(r_n)$ *is independent of* $\mathcal{F}_k$ *for all* $n > k$,

(iii) $\alpha = (\alpha_n)_{n\in\mathbb{N}}$ *is predictable,*

(iv) $B_0 > 0$ *and* $X_0^\pi = x$ *are nonrandom,*

(v) *cardinality* $\#(\Omega) < +\infty$ *with* $\mathbb{P}(\{\omega_i\}) > 0$ *for all* $\omega_i \in \Omega$,

(vi) *completeness of* $(B, S, , \pi, X^\pi)$-*market (i.e.* $\exists \mathbb{P}^* \in \mathcal{P}^* : \mathbb{P}^* \sim \mathbb{P}$) *holds, and*

(vii) *no arbitrage (i.e.* $\exists! \mathbb{P}^* \in \mathcal{P}^*$) *can be realized.*

Then, for the European option with contingent claim (f, N),

[1] *the fair price is given by*

$$c(N) = \mathbb{E}_{\mathbb{P}^*}(\mathcal{E}_N^{-1}(U)f), \quad (11.21)$$

[2] *the minimal* $(c(N), f, N)$-*hedge* $\pi^* = (\beta_n^*, \alpha_n^*)_{0\leq n\leq N}$ *exists and has the value*

$$X_n^{\pi^*} = \mathbb{E}_{\mathbb{P}^*}(\mathcal{E}_N^{-1}(U)\mathcal{E}_n(U)f | \mathcal{F}_n), \quad (11.22)$$

[3] *the coefficients of the minimal* $(c(N), f, N)$-*hedge* $\pi^* = (\beta_n^*, \alpha_n^*)_{0\leq n\leq N}$ *satisfy*

$$\alpha_n^* = \alpha_n, \quad \beta_n^* = \frac{X_{n-1}^{\pi^*} - \alpha_n^* S_{n-1}}{B_{n-1}} \quad (11.23)$$

for $0 \leq n \leq N$ *(i.e.* $(\alpha_n^*)_{0\leq n\leq N}$ *and* $(\beta_n^*)_{0\leq n\leq N}$ *can be chosen as predictable sequences).*

Proof: (Following similar ideas in parts as in Melnikov [134], Theorem 4.1, p. 33-35). Let $\pi \in H(x, f, N)$. It is obvious that $B_n > 0$ for all $0 \leq n \leq N$ under asssumption (iv).

First, the proof of claim [1]. Consider the discounted sequence $M^\pi = (M_n^\pi)_{0\leq n\leq N}$ defined by $M_n^\pi = X_n^\pi / B_n$. A simple calculation shows that

$$M_n^\pi = X_0 B_0^{-1} + \sum_{k=1}^n B_k^{-1} \alpha_k S_{k-1}(\rho_k - r_k),$$

since X_n^π possesses the representation $X_n^\pi = \mathcal{E}_n(U)(X_0 + \sum_{k=1}^n \mathcal{E}_k^{-1}(U)\Delta N_k)$ where $\Delta N_k = \alpha_k S_{k-1}(\rho_k - r_k)$ (see Lemma 28). Note that $\alpha_n S_{n-1}$ forms a predictable sequence. Recall also that $\mathbb{P}^*$ is the unique martingale-measure with $\mathbb{P}^* \sim \mathbb{P}$. Thanks to Lemma 29, the sequences $(\sum_{k=0}^n (\rho_k - r_k))_{0 \leq n \leq N}$ is a martingale with respect to $\mathbb{P}^*$. Consequently, M^π following (11.23) is a martingale with respect to $\mathbb{P}^*$. The monotonicity property of expectations of martingales implies that

$$\mathbb{E}_{\mathbb{P}^*}(M_N^\pi) = \mathbb{E}_{\mathbb{P}^*}(\frac{X_N^\pi}{B_N}) = \mathbb{E}_{\mathbb{P}^*}(\frac{X_0^\pi}{B_0}) = \frac{x}{B_0}.$$

Multiplying both sides by B_0 leads to the equivalent relation

$$x(= X_0) = \mathbb{E}_{\mathbb{P}^*}(\mathcal{E}_N^{-1}(U)X_N^\pi). \qquad (11.24)$$

Therefore, for all (x, f, N)-hedges π,

$$x \geq \mathbb{E}_{\mathbb{P}^*}(\mathcal{E}_N^{-1}(U)f) \qquad (11.25)$$

must hold, and, if this expected value exists, the minimal hedge satisfies

$$x = \mathbb{E}_{\mathbb{P}^*}(\mathcal{E}_N^{-1}(U)f). \qquad (11.26)$$

Using (11.20) gives the relation

$$c(N) = \inf\{x > 0 : H(x, f, N) \neq \emptyset\} \geq \mathbb{E}_{\mathbb{P}^*}(\mathcal{E}_N^{-1}(U)f) \qquad (11.27)$$

for the fair price $c(N)$, hence claim [1] is verified.

Second, the proof of claim [2]. We need to show the existence of minimal $(c(N), f, N)$-hedge π^*. For this purpose, set $x = \mathbb{E}_{\mathbb{P}^*}[\mathcal{E}_N^{-1}(U)f]$. Consider the martingale $M^* = (M_n^*)_{0 \leq n \leq N}$ defined by

$$M_n^* = \mathbb{E}_{\mathbb{P}^*}(B_N^{-1}f|\mathcal{F}_n)$$

for $0 \leq n \leq N$. Then, one easily recognizes that

$$M_0^* = \mathbb{E}_{\mathbb{P}^*}(B_N^{-1}f|\mathcal{F}_0) = \mathbb{E}_{\mathbb{P}^*}(B_N^{-1}f) = B_0^{-1}\mathbb{E}_{\mathbb{P}^*}(\mathcal{E}_N^{-1}f) = \frac{x}{B_0},$$

$$M_N^* = \mathbb{E}_{\mathbb{P}^*}(B_N^{-1}f|\mathcal{F}_N) = \frac{f}{B_N}.$$

Thanks to the completeness of the underlying market and resulting Theorem 11.2, we know that M^* must have the representation

$$M_n^* = M_0^* + \sum_{k=1}^n \alpha_k^* \frac{S_{k-1}}{B_k}(\rho_k - r_k)$$

for $1 \leq n \leq N$. We note that the construction of the above stated portfolio $\pi^* \in SF$ guarantees the coincidence

$$M_n^* = \frac{X_n^{\pi^*}}{B_n}$$

for all $0 \leq n \leq N$. This fact can be verified by induction on n. For example, for $n = 1$, we have $\pi_1^* = (\beta_1^*, \alpha_1^*)$ with $\alpha_1^* = \alpha_1$ and $\beta_1^* = (x - \alpha_1^*)/B_0$, and one arrives at its value $X_1^{\pi^*} = \beta_1^* B_1 + \alpha_1^* S_1$ and its discounted value

$$\begin{aligned} M_1^{\pi^*} &= \frac{X_1^{\pi^*}}{B_1} = \beta_1^* + \alpha_1^* \frac{S_1}{B_1} = \frac{x}{B_0} + \alpha_1^* \frac{S_1}{B_1} - \alpha_1^* \frac{S_0}{B_0} = \frac{x}{B_0} \\ &+ \alpha_1^* \left(\frac{S_1}{B_1} - \frac{S_0}{B_0} \right) \\ &= \frac{x}{B_0} + \alpha_1^* \frac{S_0}{B_1} \left(\frac{S_1}{S_0} - \frac{B_1}{B_0} \right) = \frac{x}{B_0} + \alpha_1^* \frac{S_0}{B_1} (\rho_1 - r_1) \\ &= M_1^* \end{aligned}$$

since $S_1/S_0 = 1 + \Delta V_1 = 1 + \rho_1$ and $B_1/B_0 = 1 + \Delta U_1 = 1 + r_1$ in linear dynamic (B, S)-markets. Continuing these calculations leads to the verification of the identities $M_n^{\pi^*} = X_n^{\pi^*}/B_n = M_n^*$ for all $1 \leq n \leq N$, hence (11.27) holds indeed. Consequently, for hedge π^*, we have

$$M_n^{\pi^*} = \frac{X_n^{\pi^*}}{B_n} = M_n^* = \mathbb{E}_{\mathbb{P}^*}(B_N^{-1} f | \mathcal{F}_n).$$

Now, multiply this result by B_n and use the $(\mathcal{F}_n, \mathcal{B}(\mathbb{R}))$-measurability of B_n to find that

$$X_n^{\pi^*} = \mathbb{E}_{\mathbb{P}^*}(B_n B_N^{-1} f | \mathcal{F}_n) = \mathbb{E}_{\mathbb{P}^*}(\mathcal{E}_N^{-1}(U) \mathcal{E}_n(U) f | \mathcal{F}_n)$$

for $0 \leq n \leq N$. In particular, the final value is found to be

$$X_N^{\pi^*} = \mathbb{E}_{\mathbb{P}^*}(f | \mathcal{F}_N) = f$$

and the initial value is

$$X_0^{\pi^*} = \mathbb{E}_{\mathbb{P}^*}(\mathcal{E}_N^{-1}(U) f | \mathcal{F}_0) = \mathbb{E}_{\mathbb{P}^*}(\mathcal{E}_N^{-1}(U) f) = x.$$

This shows the attainability and minimality of the $(c(N), f, N)$-hedge $\pi^* \in SF$. Note that we have already proven that $\pi^* \in SF$ at the end of the proof of Theorem 11.2 (take $f = x = 0$). Therefore, the proof of Theorem 11.3 is complete.

4.3 SOME PROPERTIES OF THE BINARY DYNAMIC (B, S, π, X^π)-MARKET

DEFINITION 11.10 *A (B, S, π, X^π)-market over the probability space $(\Omega, \mathcal{F}, (\mathcal{F}_n)_{0 \leq n \leq N}, \mathbb{P})$ is called binary iff there are two sequences of distinct real numbers $(a_n)_{0 \leq n \leq N}$ and $(b_n)_{0 \leq n \leq N}$ such that $a_n < b_n$, $0 < q_n = \mathbb{P}(\{\rho_n = a_n\}) < 1$ and $0 < p_n = \mathbb{P}(\{\rho_n = b_n\}) < 1$ with $p_n + q_n = 1$ for all $0 \leq n \leq N$ and $(\rho_n)_{0 \leq n \leq N}$ are independent random variables.*

Remark: Note that, in general, binary markets can possess time-dependent probabilities. An appropriate probability space for binary markets can be constructed as follows. Introduce independent random variables $\varepsilon_n : (\Omega, \mathcal{F}_n, \mathbb{P}) \to \{+1, -1\}$ with the properties that

$$\varepsilon_n = +1 \iff \rho_n = b_n, \quad \varepsilon_n = -1 \iff \rho_n = a_n.$$

Thus, we may set

$$\rho_n = \frac{a_n + b_n}{2} + \frac{b_n - a_n}{2} \varepsilon_n.$$

Define $\mathcal{F}_0^\varepsilon = \{\emptyset, \Omega\}$ since B_0 and S_0 are supposed to be nonrandom (as known certain quantities to the market participants). Also it is reasonable to assume that interest rates $r = (r_n)_{0 \leq n \leq N}$ involved in the dynamics of fairly riskless asset B are nonrandom (hence known to the participants) too. Then, the naturally induced choice for the filtration is given by the σ-algebras $\mathcal{F}_n^\varepsilon = \sigma(\varepsilon_1, \varepsilon_2, ..., \varepsilon_n)$ for $1 \leq n \leq N$. Consequently, the product probability space $(\Omega^N, \mathcal{F}_N^\varepsilon, (\mathcal{F}_n^\varepsilon)_{0 \leq n \leq N}, \bigotimes_{k=1}^N \mathbb{P}_k)$ with $\mathbb{P}_k(\varepsilon_k = +1) = p_k$ may serve as the underlying stochastic basis. Note that also $\mathcal{F}_n^\varepsilon = \sigma(\rho_1, ..., \rho_n)$ for all $1 \leq n \leq N$.

THEOREM 11.4 *The binary (B, S, π, X^π)-market with nonrandom parameters $-1 < a_n < r_n < b_n < +\infty$ is a no-arbitrage and complete financial market over the filtered probability space $(\Omega^N, \mathcal{F}_N^\varepsilon, (\mathcal{F}_n^\varepsilon)_{0 \leq n \leq N}, \bigotimes_{k=1}^N \mathbb{P}_k)$.*

Proof: Note that any binary probability measure $\mathbb{P}_k$ is completely determined by $p_k = \mathbb{P}(\rho_k = b_k)$. Choose

$$p_k^* = \frac{r_k - a_k}{b_k - a_k}$$

for all $1 \leq k \leq N$. Trivially, $0 < p_k^* < 1$ holds due to the assumption $r_k \in (a_k, b_k)$ (also $q_k^* = 1 - p_k^* = (b_k - r_k)/(b_k - a_k)$). Thus, probability measures $\mathbb{P}_k^*$ and $\bigotimes_{k=1}^N \mathbb{P}_k^*$ can be constructed out of the given data. Define $M_n = S_n/B_n$ for all $0 \leq n \leq N$. Calculate

$$\mathbb{E}_{\mathbb{P}^*}(\sum_{k=1}^n (\rho_k - r_k) - \sum_{k=1}^{n-1}(\rho_k - r_k)|\mathcal{F}_{n-1}^\varepsilon) = \mathbb{E}_{\mathbb{P}^*}(\rho_n - r_n)$$
$$= a_n(1 - p_n^*) + b_n p_n^* - r_n$$
$$= (b_n - a_n)p_n^* + a_n - r_n$$
$$= r_n - a_n + a_n - r_n$$
$$= 0,$$

hence $(\sum_{k=1}^n (\rho_k - r_k))_{0 \leq n \leq N}$ forms a $(\mathcal{F}_n^\varepsilon)_{0 \leq n \leq N}$-martingale with respect to the measure $\mathbb{P}^* = \bigotimes_{k=1}^N \mathbb{P}_k^*$. Thanks to Lemma 27, $M = (M_n)_{0 \leq n \leq N}$ is a $(\mathcal{F}_n^\varepsilon)_{0 \leq n \leq N}$-martingale under $\mathbb{P}^*$ too, i.e. $\mathbb{P}^*$ represents a martingale measure and $\mathbb{P}^* \in \mathcal{P}^*$ (equivalent by construction). Now, Theorem 11.1 implies that $SF_{arb} = \emptyset$, hence there is not any arbitrage possibility in the framework of given binary (B, S, π, X^π)-market model. Moreover, its completeness follows from the martingale-representation Lemma 30 below. Thus, the proof of Theorem 11.4 is complete.

LEMMA 30 *Let $(\Omega_n, \mathcal{F}_n, \mathbb{P}_n)$ be a sequence of probability spaces and $\rho = (\rho_n)_{0 \leq n \leq N}$ be a sequence of independent random variables with non-random parameters $-1 < a_n < r_n < b_n < +\infty$, $p_n + q_n = 1$, $0 < q_n = \mathbb{P}_n(\{\rho_n = a_n\}) < 1$ and $0 < p_n = \mathbb{P}_n(\{\rho_n = b_n\}) < 1$ where $p_n = (r_n - a_n)/(b_n - a_n)$ for $1 \leq n \leq N$. Define $\mathcal{F}_n = \sigma(\rho_1, ..., \rho_n)$ and $\mathcal{F}_0 = \{\emptyset, \Omega\}$. Then, any arbitrary $(\mathcal{F}_n)_{0 \leq n \leq N}$-martingale $M = (M_n)_{0 \leq n \leq N}$ with $\mathbb{E}(M_0) = 0$ possesses a discrete integral representation of the form*

$$M_n = \sum_{k=1}^n \alpha_k \Delta m_k \qquad (11.28)$$

where $(\alpha_k)_{1 \leq k \leq N}$ is a predictable sequence and $m = (m_k)_{1 \leq k \leq N}$ defined by $m_n = \sum_{k=1}^n (\rho_k - r_k)$ for $1 \leq n \leq N$ and $m_0 = 0$ is the underlying martingale with respect to $(\mathcal{F}_n)_{0 \leq n \leq N}$.

Proof: First, note that M_n is $(\mathcal{F}_n, \mathcal{B}(\mathbb{R}))$-measurable. Therefore, there must exist Borel-measurable functions $f_n : \{a_1, b_1\} \times \{a_2, b_2\} \times ... \times \{a_n, b_n\} \to \mathbb{R}$ such that $M_n(\omega) = f_n(\rho_1(\omega), \rho_2(\omega), ..., \rho_n(\omega))$ for all $\omega \in \Omega$ and $1 \leq n \leq N$. This implies that

$$\Delta M_n = \alpha_n \Delta m_n \qquad (11.29)$$

holds iff

$$f_n(\rho_1, \rho_2, ..., \rho_n) - f_{n-1}(\rho_1, \rho_2, ..., \rho_{n-1}) = \alpha_n \Delta m_n = \alpha_n(\rho_n - r_n).$$

This is equivalent to the following system of equations

$$f_n(\rho_1, \rho_2, ..., \rho_{n-1}, b_n) - f_{n-1}(\rho_1, \rho_2, ..., \rho_{n-1}) = \alpha_n(b_n - r_n),$$
$$f_n(\rho_1, \rho_2, ..., \rho_{n-1}, a_n) - f_{n-1}(\rho_1, \rho_2, ..., \rho_{n-1}) = \alpha_n(a_n - r_n).$$

Solving the latter system with respect to α_n yields the identities

$$\alpha_n = \frac{f_n(\rho_1, \rho_2, ..., \rho_{n-1}, b_n) - f_{n-1}(\rho_1, \rho_2, ..., \rho_{n-1})}{b_n - r_n},$$
$$= \frac{f_n(\rho_1, \rho_2, ..., \rho_{n-1}, a_n) - f_{n-1}(\rho_1, \rho_2, ..., \rho_{n-1})}{a_n - r_n}, \quad (11.30)$$

which also follow directly from the martingale property of M

$$\mathbb{E}(M_n - M_{n-1} | \mathcal{F}_{n-1})$$
$$= \mathbb{E}(f_n(\rho_1, \rho_2, ..., \rho_n) - f_{n-1}(\rho_1, \rho_2, ..., \rho_{n-1}) | \mathcal{F}_{n-1})$$
$$= 0$$

which can be equivalently rewritten as

$$p_n f_n(\rho_1, ..., \rho_{n-1}, b_n) + (1 - p_n) f_n(\rho_1, ..., \rho_{n-1}, a_n)$$
$$= p_n f_{n-1}(\rho_1, ..., \rho_{n-1}) + (1 - p_n) f_{n-1}(\rho_1, ..., \rho_{n-1})$$

or, in another form,

$$\frac{f_n(\rho_1, ..., \rho_{n-1}, b_n) - f_{n-1}(\rho_1, ..., \rho_{n-1})}{1 - p_n}$$
$$= \frac{f_{n-1}(\rho_1, ..., \rho_{n-1}) - f_n(\rho_1, ..., \rho_{n-1}, a_n)}{p_n}$$

by taking the choice $p_n = (r_n - a_n)/(b_n - a_n)$. Backwards, we may also conlude the validity of (11.29) from (11.30), and hence the discrete integral representation (11.28) is confirmed. This completes the proof of Lemma 30.

5. γ-PRICING AND γ-HEDGING

So far we have considered claims which are attainable with probability one. It is natural to ask what happens for claims which are only attainable with some positive probability. This leads to an essentially weaker concept and requires less initial capital for investments. For simplicity, we suppose that we have a complete (B, S)-market with nonrandom interest rates $r_n > -1$, nonrandom initial values $X_0^\pi = x$, nonrandom initial riskless assets B_0 which are normalized to $B_0 = 1$ and contingent claims (f, N) such that $\mathbb{E}_{\mathbb{P}^*}(f(S_N)) > 0$. Define

$$SF(f, N) = \{\pi \in SF : X_N^\pi > f - \mathbb{E}_{\mathbb{P}^*}(f)\}.$$

In analogy to statistics, introduce the level $\gamma \in (0,1)$ of significance of contingent claims (f, N).

DEFINITION 11.11 *A portfolio $\pi \in SF(f, N)$ is called an $\gamma - (x, f, N)$-hedge for the probability measure $\mathbb{P}^*$ iff*

$$\mathbb{P}^*(\{X_N^\pi \geq f\}) \geq 1 - \gamma. \qquad (11.31)$$

The set $H(x, f, N, \gamma) = \{\pi \in SF(f, N) : \pi\gamma - (x, f, N)\text{-hedge}\}$ is called the set of $\gamma - (x, f, N)$-hedges. The γ-price $C(N, \gamma)$ of the contingent claim (f, N) is defined by

$$C(N, \gamma) = \inf\{x > 0 : H(x, f, \gamma) \neq \emptyset\}.$$

THEOREM 11.5 *Under the forementioned conditions and for the martingale-measure $\mathbb{P}^* \sim \mathbb{P}$, we have*

$$\forall \pi \in SF(f, N) : \mathbb{P}^*(\{X_N^\pi \geq f\}) \leq \frac{X_0^\pi}{C(N)}, \qquad (11.32)$$

where $C(N) = \mathbb{E}_{\mathbb{P}^*}(\mathcal{E}_N^{-1} f)$ is the fair price of the European option with claim (f, N) with $f \geq 0$.

Proof: Using Chebyshev's inequality, the definition of $SF(f, N)$, the nonrandomness of r_n and the martingale property of X^π/B under $\mathbb{P}^*$ leads to

$$\begin{aligned}
\mathbb{P}^*(\{X_N^\pi \geq f\}) &= \mathbb{P}^*(\{X_N^\pi - f + \mathbb{E}_{\mathbb{P}^*}(f) \geq \mathbb{E}_{\mathbb{P}^*}(f)\}) \\
&\leq \frac{\mathbb{E}_{\mathbb{P}^*}\left(X_N^\pi - f + \mathbb{E}_{\mathbb{P}^*}(f)\right)}{\mathbb{E}_{\mathbb{P}^*}(f)} \\
&= \frac{\mathbb{E}_{\mathbb{P}^*}(X_N^\pi)}{\mathbb{E}_{\mathbb{P}^*}(f)} \\
&= \frac{\mathbb{E}_{\mathbb{P}^*}(\mathcal{E}_N^{-1} X_N^\pi)}{\mathbb{E}_{\mathbb{P}^*}(\mathcal{E}_N^{-1} f)} \\
&= \frac{X_0^\pi}{C(N)},
\end{aligned}$$

hence the assertion of Theorem 11.5 is rather obvious.

Remark: As a consequence, a necessary condition for $\pi \in H(x, f, N, \gamma)$ is

$$1 - \gamma \leq \mathbb{P}^*(\{X_N^\pi \geq f\}) \leq \frac{x}{C(N)} \quad (11.33)$$

as a determining relation between significance-level γ and initial capital x.

Following ideas from the proof of Theorem 11.2, we are able to construct a γ-hedge π. For this purpose, define the density

$$Z_N = \left.\frac{d\mathbb{P}}{d\mathbb{P}^*}\right|_{\mathcal{F}_N}$$

with $Z_N > 0$. Clearly, there exists a quantity $\lambda = \lambda(\gamma)$ such that

$$\mathbb{P}^*(\{Z_N \geq \lambda(\gamma)\}) = 1 - \gamma \quad (11.34)$$

or, equivalently, $\mathbb{P}^*(\{Z_N < \lambda(\gamma)\} = \gamma\})$. For simplicity, suppose that $\lambda \geq 1$. Consider the following $(\mathcal{F}_n)_{0 \leq n \leq N}$-martingales defined by

$$M_n^\gamma = \mathbb{E}_{\mathbb{P}^*}(\mathbf{1}_{\{Z_N < \lambda(\gamma)\}}|\mathcal{F}_n), \quad M_n^c = \mathbb{E}_{\mathbb{P}^*}(\mathcal{E}_N^{-1}(U)f|\mathcal{F}_n).$$

Recall that the market (B, S) is complete (as we assumed it). Hence the martingales $M^\gamma = (M_n^\gamma)_{0 \leq n \leq N}$ and $M^c = (M_n^c)_{0 \leq n \leq N}$ admit the representations

$$M_n^\gamma = \gamma + \sum_{k=1}^n \varphi_k \mathcal{E}_k^{-1}(U) S_{k-1}(\rho_k - r_k),$$

$$M_n^c = C(N) + \sum_{k=1}^n \alpha_k^* \mathcal{E}_k^{-1}(U) S_{k-1}(\rho_k - r_k)$$

with predictable sequences $(\varphi_k)_{0 \leq k \leq N}$ and $(\alpha_k^*)_{0 \leq k \leq N}$, where $M_0^\gamma = \gamma$ and $M_0^c = C(N)$.

THEOREM 11.6 *Under the above assumptions and for nonrandom initial value* $X_0^\pi = x = (1-\gamma)C(N)$, *the portfolio* $\pi_\gamma = (\beta_n^\gamma, \alpha_n^\gamma)_{0 \leq n \leq N}$ *defined by*

$$\alpha_n^\gamma = \alpha_n^* - \varphi_n C(N), \quad \beta_n^\gamma = \frac{X_{n-1}^{\pi_\gamma} - \alpha_n^\gamma S_{n-1}}{\mathcal{E}_{n-1}(U)} \quad (11.35)$$

is a $\gamma - ((1-\gamma)C(N), f, N)$-hedge.

Proof: From section 2, recall the representation
$$X_n^{\pi\gamma} = \mathcal{E}_n(U)\left(X_0^{\pi\gamma} + \sum_{k=1}^{n} \mathcal{E}_k^{-1}(U)\Delta N_k\right).$$

In view of the construction (11.35), we conclude that

$$\mathcal{E}_n^{-1}(U)X_n^{\pi\gamma} = (1-\gamma)C(N) + \sum_{k=1}^{n} \mathcal{E}_k^{-1}(U)\alpha_k^\gamma S_{k-1}(\rho_k - r_k)$$

$$= (1-\gamma)C(N) + \sum_{k=1}^{n} \mathcal{E}_k^{-1}(U)\alpha_k^* S_{k-1}(\rho_k - r_k)$$

$$- C(N)\sum_{k=1}^{n} \mathcal{E}_k^{-1}(U)\varphi_k S_{k-1}(\rho_k - r_k)$$

$$= C(N) + \sum_{k=1}^{n} \mathcal{E}_k^{-1}(U)\alpha_k^* S_{k-1}(\rho_k - r_k) - \gamma C(N)$$

$$- C(N)\sum_{k=1}^{n} \mathcal{E}_k^{-1}(U)\varphi_k S_{k-1}(\rho_k - r_k)$$

$$= M_n^c - C(N)M_n^\gamma = \mathbb{E}_{\mathbb{P}^*}(\mathcal{E}_N^{-1}(U)f|\mathcal{F}_n) - C(N)\mathbb{E}_{\mathbb{P}^*}(\mathbf{1}_{\{Z_N < \lambda(\gamma)\}}|\mathcal{F}_n).$$

Now, just set $n = N$ in these computations. Thus, one obtains the identity

$$\mathcal{E}_N^{-1}(U)X_N^{\pi\gamma} = \mathcal{E}_N^{-1}(U)f(S_N) - C(N)\mathbf{1}_{\{Z_N < \lambda(\gamma)\}}.$$

Therefore, one can estimate

$$\mathcal{E}_N^{-1}(U)X_N^{\pi\gamma} \geq \mathcal{E}_N^{-1}(U)f(S_N) - C(N),$$

hence $\pi_\gamma \in SF(f, N)$. Furthermore, from (11.34), we know that

$$\mathbb{P}^*(\{X_N^{\pi\gamma} \geq f\}) = \mathbb{P}^*(\{f - C(N)\mathcal{E}_N(U)\mathbf{1}_{\{Z_N < \lambda(\gamma)\}} \geq f\})$$
$$= \mathbb{P}^*(\{\mathbf{1}_{\{Z_N < \lambda(\gamma)\}} \leq 0\}) \qquad (11.36)$$
$$= \mathbb{P}^*(\{\mathbf{1}_{\{Z_N < \lambda(\gamma)\}} = 0\}) \qquad (11.37)$$
$$= \mathbb{P}^*(\{Z_N \geq \lambda(\gamma)\}) \qquad (11.38)$$
$$= 1 - \gamma. \qquad (11.39)$$

Finally, from (11.34) and (11.37), we arrive at

$$\mathbb{P}(\{X_N^{\pi\gamma} \geq f\}) = \mathbb{E}_{\mathbb{P}}(\mathbf{1}_{\{X_N^{\pi\gamma} \geq f\}})$$

$$\begin{aligned}
&= \mathbb{E}_{\mathbb{P}^*}(\mathbf{1}_{\{X_N^{\pi_\gamma} \geq f\}} Z_N) \\
&\geq \mathbb{E}_{\mathbb{P}^*}(\mathbf{1}_{\{X_N^{\pi_\gamma} \geq f\}} \mathbf{1}_{\{Z_N \geq \lambda(\gamma)\}} Z_N) \\
&\geq \lambda(\gamma) \mathbb{E}_{\mathbb{P}^*}(\mathbf{1}_{\{X_N^{\pi_\gamma} \geq f\}}) \\
&= \lambda(\gamma) \mathbb{P}^*(\{X_N^{\pi_\gamma} \geq f\}) \\
&= \lambda(\gamma)(1-\gamma) \\
&\geq (1-\gamma)
\end{aligned}$$

if $\lambda(\gamma) \geq 1$. Consequently, π_γ is a $\gamma - ((1-\gamma)C(N), f, N)$-hedge, and the proof of Theorem 11.5 is complete.

Remark: So it is possible to hedge contingent claims with a specified probability $1 - \gamma$. Moreover, the initial capital can be reduced to $(1-\gamma)C(N)$ by the amount of $\gamma C(N)$ with allowing a risk probability γ that the accepted claim cannot totally be repaid.

6. ASYMPTOTIC BEHAVIOR OF BINARY (B, S)-MARKETS

6.1 A FEW WORDS ON RELATION TO LIMIT-MODELS

Employing the heuristic argument that stock prices are either rising or falling at any moment of time, Cox, Ross, and Rubinstein [27] proposed regarding these changes as discrete and introduced a binomial model of financial markets. They showed that the binomial model has a Brownian motion as a limit, and that the formula obtained for a fair price converges to the classical Black-Scholes formula. In some studies of financial time series [130, 171] it has been demonstrated that the stock market prices exhibit the so-called long-range-dependence property. Therefore, it has been proposed by Sottinen and Valkeila [175, 176] to replace the Brownian motion in the classical Black-Scholes pricing model by the fractional Brownian motion. Sottinen's construction is based on the path properties of some "disturbed" random walk.

We consider a discrete time approximation of the Black-Scholes model with a dynamic Cox-Ross-Rubinstein's model based on dynamic random walks.

This section aims at proving the convergence of the binary dynamic binomial model to the continuous Black-Scholes model. So consistency of modeling with discrete time dynamic (B, S)-markets and their continuous time counterparts is verified. Especially, we shall prove that, for every $e \in E$, the sequence $\left(B_{[nt]}^{(n)}, S_{[nt]}^{(n)}\right)_{t \in [0,T]}$ converges weakly in the

Skorohod space $\mathcal{D} = \mathcal{D}([0,T], \mathbb{R}^2)$ to $(B_t, S_t)_{t \in [0,T]}$ as n goes to infinity. For this purpose, the binary univariate dynamic (B, S)-market is supposed to possess the probabilities

$$\mathbb{P}(\rho_k = z) = \begin{cases} g(T^k e) & \text{if } z = b \\ 1 - g(T^k e) & \text{if } z = a \\ 0 & \text{otherwise} \end{cases}$$

where $e \in E$, $a < b \in \mathbb{R}$.

6.2 CONVERGENCE OF BINARY DYNAMIC (B, S)-MARKETS TO THE CONTINUOUS BLACK-SCHOLES MODEL

In the classical Black-Scholes pricing model two assets are traded continuously over the time interval $[0, T]$. Let us denote by B the continuous time risk-less asset, or bond, and by S the continuous time risky asset, or stock. The dynamics of the continuous time assets are given by

$$dB_t = rB_t dt, \quad B_0 > 0 \tag{11.40}$$

and

$$dS_t = S_t(-\frac{\sigma^2}{2}dt + \sigma dW_t), \quad S_0 > 0. \tag{11.41}$$

Here r is a deterministic interest rate, $\sigma > 0$ is a constant and $(W_t)_{t \geq 0}$ is the classical one-dimensional Brownian motion. We consider a discrete time approximation of the Black-Scholes model with the following sequence of dynamic Cox-Ross-Rubinstein's model: The assets are traded on time points $0 = t_0^{(n)} < t_1^{(n)} < \ldots < t_n^{(n)} = T$. The dynamics are

$$B_k^{(n)} = \left(1 + r_k^{(n)}\right) B_{k-1}^{(n)}, \quad B_0^{(n)} = B_0$$

and

$$S_k^{(n)} = \left(1 + \rho_k^{(n)}\right) S_{k-1}^{(n)}, \quad S_0^{(n)} = S_0.$$

Here, $B_k^{(n)}$ and $S_k^{(n)}$ are the values of the bond and stock, respectively, on $[t_k^{(n)}, t_{k+1}^{(n)})$. We assume that for each k,

$$r_k^{(n)} = \frac{r}{n}$$

where r is the constant interest rate appearing in (11.40). The process $\rho^{(n)} = (\rho_k^{(n)})_{k \in \mathbb{N}}$ is defined for n fixed by

$$1 + \rho_k^{(n)} = \exp(X_k^{(n)}), k \in \mathbb{N}$$

where $(X_k^{(n)})_{k\in\mathbb{N}}$ is a sequence of independent random variables with distribution

$$\mathbb{P}(X_k^{(n)} = \frac{\sigma}{\sqrt{n}}) = g_n(T^k e) = 1 - \mathbb{P}(X_k^{(n)} = -\frac{\sigma}{\sqrt{n}}),$$

where $D = (E, \mathcal{A}, \mu, T)$ is a dynamical system, the functions g_n are defined on E with values in $[0, 1]$ and e is a point of E. We assume that the functions g_n satisfy the two hypotheses

$$(H_1) \quad : \quad \sup_{t\in[0,T]; x\in E} \left| \frac{1}{\sqrt{n}} \sum_{k=1}^{[nt]} (2g_n(T^k x) - 1) + \frac{\sigma t}{2} \right| = o(1),$$

$$(H_2) \quad : \quad \sup_{x\in E} \left| \frac{1}{n} \sum_{k=1}^{[nt]} [4g_n(T^k x)(1 - g_n(T^k x))] - t \right| = o(1).$$

6.3 A PRELIMINARY RESULT

THEOREM 11.7 *For every $e \in E$, the sequence of processes*

$$(\sum_{k=1}^{[nt]} X_k^{(n)})_{t\in[0,T]}$$

weakly converges in the Skorohod space $\mathcal{D}([0, T])$ (see [15]) to the real Brownian motion with mean $-\frac{\sigma^2 t}{2}$ and variance $\sigma^2 t$.

Proof: Introduce the characteristic function of $\sum_{k=1}^{[nt]} X_k^{(n)}$ by

$$\phi_n(u) = \mathbb{E}(\exp(iu \sum_{k=1}^{[nt]} X_k^{(n)})).$$

By independence of the random variables $X_k^{(n)}, k \geq 1$,

$$\phi_n(u) = \prod_{k=1}^{[nt]} \mathbb{E}(\exp(iuX_k^{(n)})) = \prod_{k=1}^{[nt]} Q_{k,n}(u)$$

where

$$Q_{k,n}(u) = \cos(\frac{u\sigma}{\sqrt{n}}) + i(2g_n(T^k e) - 1)\sin(\frac{u\sigma}{\sqrt{n}}).$$

A direct calculation gives

$$|Q_{k,n}(u)|^2 = 1 - 4g_n(T^k e)(1 - g_n(T^k e))\sin^2(\frac{u\sigma}{\sqrt{n}})$$

$$= 1 - 4g_n(T^k e)(1 - g_n(T^k e))\frac{u^2 \sigma^2}{n} + \mathcal{O}(n^{-2})$$

and then

$$|\phi_n(u)| = \prod_{k=1}^{[nt]} |Q_{k,n}(u)|$$

$$= \exp\left[\frac{1}{2}\sum_{k=1}^{[nt]} \log[1 - 4g_n(T^k e)(1 - g_n(T^k e))\frac{u^2\sigma^2}{n} + \mathcal{O}(n^{-2})]\right]$$

$$= \exp\left(-\frac{\sigma^2 t}{2}u^2 + o(1)\right)$$

using hypothesis (H_2).

The imaginary part of the characteristic function can be rewritten as

$$\prod_{k=1}^{[nt]} \exp\left(i \arctan\left(\frac{(2g_n(T^k e) - 1)\sin(\frac{u\sigma}{\sqrt{n}})}{\cos(\frac{u\sigma}{\sqrt{n}})}\right)\right)$$

$$= \exp\left(i \sum_{k=1}^{[nt]} (2g_n(T^k e) - 1)\frac{u\sigma}{\sqrt{n}} + o(1)\right) = \exp(-\frac{iu\sigma^2 t}{2}) + o(1)$$

using hypothesis (H_1). The convergence of the finite dimensional distributions of processes $(\sum_{k=1}^{[nt]} X_k^{(n)})_t$ to the one of the one-dimensional Brownian motion with mean $-\frac{\sigma^2 t}{2}$ and variance $\sigma^2 t$ is obtained in a classical way using the independence of the increments.

It remains to prove the tightness in $\mathcal{D}([0,T])$ of the sequence $(\sum_{k=1}^{[nt]} X_k^{(n)})_{t\in[0,T]}$. The classical criterion: Theorem 15.6 in Billingsley ([15]) is used. Let $0 \le t_1 < t < t_2 \le T$, by independence of the random variables $X_k^{(n)}$,

$$\mathbb{E}(|\sum_{k=1}^{[nt]} X_k^{(n)} - \sum_{k=1}^{[nt_1]} X_k^{(n)}|^2 \cdot |\sum_{k=1}^{[nt_2]} X_k^{(n)} - \sum_{k=1}^{[nt]} X_k^{(n)}|^2) \qquad (11.42)$$

$$= \mathbb{E}(|\sum_{k=1}^{[nt]} X_k^{(n)} - \sum_{k=1}^{[nt_1]} X_k^{(n)}|^2)\mathbb{E}(|\sum_{k=1}^{[nt_2]} X_k^{(n)} - \sum_{k=1}^{[nt]} X_k^{(n)}|^2).$$

Now,

$$\mathbb{E}(|\sum_{k=1}^{[nt]} X_k^{(n)} - \sum_{k=1}^{[nt_1]} X_k^{(n)}|^2) = \mathbb{E}((\sum_{k=[nt_1]+1}^{[nt]} X_k^{(n)})^2)$$

$$= \frac{\sigma^2}{n} \sum_{k=[nt_1]+1}^{[nt]} (4g_n(T^k e)(1 - g_n(T^k e)))$$

$$+ \frac{\sigma^2}{n} \left(\sum_{k=[nt_1]+1}^{[nt]} (2g_n(T^k e) - 1) \right)^2$$

Using hypothesis (H_1), there exists a constant $C_1 > 0$ such that for every n,

$$\left| \sum_{k=1}^{[nt]-[nt_1]} (g_n(T^{[nt_1]+k} e) - \frac{1}{2}) \right|^2 \leq C_1 n(t_2 - t_1)^2.$$

And, since for every k, n and $e \in E$, $0 \leq 4g_n(T^k e)(1 - g_n(T^k e)) \leq 1$, we have

$$\sum_{k=[nt_1]+1}^{[nt]} 4g_n(T^k e)(1 - g_n)(T^k e) \leq [nt] - [nt_1] \leq [nt_2] - [nt_1].$$

If $t_2 - t_1 \geq \frac{1}{n}$, $[nt_2] - [nt_1] \leq 2n(t_2 - t_1)$ so there exists $C > 0$ such that

$$\mathbb{E}(|\sum_{k=1}^{[nt]} X_k^{(n)} - \sum_{k=1}^{[nt_1]} X_k^{(n)}|^2 \cdot |\sum_{k=1}^{[nt_2]} X_k^{(n)} - \sum_{k=1}^{[nt]} X_k^{(n)}|^2) \leq C(t_2 - t_1)^2.$$

If $t_2 - t_1 < \frac{1}{n}$, either t and t_1 belong in the same interval $[\frac{(i-1)}{n}, \frac{i}{n})$ or t and t_2 do. In both cases, the left hand side of (11.42) vanishes. The tightness follows from Billingsley's Theorem 15.6 ([15]).

6.4 THE MAIN CONVERGENCE RESULT.

We now are able to prove that the sequence of dynamic Cox-Ross-Rubinstein's models converges to the continuous Black-Scholes model.

THEOREM 11.8 *For every $e \in E$, the sequence $\left(B_{[nt]}^{(n)}, S_{[nt]}^{(n)} \right)_{t \in [0,T]}$ converges weakly in $\mathcal{D}([0,T], \mathbb{R}^2)$ to $(B_t, S_t)_{t \in [0,T]}$ as n tends to infinity.*

Proof: By definition,

$$\log(S_{[nt]}^{(n)}) = \log(S_0^{(n)}) + \sum_{k=1}^{[nt]} \log(1 + \rho_k^{(n)}) = \log(S_0^{(n)}) + \sum_{k=1}^{[nt]} X_k^{(n)}$$

From Theorem 11.7, the sequence of processes

$$\left(\sum_{k=1}^{[nt]} X_k^{(n)}\right)_{t \in [0,T]}$$

weakly converges in the Skorohod space $\mathcal{D}([0,T])$ to the one-dimensional Brownian motion with mean $-\frac{\sigma^2}{2}t$ and variance $\sigma^2 t$. Since the functional exponential is continuous in the Skorohod space, we deduce that $(S_{[nt]}^{(n)})_{t \in [0,T]}$ weakly converges to the stock price

$$S_t = S_0 \exp\left(\frac{-\sigma^2 t}{2} + \sigma W_t\right)$$

which is the unique solution of (11.41).
The theorem is proved since the deterministic sequence

$$B_n^{([nt])} = \left(1 + \frac{r}{n}\right)^{[nt]} B_0$$

converges to $B_t = B_0 \exp(rt)$, the solution of equation (11.40).

6.5 PARTICULAR EXAMPLES
1. **Cox-Ross-Rubinstein Theorem:**
   Let us fix $n \geq 1$ and for every $k \in \mathbb{N}$, we choose

$$g_n = g_n(T^k e) = \frac{r_n - a_n}{b_n - a_n}$$

with $r_n = r/n$, $1 + a_n = (1 + r_n)\exp(-\sigma/\sqrt{n})$, $1 + b_n = (1 + r_n)\exp(\sigma/\sqrt{n})$ with $-1 < a_n < r_n < b_n$. For each n fixed, this dynamic (B,S)-market is the classical Cox-Ross-Rubinstein's model. For these particular functions g_n, hypotheses (H_1) and (H_2) are clearly satisfied and Theorem 11.8 is the well-known Cox-Ross- Rubinstein's result (see [27]) which asserts that a discrete time approximation of the Black-Scholes model is given by a sequence of Cox-Ross-Rubinstein's models. It is worth noting that in this particular case, with the notations $\rho_n^{(n)} = \rho_n, n \geq 1$, the sequence of random variables

$$\left(\sum_{k=1}^{n}(\rho_k - r_k)\right)_{k \geq 1}$$

is a martingale with respect to the natural filtration $\mathcal{F}_n = \mathcal{F}_n^S = \sigma(\rho_1, \ldots, \rho_n), n \geq 1$. So, from Lemma 29, we deduce that this particular measure is risk-neutral. However, in general, i.e. when the functions g_n are not all constant, the above sequence does not have to be a martingale.

2. **A perturbated model:**
   We add to the previous model dynamical perturbations as follows. Let g be a bounded function defined on E with values in $\mathbb{R}$. We denote by $||g||_\infty$ the supremum norm of g. For every $n \geq 1$, the functions giving the distributions of the dynamic random variables $\rho_k^{(n)}$ (or $X_k^{(n)}$) are chosen as

$$g_n(e) = \frac{r_n(e) - a_n(e)}{b_n(e) - a_n(e)}$$

where for every $e \in E$,

$$a_n(e) = (1 + r_n(e))\exp(-\frac{g(e)}{n^\beta}) - 1$$

and

$$b_n(e) = (1 + r_n(e))\exp(\frac{g(e)}{\sqrt{n}}) - 1.$$

From this definition, it clearly follows that for every $n \geq 1$, for every $e \in E$, $-1 < a_n(e) < r_n(e) < b_n(e)$ and hence, each function g_n take its values in the interval $[0, 1]$.
Let us assume that

$$(H_3): \quad \sup_{t \in [0,T]; x \in E} \left|\frac{1}{n}\sum_{k=1}^{[nt]} g(T^k x) - \sigma t\right| = o(1).$$

Large classes of functions g for which this kind of convergence holds will be given in the next section. For this particular model which is a modification of the so-called Cox-Ross-Rubinstein model, Theorem 11.8 holds. To prove it we have to show that hypotheses (H_1) and (H_2) hold. Let us define for every $t \in [0, T]$, $x \in E$,

$$M_{n,t}(x) = \frac{1}{\sqrt{n}}\sum_{k=1}^{[nt]}(2g_n(T^k x) - 1).$$

By definition of the g_n's,

$$\begin{aligned}M_{n,t}(x) &= \frac{1}{\sqrt{n}}\sum_{k=1}^{[nt]}\left(\frac{2r_n(T^kx)-a_n(T^kx)-b_n(T^kx)}{b_n(T^kx)-a_n(T^kx)}\right)\\ &= \frac{1}{\sqrt{n}}\sum_{k=1}^{[nt]}\left(\frac{(1+r_n(T^kx))(1-\cosh(\frac{g(T^kx)}{\sqrt{n}}))}{(1+r_n(T^kx))\sinh(\frac{g(T^kx)}{\sqrt{n}})}\right)\\ &= -\frac{1}{\sqrt{n}}\sum_{k=1}^{[nt]}\tanh\left(\frac{g(T^kx)}{2\sqrt{n}}\right)\\ &= -\frac{1}{\sqrt{n}}\sum_{k=1}^{[nt]}\left[\tanh\left(\frac{g(T^kx)}{2\sqrt{n}}\right)-\frac{g(T^kx)}{2\sqrt{n}}\right]-\frac{1}{2n}\sum_{k=1}^{[nt]}g(T^kx)\end{aligned}$$

Now, on every compact $[-M,M]$, we have $|\tanh(x)-x|\leq C|x|^3$, thus

$$\sup_{t\geq 0;x\in E}\left|\frac{1}{\sqrt{n}}\sum_{k=1}^{[nt]}\left[\tanh\left(\frac{g(T^kx)}{2\sqrt{n}}\right)-\frac{g(T^kx)}{2\sqrt{n}}\right]\right|\leq C\frac{[nT]\|g\|_\infty^3}{n^2}=o(1).$$

Then, hypothesis (H_1) follows from hypothesis (H_3).
Moreover, hypothesis (H_2) can be proved when hypothesis (H_3) is satisfied. Straightforward computations give

$$\begin{aligned}&\frac{1}{n}\sum_{k=1}^{[nt]}[4g_n(T^kx)(1-g_n(T^kx))]\\ &=\frac{4}{n}\sum_{k=1}^{[nt]}\frac{(r_n(T^kx)-a_n(T^kx))(b_n(T^kx)-r_n(T^kx))}{(b_n(T^kx)-a_n(T^kx))^2}\\ &=\frac{2}{n}\sum_{k=1}^{[nt]}\frac{(\cosh(\frac{g(T^kx)}{\sqrt{n}})-1)}{\sinh^2(\frac{g(T^kx)}{\sqrt{n}})}\\ &=\frac{[nt]}{n}-\frac{1}{n}\sum_{k=1}^{[nt]}\tanh^2\left(\frac{g(T^kx)}{2\sqrt{n}}\right)\end{aligned}$$

Then, from hypothesis (H_3), we get that hypothesis (H_2) is satisfied:

$$\begin{aligned}&\sup_{x\in E}\left|\frac{1}{n}\sum_{k=1}^{[nt]}[4g_n(T^kx)(1-g_n(T^kx))]-t\right|\\ &\leq\left|\frac{[nt]}{n}-t\right|+\frac{1}{n}\sup_{x\in E}\sum_{k=1}^{[nt]}\tanh^2(\frac{g(T^kx)}{2\sqrt{n}})\end{aligned}$$

$$\leq \left|\frac{[nt]}{n} - t\right| + \frac{T}{4n}||g||_\infty^2 = o(1)$$

since $|\tanh(x)| \leq |x|$ for every x.

References

This chapter is a preliminary detailed version of [75].

Appendix A
Ergodic theory

1. SOME DEFINITIONS AND BASIC THEOREMS

Let $(E, \mathcal{A}, \mu)$ be a complete probability space, that is a set E endowed with a σ-algebra $\mathcal{A}$ of measurable sets of E, a probability measure μ on $\mathcal{A}$ and $\mathcal{A}$ is assumed to contain all subsets of sets of measure 0.
Let $T : E \to E$ be a one-to-one onto map such that T and T^{-1} are both measurable.
A map T is called a measure-preserving transformation if

$$\mu(T^{-1}A) = \mu(A), \quad \forall\, A \in \mathcal{A}.$$

We call $(E, \mathcal{A}, \mu, T)$ a dynamical system.
Let x be a point of E, then the orbit of this point $(T^n x)_{n \in \mathbb{Z}}$ represents the complete history of the system from the infinite past to the infinite future. If $f : E \to \mathbb{R}$ is a measurable function representing some measurement made on the system; $f(x), f(Tx), f(T^2 x), \ldots$ can be seen as the values of some physically interesting variable measured at time $t = 0, 1, 2, \ldots$ when the system starts from the initial point x.
A basic question of ergodic theory is that of the convergence of the long-term averages

$$\frac{1}{n} \sum_{k=0}^{n-1} f(T^k x)$$

when the number of observations n is large.
This problem was originally raised by the physicist L. Boltzmann in 1885 who wanted to establish the "ergodic hypothesis", that is, the time mean

$$\frac{1}{n} \sum_{k=0}^{n-1} f(T^k x)$$

coincides almost everywhere with the space mean

$$\int_E f(x)d\mu(x)$$

for n large when the orbits of the dynamical system penetrate all corners of the space E.

Under this ergodic hypothesis, the physicists could approximate the time means (impossible to get in practice) by the space mean.

Rigourous mathematical results arise in 1931 with Von Neumann's Theorem (also known as the Mean Ergodic Theorem) and Birkoff's Theorem (or Pointwise Ergodic Theorem). The first one is about the general convergence in the mean square (L^2) sense and the second one the almost everywhere convergence.

THEOREM A.1 *(Von Neumann 1931)* Let $(E, \mathcal{A}, \mu)$ be a σ-finite measure space, $T : E \to E$ a measure-preserving transformation. Then, for every $f \in L^2(\mu)$, the average

$$\frac{1}{n}\sum_{k=0}^{n-1} f \circ T^k$$

converges in $L^2(\mu)$ to a function $\bar{f} \in L^2(\mu)$.

DEFINITION A.12 *A set $A \in \mathcal{A}$ is called invariant if $\mu(T^{-1}A \,\Delta\, A) = 0$. The invariant sets form a sub-σ-algebra. The transformation T is called ergodic if every invariant set has measure 0 or 1 or equivalently if the σ-algebra is trivial.*

THEOREM A.2 *(Birkhoff 1931)* Let $(E, \mathcal{A}, \mu)$ be a probability space, $T : E \to E$ be a measure-preserving transformation and $f \in L^1(E, \mathcal{A}, \mu)$. Then,

$$\lim_{n \to \infty} \frac{1}{n}\sum_{k=0}^{n-1} f \circ T^k = \mathbb{E}(f|\mathcal{I}) \quad \mu - a.e., \quad L^1$$

where $\mathbb{E}(f|\mathcal{I})$ means the conditional expectation of the function f with respect to the invariant σ-algebra $\mathcal{I}$.

Equality, for large n, between the time mean and the space mean then holds when the dynamical system is ergodic.

THEOREM A.3 *Assume that E is a compact metric space. The following assertions are equivalent:*
1- The dynamical system $(E, \mathcal{A}, \mu, T)$ is uniquely ergodic i.e. μ is the

unique probability measure which is preserved by the transformation T.
2- For every continuous function f on E, the sequence of averages

$$\frac{1}{n}\sum_{k=0}^{n-1} f \circ T^k$$

uniformly converges to $\mathbb{E}(f) = \int_E f\, d\mu$ as n goes to infinity.

The proof can be found in [149].

2. EXAMPLES OF DYNAMICAL SYSTEMS
2.1 THE ROTATION ON THE TORUS

Given $\alpha \in \mathbb{R}^r$, we consider the map

$$T_\alpha : \mathbb{T}^r \to \mathbb{T}^r$$

$$x \mapsto x + \alpha \bmod 1 = (x_1 + \alpha_1 \bmod 1, \ldots, x_r + \alpha_r \bmod 1)$$

It is clear that T_α preserves Lebesgue measure on $\mathbb{T}^r$. The angle $\alpha = (\alpha_1, \ldots, \alpha_r)$ is said irrational if $1, \alpha_1, \ldots, \alpha_r$ are linearly independent on $\mathbb{Q}$. It can be proved that T_α is uniquely ergodic if and only if α is irrational (see for instance [149]). In this case, for every continuous function f on $\mathbb{T}^r$,

$$\lim_{n\to\infty} \frac{1}{n} \sum_{k=0}^{n-1} f(x + k\alpha) = \int_{\mathbb{T}^r} f(x)dx, \quad \text{uniformly in } x.$$

More details are given in Appendix B about this particular and very interesting system.

2.2 STATIONARY STOCHASTIC PROCESSES

Let $(\Omega, \mathcal{F}, \mathbb{P})$ be a probability space and $f = (f_n)_{n\in\mathbb{Z}}$ be a stationary sequence of measurable functions defined on Ω, in the sense that for every $k \geq 1$, for every $i_1 \in \mathbb{Z}, \ldots, i_k \in \mathbb{Z}$, for every Borel subsets $B_1, B_2, \ldots, B_k$ of $\mathbb{R}$ and for every $l \in \mathbb{Z}$,

$$\mathbb{P}(\{w; f_{i_1}(w) \in B_1, \ldots, f_{i_k}(w) \in B_k\})$$

$$= \mathbb{P}(\{w; f_{i_1+l}(w) \in B_1, \ldots, f_{i_k+l}(w) \in B_k\}).$$

Let $\mathbb{R}^{\mathbb{Z}}$ and μ be the probability measure defined on the Borel subsets B of $\mathbb{R}^{\mathbb{Z}}$ by

$$\mu(B) = \mathbb{P}(\{w; f(w) \in B\}).$$

Let $\sigma : \mathbb{R}^{\mathbb{Z}} \to \mathbb{R}^{\mathbb{Z}}$ be the shift transformation defined by

$$\forall n \in \mathbb{Z},\ (\sigma x)_n = x_{n+1}.$$

The stationarity of the sequence $(f_n)_{n\in\mathbb{Z}}$ implies that the measure μ is shift-invariant on the cylinder sets and hence on all sets of the Borel field $\mathcal{B}$ of $\mathbb{R}^{\mathbb{Z}}$, so we have constructed a measure-preserving system $(\mathbb{R}^{\mathbb{Z}}, \mathcal{B}, \mu, \sigma)$. Denote by $\Pi_n : \mathbb{R}^{\mathbb{Z}} \to \mathbb{R}$ the projection onto the nth coordinate: $\Pi_n(x) = x_n$ then $(\Pi_n)_n$ has the same joint distributions on $\mathbb{R}^{\mathbb{Z}}$ as $(f_n)_n$, on Ω.

This simple and classical construction proves that every stationary stochastic process can be viewed as a dynamical system in the setting of ergodic theory. Much more examples can be found in [149].

Appendix B
Some Results on Diophantine Approximations

Let us consider the dynamical system $(\mathbb{T}^r, \mathcal{B}(\mathbb{T}^r), \lambda, T_\alpha)$ where λ is the Lebesgue measure on the torus $\mathbb{T}^r$ and T_α is the irrational rotation over $\mathbb{T}^r$ defined by $x \to x + \alpha \mod 1$. It is well known that under these conditions this dynamical system is ergodic and for every $f \in L^1(\lambda)$, for almost every $x \in \mathbb{T}^r$,

$$M_n = \frac{1}{n} \sum_{l=1}^{n} f(T_\alpha^l x) - \int_{\mathbb{T}^r} f(t)dt \underset{n \to \infty}{\to} 0$$

When f is with bounded variation, this result holds for every $x \in \mathbb{T}^r$ and it is possible to determine the speed of convergence of the sequence M_n to 0 in terms of arithmetic properties of the irrational vector α. When $r = 1$, for all irrational badly approximated by rationals, Denjoy-Koksma's inequality gives us a majorization of M_n uniformly in x for n large enough. But when $r \geq 2$, Denjoy-Koksma's inequality does not hold (see Yoccoz [188]) and the method of low discrepancy sequences has to be used.

A. Case of one-dimensional torus

Let α be an irrational. We call a rational $\frac{p}{q}$ with p, q relatively prime such that $|\alpha - \frac{p}{q}| < \frac{1}{q^2}$, a rational approximation of α. When α has the continued fraction expansion $\alpha = [\alpha] + [a_1, \ldots, a_n, \ldots]$, the n-th principal convergent of α is $\frac{p_n}{q_n}$ where, $\forall n \geq 2$,

$$p_n = a_n p_{n-1} + p_{n-2}$$

$$q_n = a_n q_{n-1} + q_{n-2};$$

the recurrence is given by defining the values of p_0, p_1 and q_0, q_1.

Denjoy-Koksma's inequality *Let $f : \mathbb{R} \to [0,1]$ be a function with bounded variation $V(f)$ and $\frac{p}{q}$ a rational approximation of α. Then, for every $x \in \mathbb{T}^1$,*

$$\left| \sum_{l=1}^{q} f(T_\alpha^l x) - q \int_{\mathbb{T}^1} f(t)dt \right| \leq V(f).$$

PROPOSITION 14 *Let f be a function with bounded variation $V(f)$. For every irrational α such that the inequality $a_m < m^{1+\epsilon}$, where $\epsilon > 0$, is satisfied eventually for all m,*

$$\sup_{x \in \mathbb{T}^1} \left| \sum_{l=1}^{n} \left(f(T_\alpha^l x) - \int_{\mathbb{T}^1} f(t)dt \right) \right| = \mathcal{O}(\log^{2+\epsilon} n).$$

Proof:
The sequence of integers $(q_i)_{i \geq 1}$ being strictly increasing, for a given $n \geq 1$, there exists $m_n \geq 0$ such that

$$q_{m_n} \leq n < q_{m_n+1}.$$

By Euclidean division, we have $n = b_{m_n} q_{m_n} + n_{m_n-1}$ with $0 \leq n_{m_n-1} < q_{m_n}$. We can use the usual relations

$$q_0 = 1, q_1 = a_1$$

$$q_n = a_n q_{n-1} + q_{n-2}, n \geq 2. \tag{B.1}$$

We obtain that $(a_{m_n+1} + 1)q_{m_n} > q_{m_n+1} > n$ and so $b_{m_n} \leq a_{m_n+1}$. If $m_n > 0$, we may write $n_{m_n-1} = b_{m_n-1} q_{m_n-1} + n_{m_n-2}$ with $0 \leq n_{m_n-2} < q_{m_n-1}$. Again, we find $b_{m_n-1} \leq a_{m_n}$. Continuing in this manner, we arrive at a representation for n of the form

$$n = \sum_{i=0}^{m_n} b_i q_i$$

with $0 \leq b_i \leq a_{i+1}$ for $0 \leq i \leq m_n$ and $b_{m_n} \geq 1$. Using Denjoy-Koksma's inequality, we get

$$\left| \sum_{l=1}^{n} f(T_\alpha^l x) - n \int_{\mathbb{T}^1} f(x)dx \right| \leq V(f) \sum_{i=0}^{m_n} b_i$$

$$\leq V(f) \sum_{i=0}^{m_n} a_{i+1}.$$

Appendix B: Some Results on Diophantine Approximations

By hypothesis, there exists $m_0 \geq 1$ such that,
$$a_m < m^{1+\epsilon}, \forall m \geq m_0.$$

Let n be such that $m_n > m_0$. Thus,
$$\left|\sum_{l=1}^{n} f(T_\alpha^l x) - n \int_{\mathbb{T}^1} f(t)dt\right| \leq V(f)\left(\sum_{i=0}^{m_0-1} a_{i+1} + (m_n+1)^{2+\epsilon}\right).$$

We need to know the asymptotic behavior of m_n. When α is the golden ratio, $a_n = 1$, $\forall n \geq 1$ and the relation (B.1) implies that $q_n \sim \frac{1}{\sqrt{5}}\alpha^{n+1}$. Let α' be another irrational; its partial quotients a'_n satisfy necessarily $a'_n \geq 1$. Using the relation (B.1), we see that $q'_n \geq q_n, \forall n \geq 1$. Therefore, $m_n = \mathcal{O}(\log n)$ and the proposition is proved.

B. Generalization to r-dimensional torus

We recall some definitions and well known results from the method of low discrepancy sequences in dimension $r \geq 1$.

Suppose we are given a function $f(x) = f(x^{(1)}, \ldots, x^{(r)})$ with $r \geq 1$. By a partition P of $[0,1]^r$, we mean a set of r finite sequences $\eta_0^{(j)}, \eta_1^{(j)}, \ldots, \eta_{m_j}^{(j)}$ ($j = 1, \ldots, r$), with $0 = \eta_0^{(j)} \leq \eta_1^{(j)} \leq \ldots \leq \eta_{m_j}^{(j)} = 1$ for $j = 1, \ldots, r$. In connection with such a partition, we define, for $j = 1, \ldots, r$ an operator Δ_j by

$$\Delta_j f(x^{(1)}, \ldots, x^{(j-1)}, \eta_i^{(j)}, x^{(j+1)}, \ldots, x^{(r)}) = f(x^{(1)}, \ldots, x^{(j-1)}, \eta_{i+1}^{(j)},$$
$$x^{(j+1)}, \ldots, x^{(r)}) - f(x^{(1)}, \ldots, x^{(j-1)}, \eta_i^{(j)}, x^{(j+1)}, \ldots, x^{(r)}),$$

for $0 \leq i < m_j$.

DEFINITION B.13 1. For a function f on $[0,1]^r$, we set
$$V^{(r)}(f) = \sup_P \sum_{i_1=0}^{m_1-1} \cdots \sum_{i_r=0}^{m_r-1} |\Delta_{1,\ldots,r} f(\eta_{i_1}^{(1)}, \ldots, \eta_{i_r}^{(r)})|,$$

where the supremum is extended over all partitions P of $[0,1]^r$. If $V^{(r)}(f)$ is finite, then f is said to be of bounded variation on $[0,1]^r$ in the sense of Vitali.

2. For $1 \leq p \leq r$ and $1 \leq i_1 < i_2 < \ldots < i_p \leq r$, we denote by $V^{(p)}(f; i_1, \ldots, i_p)$ the p-dimensional variation in the sense of Vitali of the restriction of f to

$$E_{i_1\ldots i_p}^r = \{(t_1,\ldots,t_r) \in [0,1]^r; t_j = 1 \text{ whenever } j \text{ is none of the } i_r, 1 \leq r \leq p\}.$$

If all the variations $V^{(p)}(f; i_1, \ldots, i_p)$ are finite, the function f is said to be of bounded variation on $[0,1]^r$ in the sense of Hardy and Krause.

Let $x_1, \ldots, x_n$ be a finite sequence of points in $[0,1]^r$ with $x_l = (x_{l_1}, \ldots, x_{l_r})$ for $1 \leq l \leq n$. We introduce the function

$$R_n(t_1, \ldots, t_r) = \frac{A(t_1, \ldots, t_r; n)}{n} - t_1 \ldots t_r$$

for $(t_1, \ldots, t_r) \in [0,1]^r$, where $A(t_1, \ldots, t_r; n)$ denotes the number of elements $x_l, 1 \leq l \leq n$, for which $x_{l_i} < t_i$ for $1 \leq i \leq r$.

DEFINITION B.14 *The discrepancy D_n^* of the sequence $x_1, \ldots, x_n$ in $[0,1]^r$ is defined to be*

$$D_n^* = \sup_{(t_1,\ldots,t_r) \in [0,1]^r} |R_n(t_1, \ldots, t_r)|.$$

For a real number t, let $\|t\|$ denote its distance to the nearest integer, namely,

$$\begin{aligned}\|t\| &= \inf_{n \in \mathbb{Z}} |t-n| \\ &= \inf(\{t\}, 1-\{t\})\end{aligned}$$

where $\{t\}$ is the fractional part of t.

DEFINITION B.15 *For a real number η, a r-tuple $\alpha = (\alpha_1, \ldots, \alpha_r)$ of irrationals is said to be of type η if η is the infimum of all numbers σ for which there exists a positive constant $c = c(\sigma; \alpha_1, \ldots, \alpha_r)$ such that*

$$r^\sigma(h)\| < h, \alpha > \| \geq c$$

holds for all $h \neq 0$ in $\mathbb{Z}^r$, where $r(h) = \prod_{i=1}^r \max(1, |h_i|)$ and $< \cdot, \cdot >$ denotes the standard inner product in $\mathbb{R}^r$.

The type η of α is also equal to

$$\sup\{\gamma : \inf_{h \in (\mathbb{Z}^r)^*} r^\gamma(h)\| < h, \alpha > \| = 0\}.$$

We always have $\eta \geq 1$ (see [143]). Now we give a result (see [108]) which yields the asymptotic behavior of the discrepancy of the sequence $w = (x_1 + l\alpha_1, \ldots, x_r + l\alpha_r), l = 1, 2, \ldots$ as a function of the mutual irrationality of the components of α.

Appendix B: Some Results on Diophantine Approximations

PROPOSITION 15 *Let $\alpha = (\alpha_1, \ldots, \alpha_r)$ be an irrational vector. Suppose there exists $\eta \geq 1$ and $c > 0$ such that*

$$r^\eta(h) \| <h, \alpha> \| \geq c$$

for all $h \neq 0$ in $\mathbb{Z}^r$. Then, for every $x \in [0,1]^r$, the discrepancy of the sequence $w = (x_1 + l\alpha_1, \ldots, x_r + l\alpha_r), l = 1, 2, \ldots$ satisfies $D_n^*(w) = \mathcal{O}(n^{-1} \log^{r+1} n)$ for $\eta = 1$ and $D_n^*(w) = \mathcal{O}(n^{-\frac{1}{((\eta-1)r+1)}} \log n)$ for $\eta > 1$.

The proof is based on the Erdös-Turán-Koksma's theorem: For $h \in \mathbb{Z}^r$, define $p(h) = \max_{1 \leq j \leq r} |h_j|$. Let $x_1, \ldots, x_n$ be a finite sequence of points in $\mathbb{R}^r$. Then, for any positive integer m, we have

$$D_n^* \leq C_r \left(\frac{1}{m} + \sum_{0 \leq p(h) \leq m} \frac{1}{r(h)} \left| \frac{1}{n} \sum_{l=1}^n e^{2\pi i <h, x_l>} \right| \right)$$

where C_r only depends on the dimension r. This theorem combined with the results of [108] (p.131) gives us the result.

THEOREM B.1 (HLAWKA, ZAREMBA) *Let f be of bounded variation on $[0,1]^r$ in the sense of Hardy and Krause, and let ω be a finite sequence of points $x_1, \ldots, x_n$ in $[0,1]^r$. Then, we have*

$$\left| \frac{1}{n} \sum_{l=1}^n f(x_l) - \int_{\mathbb{T}^r} f(t) dt \right| \leq \sum_{p=1}^r \sum_{1 \leq i_1 < i_2 < \ldots < i_p \leq r} V^{(p)}(f; i_1, \ldots, i_p) D_n^*(\omega_{i_1 \ldots i_p}),$$

where $D_n^*(\omega_{i_1 \ldots i_p})$ is the discrepancy in $E_{i_1 \ldots i_p}^r$ of the sequence $\omega_{i_1 \ldots i_p}$ obtained by projecting ω onto $E_{i_1 \ldots i_p}^r$.

PROPOSITION 16 *Let f be a function with bounded variation in the sense of Hardy and Krause, and α an irrational vector of type η, then*

$$\sup_{x \in \mathbb{T}^r} \left| \sum_{l=1}^n (f(T_\alpha^l x) - \int_{\mathbb{T}^r} f(t) dt) \right| = \begin{cases} \mathcal{O}(\log^{r+1} n) & \text{if } \eta = 1 \\ \mathcal{O}(n^{1 - \frac{1}{((\eta-1)r+1)}} \log n) & \text{if } \eta > 1. \end{cases}$$

Proof:
Let η' be such that $\eta \leq \eta' < 1 + \frac{1}{r}$. There exists $c > 0$ such that

$$r^{\eta'}(h) \| <h, \alpha> \| \geq c$$

holds for all $h \neq 0$ in $\mathbb{Z}^r$. Suppose we are given a p-tuple $\alpha_p = (\alpha_{i_1}, \ldots, \alpha_{i_p}), 1 \leq p \leq r$, of α, then

$$r^{\eta'}(h) \| <h, \alpha_p> \| \geq c$$

holds for all $h \neq 0$ in $\mathbb{Z}^p, 1 \leq p \leq r$. Thus, every p-tuple, $1 \leq p \leq r$, is of type δ such that $1 \leq \delta \leq \eta$ and $(\alpha_{i_1}, \ldots, \alpha_{i_p})$ is an irrational vector. For every $p, 1 \leq p \leq r$, we define $w_{i_1\ldots i_p}$ by the projection of w on $E^r_{i_1\ldots i_p}$. From the previous proposition, we have for every $p, 1 \leq p \leq r$,

$$\begin{cases} nD_n^*(w_{i_1\ldots i_p}) &= \mathcal{O}(\log^{p+1} n) \text{ if } \delta = 1 \\ nD_n^*(w_{i_1\ldots i_p}) &= \mathcal{O}(n^{1-\frac{1}{((\delta-1)p+1)}} \log n) \text{ if } 1 < \delta \leq \eta. \end{cases}$$

Now, $\forall p = 1, \ldots, r$,

$$0 \leq 1 - \frac{1}{(\delta-1)p+1} \leq 1 - \frac{1}{(\eta-1)r+1} \leq 1.$$

Therefore, using Hlawka-Zaremba's theorem, we obtain Proposition 16.

Appendix C
Skorohod metric

Let $\mathcal{D}$ be the space of functions $x = (x_t)$, $t \in [0,1]$, that are continuous on the right ($x_t = x_{t+}$ for all $t \in [0,1]$) and have limits from the left (at every $t > 0$) and $\mathcal{B}(\mathcal{D})$ the σ-algebra generated by the cylinder sets. Skorohod introduced a metric $d(x, y)$ on $\mathcal{D}$ such that the σ-algebra generated by the open sets will coincide with $\mathcal{B}(\mathcal{D})$. Skorohod's metric is defined as follows:

$$d(x,y) = \inf\{\epsilon > 0 : \exists \lambda \in \Lambda, \sup_t |x_t - y_{\lambda(t)}| + \sup_t |t - \lambda(t)| \leq \epsilon\}$$

where Λ is the set of strictly increasing functions $\lambda = \lambda(t)$ that are continuous on $[0,1]$ and have $\lambda(0) = 0$, $\lambda(1) = 1$.

DEFINITION C.16 *The Skorohod space $\mathcal{D}([0, \infty))$ is the set of all right continuous and with left limit functions (RCLL) from $[0, \infty)$ in $\mathbb{R}$.*

Cylinder sets are defined as follows: We consider (for illustration) the space $\mathbb{R}^\infty$ of ordered sequences of numbers,

$$x = (x_1, x_2, \ldots), -\infty < x_k < \infty, k = 1, 2, \ldots$$

Let I_k and B_k denote, respectively, the intervals $(a_k, b_k]$ and the Borel subsets of the kth line (with coordinate x_k).
Cylinder sets are defined by:

$$\mathcal{F}(I_1 \times \ldots \times I_n) = \{x : x = (x_1, x_2, \ldots), x_1 \in I_1, \ldots, x_n \in I_n\},$$

$$\mathcal{F}(B_1 \times \ldots \times B_n) = \{x : x = (x_1, x_2, \ldots), x_1 \in B_1, \ldots, x_n \in B_n\},$$

$$\mathcal{F}(B^n) = \{x : (x_1, \ldots, x_n) \in B^n\},$$

where B^n is a Borel set in $\mathcal{B}(\mathbb{R}^n)$. Each cylinder $\mathcal{F}(B_1 \times \ldots \times B_n)$, or $\mathcal{F}(B^n)$ can also be thought of as a cylinder with base in $\mathbb{R}^{n+1}$, $\mathbb{R}^{n+2}$,..., since
$$\mathcal{F}(B_1 \times \ldots \times B_n) = \mathcal{F}(B_1 \times \ldots \times B_n \times \mathbb{R}),$$
$$\mathcal{F}(B^n) = \mathcal{F}(B^{n+1}),$$
where $B^{n+1} = B^n \times \mathbb{R}$.
It follows that both systems of cylinders $\mathcal{F}(B_1 \times \ldots \times B_n)$ and $\mathcal{F}(B^n)$ are algebras.

Reference
The contain of this appendix is taken from [170].

Appendix D
Fourier series

Let f be a real function defined on $[0, 2L]$ and differentiable then this function can be expanded into a Fourier series for every $x \in]0, 2L[$ as

$$f(x) = \frac{a_0}{2} + \sum_{n=1}^{\infty} a_n \cos\left(\frac{n\pi x}{L}\right) + \sum_{n=1}^{\infty} b_n \sin\left(\frac{n\pi x}{L}\right),$$

with

$$a_n = \frac{1}{L} \int_0^{2L} f(x) \cos\left(\frac{n\pi x}{L}\right) dx$$

and

$$b_n = \frac{1}{L} \int_0^{2L} f(x) \sin\left(\frac{n\pi x}{L}\right) dx.$$

- Let us establish the formula

$$\sum_{n \geq 1} \frac{n \sin(nx)}{a^2 + n^2} = \frac{\pi}{2} \frac{\sinh(a(\pi - x))}{\sinh(a\pi)}, \quad x \in]0, 2\pi[.$$

Let $L = \pi$ and

$$f(x) = \frac{\pi}{2} \frac{\sinh(a(\pi - x))}{\sinh(a\pi)}.$$

Since, for every $n \geq 0$,

$$\begin{aligned}
a_n &= \frac{1}{\pi} \int_0^{2\pi} \frac{\pi}{2} \frac{\sinh(a(\pi - x))}{\sinh(a\pi)} \cos(nx) \, dx \\
&= \frac{1}{2 \sinh(a\pi)} \int_0^{2\pi} \sinh(a(\pi - x)) \cos(nx) \, dx \\
&= \frac{1}{2 \sinh(a\pi)} \int_{-\pi}^{\pi} \sinh(ax) \cos(n(\pi - x)) \, dx
\end{aligned}$$

243

$$= \frac{(-1)^n}{2\sinh(a\pi)} \int_{-\pi}^{\pi} \sinh(ax)\cos(nx)dx = 0,$$

we only have

$$\frac{\pi}{2}\frac{\sinh(a(\pi-x))}{\sinh(a\pi)} = \sum_{n=1}^{\infty} b_n \sin(nx)$$

where

$$b_n = \frac{1}{2\sinh(a\pi)} \int_0^{2\pi} \sinh(a(\pi-x))\sin(nx)\,dx.$$

Now, integrating by parts twice, we get

$$b_n = \frac{n}{2\sinh(a\pi)} \int_0^{2\pi} \frac{\cosh(a(\pi-x))}{a}\cos(nx)\,dx$$

$$= \frac{n}{a^2} - \frac{n^2}{a^2}b_n.$$

Thus,

$$b_n = \frac{n}{a^2+n^2}$$

and the formula follows.

- Let us prove that for $x \in \,]0,1[$,

$$\sum_{n=1}^{\infty} \frac{\sin(2n\pi x)}{n} = -\pi(x-\frac{1}{2}).$$

Let $L = \frac{1}{2}$ and $f(x) = -\pi(x-\frac{1}{2})$. For every $n \geq 0$,

$$a_n = -2\pi \int_0^1 (x-\frac{1}{2})\cos(2n\pi x)\,dx$$

$$= 2\pi(-1)^{n+1} \int_{-\frac{1}{2}}^{\frac{1}{2}} x\cos(2n\pi x)\,dx = 0$$

and, for every $n \geq 1$,

$$b_n = -2\pi \int_0^1 (x-\frac{1}{2})\sin(2n\pi x)\,dx$$

$$= -2\pi \left(\left[-\left(x-\frac{1}{2}\right)\frac{\cos(2n\pi x)}{2n\pi} \right]_0^1 + \int_0^1 \frac{\cos(2n\pi x)}{2n\pi}\,dx \right)$$

$$= \frac{1}{n}.$$

Consequently, the function f can be expanded into the Fourier series

$$\sum_{n=1}^{\infty} \frac{\sin(2n\pi x)}{n}, \quad x \in \,]0,1[.$$

Appendix E
Hilbert spaces, representations, *-algebras, von Neumann algebras

1. HILBERT SPACES

DEFINITION E.17 *On a complex vector space $\mathcal{V}$ an inner product is a mapping $<.,.>: \mathcal{V} \times \mathcal{V} \to \mathbb{C}$ such that, for $u, v, w \in \mathcal{V}, \alpha, \beta \in \mathbb{C}$:*

- $<\alpha u + v, \beta w> = \alpha <u,w> + \bar{\beta} <v,w>$,
- $<u,v> = \overline{<v,u>}$,
- $<u,u> \geq 0$ *and equal if and only if $u = 0$.*

DEFINITION E.18 *A complex vector space equipped with an inner product such that the norm defined by*

$$\| v \| = \sqrt{<v,v>}$$

turns it into a complete metric space is called Hilbert space.

Let us recall Schwartz' inequality

$$|<u,v>| \leq \| u \| \| v \|$$

To define an inner product on operators, one uses the trace. Setting

$$<X,Y> = \text{tr}(XY^*)$$

gives a positive-definite inner product for operators on a finite-dimensional vector space (and, suitably interpreted and restricted, for infinite-dimensional spaces as well).

It is also interesting to consider the bilinear form $\text{tr}(XY)$. In the theory

of Lie algebras, this leads to the Killing form.
Properties of the trace yield properties of this inner product.

PROPOSITION 17 *The inner product is consistent with the adjoint operation:*
$< AX, Y > = < X, A^*Y >$ *and* $< X, YA > = < XA^*, Y >$

2. LIE ALGEBRAS AND REPRESENTATIONS

Consider an algebra e.g. $n \times n$ matrices of real numbers. First, it is a vector space having operations of addition and multiplication by scalars. Second, it is a ring having an associative multiplication compatible with the addition operation, i.e., with distributive laws. We are interested in Lie algebras.

DEFINITION E.19 *A lie algebra is a vector space with a non-associative multiplication - the Lie bracket, denoted $[A, B]$ - satisfying:*

$$[A, B] = -[B, A]$$

$$[A, [B, C]] + [B, [C, A]] + [C, [A, B]] = 0$$

i.e., antisymmetry and the Jacobi identity.

We consider representations. Namely, where the algebra acts as linear operators on a vector space. For a Lie algebra, the abstract bracket becomes the commutator

$$[A, B] = AB - BA.$$

Consider $n \times n$ matrices of real or complex numbers, denoted here by $\mathcal{M}$. Let $Lin(\mathcal{M})$ denote the linear transformations of $\mathcal{M}$. We can define mappings from $\mathcal{M}$ into $Lin(\mathcal{M})$ by

$$r_A X = XA$$

$$l_A X = AX$$

Notice that l is a homomorphism: $l_{AB} = l_A l_B$, hence l gives a representation; while r is an anti-homomorphism: $r_{AB} = r_B r_A$, the order of multiplication is reversed. The representation l is called the left regular representation of $\mathcal{M}$. The algebra generated by these operators $\{l_A, r_A : A \in \mathcal{M}\}$ has nice properties, namely:

PROPOSITION 18 *For all $A, B \in \mathcal{M}$, l_A and r_B commute i.e $[l_A, r_B] = 0$.*

See [48] for more details on representations of Lie groups.

3. *-ALGEBRAS AND VON NEUMANN ALGEBRAS

Let $\mathcal{H}$ be a (complex) Hilbert space, $\mathcal{L}(\mathcal{H})$ the set of continuous linear applications from $\mathcal{H}$ into $\mathcal{H}$. In $\mathcal{L}(\mathcal{H})$ we have an operation called ajonction. If $S \in \mathcal{L}(\mathcal{H})$, we denote by S^* the adjoint of S. We get:

$$(S+T)^* = S^* + T^*, (\lambda S)^* = \bar{\lambda} S^*, (ST)^* = T^* S^*, S^{**} = S$$

λ stands for a complex number and $\bar{\lambda}$ is its complex conjugate.
This means that $\mathcal{L}(\mathcal{H})$ is a *-algebra.
Each subalgebra of $\mathcal{L}(\mathcal{H})$ stable for the ajonction operation is called *-subalgebra of $\mathcal{L}(H)$.
Let $\mathcal{A}$ be a *-subalgebra of $\mathcal{L}(\mathcal{H})$ and $\mathcal{A}'$ the commutator of $\mathcal{A}$ (i.e. the set of elements of $\mathcal{L}(\mathcal{H})$ which permute with the elements of $\mathcal{A}$). According to [41], if $\mathcal{A}$ is stable for the addition operation, then $\mathcal{A}'$ is also a *-subalgebra of $\mathcal{L}(\mathcal{H})$. We denote by $\mathcal{A}''$ the bicommutator of $\mathcal{A}$ (i.e. the commutator of $\mathcal{A}'$).

DEFINITION E.20 *A von Neumann algebra in $\mathcal{H}$ is a *-subalgebra $\mathcal{A}$ of $\mathcal{L}(\mathcal{H})$ such that $\mathcal{A} = \mathcal{A}''$.*

Examples:
$\mathcal{L}(\mathcal{H})$ is a von Neumann algebra.
The set of scalar operators of $\mathcal{H}$ is a von Neumann algebra.
If $\mathcal{B}$ is a subset of $\mathcal{L}(\mathcal{H})$ which is stable for the addition operation, then $\mathcal{B}$ is a von Neumann algebra.

References

[1] ACCARDI, L., FRIGERIO A. and LEWIS, J.T. Quantum stochastic processes. Publ. R.I.M.S, Kyoto (1982), 97-133.

[2] ARNOLD, L. Stochastic Differential Equations: Theory and Applications Wiley, New York, 1974.

[3] K. BACK and S.R. PLISKA On the fundamental theorem of asset pricing with an infinite state space. J. Math. Econom. 20 (1991), no. 1, pp. 1-18.

[4] BAKER, A. On some diophantine inequalities involving the exponential function. *Canad. J. Math.* (1965), 17, 616–626.

[5] BENSOUSSAN, A., LIONS, J.L., PAPANICOLAOU, G. *Asymptotic analysis for periodic structures* (1978) North-Holland, Amsterdam.

[6] BERGELSON, V. Weakly mixing PET. *Ergodic Theory Dynam. Systems* (1987), vol 7, no. 3, 337–349.

[7] BERGÉ, P., POMEAU, Y. and VIDAL, CH. L'ordre dans le chaos (collection enseignement des sciences), Hermann, Paris (1992).

[8] BERTHUET, R. Loi du logarithme itéré pour certaines intégrales stochastiques. *C. R. Acad. Sci. Paris Sér.* (1979), 289, no. 16, 813 -815

[9] BERTHUET, R. Etude de processus généralisant l'aire de Lévy. (French) [Study of a process generalizing the Levy stochastic area] *Probab. Theory Related Fields* (1986), 73, no. 3, 463 – 480.

[10] BIANE, P. Marches de Bernoulli quantiques. *Lect. Notes Math. 1426* (1990), 329-344, Springer Verlag.

[11] BIANE, P. Some properties of quantum Bernoulli random walks. *Proceedings of Quantum Probabilities 6* (1989), 193-203.

[12] BIANE, P. Quantum random walk on the dual of $SU(n)$. *Proba. Th. Rel. Fields* (1991), 89, 117–129,Springer Verlag.

[13] BIANE, P. Calcul stochastique non-commutatif. *Lect. Notes Math. 1608* (1994), 1-93, Springer Verlag.

[14] BIANE, P. Théorème de Ney-Spitzer sur le dual de $SU(2)$. *Transactions of the American Mathematical Society* (1994), 345, 1, 179–194.

[15] BILLINGSLEY, P. Convergence of probability measures. Wiley, New York (1968).

[16] BOLDRIGHINI, C., MINLOS, R. A., and PELLEGRINOTTI, A. Almost-sure central limit theorem for a Markov model of random walk in dynamical random environment. *Probab. Theory Related Fields* (1997), Vol. **109**, No 2, 245-273.

[17] BOLTHAUSEN, E. A central limit theorem for two-dimensional random walks in random sceneries. *Annals of Probability.* (1989), 17, no.1, 108–115.

[18] BOURGAIN, J. Almost sure convergence and bounded entropy. *Israel Journal of Mathematics.* (1988), 63, 79–95.

[19] BOURGAIN, J. On the pointwise ergodic theorem on L^p for arithmetic sets. *Israel Journal of Mathematics.* (1988), 61, 73–84.

[20] BOURGAIN, J. Problems of almost everywhere convergence related to Harmonic Analysis and Number Theory. *Israel Journal of Mathematics.* (1990), 71, 97–127.

[21] BUFFET, E. and PULE, J. A model of continuous polymers with random charges. *J. Math. Phys.* (1997), 38, 10, 5143 – 5152.

[22] CABUS, P. and GUILLOTIN-PLANTARD, N. Functional limit theorems for U-statistics indexed by a random walk. *Stochastic Process. Appl.* (2002), 101, 1, 143 – 160.

[23] CAMPANINO, M. and PÉTRITIS, D. Random walks in randomly oriented lattices. *Markov Processes and Related Fields* (2003), 3, 391-412.

[24] CAMPANINO, M. and PÉTRITIS, D. On the physical relevance of random walks: an example of random walks on randomly oriented

lattices. *Random walks and geometry*, 393-411, Walter de Gruyter (2004).

[25] COMETS, F., DELARUE, F., and SCHOTT, R. Distributed algorithms in an ergodic Markovian environment *Prépublication Institut Elie Cartan*, 31, 2005.

[26] COMETS, F., GANTERT, N., and ZEITOUNI, O. Quenched, annealed and functional large deviations for one-dimensional random walk in random environment. *Probab. Theory Related Fields* (2000), 118, 1, 65-114.

[27] COX, J.C., ROSS, R.A. and M. RUBINSTEIN Option pricing: a simplified approach. *J. Finan. Econ.* (1979), 7, 3, 229-263.

[28] COX, J.C. and M. RUBINSTEIN Option Markets. Prentice-Hall (1985).

[29] CREPEL, P. and RAUGI, A. Théorème central limite sur les groupes nilpotents. *Ann. Inst. H. Poincaré. Statist.* (1978), 14, 145-162.

[30] CREPEL, P. and ROYNETTE, B. Une loi du logarithme itéré pour le groupe d'Heisenberg. *Z. Wahrscheinlichkeitstheorie und verw. Geb.* (1977), 14, 217-229.

[31] CSÁKI, E., KÖNIG, W. and SHI, Z. An embedding for the Kesten-Spitzer random walk in random scenery. *Stochastic Process. Appl.* (1999), 82, 2, 283-292.

[32] DACUNHA-CASTELLE, D. and DUFLO, M. Probabilités et statistiques 2. Problèmes à temps mobile, Masson, Paris (1983).

[33] DALANG, R.C., MORTON, A. and WILLINGER, W. Equivalent martingale-measures and no-arbitrage in stochastic securities market models. *Stochastics Stochastics Rep.* (1990), 29, 2, 185-201.

[34] DELARUE, F. Recurrence and Transience of a Reflected Diffusion in the Square. Preprint of the LPMA, Universities Paris VI and VII, 2005.

[35] DELBAEN, F. and SCHACHERMEYER, W. A general version of the fundamental theorem of asset pricing. *Math. Ann.* (1994), 300, 3, 463-520.

[36] DEL JUNCO, A. and ROSENBLATT, J Counterexamples in ergodic theory and number theory. *Math. Ann.* (1979), 245, 185–197.

[37] DEMBO, A. and ZEITOUNI, O. Large Deviations Techniques and Applications. Springer Verlag, (1998).

[38] DESCLOUX, J. and TOLLEY, M. An accurate algorithm for computing the eigenvalues of polygonal membrane. *Comp. Meth. Appl. Mech. and Engineering* (1983), 39, 37-53.

[39] DINWOODIE, I. H. and ZABELL, S. L. Large deviations for exchangeable random vectors. *Ann. Probab.* (1992), 20, 3, 1147-1166.

[40] DIXMIER, J. Les C^*-algèbres et leurs représentations. (Gauthier-Villars (1964).

[41] DIXMIER, J. Les algèbres d'opérateurs dans l'espace hilbertien. Gauthier-Villars (1957).

[42] DOMBRY, C., GUILLOTIN-PLANTARD, N., PINÇON, B. and SCHOTT, R. Data structures with dynamical random transitions. *Random Structures and Algorithms* (in press).

[43] DOOB, J.L. Stochastic Processes. Wilcy, New York (1953).

[44] DRMOTA, M. and TICHY, R. Sequences, discrepancies and applications. *Lect. Notes Math. 1651.* Springer-Verlag (1997).

[45] DURAND, S. and SCHNEIDER, D. Théorèmes ergodiques et processus m-dépendants. *Ergodic Theory and Dynamical Systems* (2002).

[46] ELLIS C. A.. Probabilistic Model of Computer Deadlock. *IEEE Trans. on ASSP* (1977), 12, 43-60.

[47] EYMARD, P. and ROYNETTE, B. Marches aléatoires sur le dual de SU(2). *Lect. Notes Math. 497* (1975), 108-152, Springer-Verlag.

[48] FEINSILVER, P. and SCHOTT, R. Algebraic Structures and Operator Calculus. vol. I, Kluwer Academic Publishers (1993).

[49] FELLER, W. Introduction to Probability Theory and its Applications. vol. I, Wiley, New York (1971).

[50] FELLER, W. Introduction to Probability Theory and its Applications. vol. II, Wiley, New York (1971).

[51] FERNÁNDEZ, R., FRÖHLICH, J. and SOKAL, A.D. Random-Walks, Critical Phenomena, and Triviality in Quantum Field Theory. Springer-Verlag (Texts and Monographs in Physics), 1992.

[52] FERNIQUE, X. Un exemple illustrant l'emploi des méthodes gaussiennes (french)[An example illustrating the use of Gaussian

methods]. *Proceedings of the Conference in honor to J.P. Kahane (Orsay, 1993* J. Fourier Anal. Appl. (1995), Special Issue, 209-213.

[53] FERNIQUE, X. Une majoration des fonctions aléatoires gaussiennes à valeurs vectorielles. *C. R. Acad. Sci. Paris* (1985), 300, Série I, 315–318.

[54] FERNIQUE, X. Fonctions aléatoires à valeurs dans les espaces Lusiniens (french) [Random functions with values in Luzin spaces]. *Expositiones Math.* (1990), 8, 4, 289-364.

[55] FLAJOLET, P. The Evolution of Two Stacks in Bounded Space and Random Walks in a Triangle. *Proceedings of FCT'86.* (1986), LNCS, 233, 425-340, Springer Verlag.

[56] FLAJOLET, P., FRANÇON, J., and VUILLEMIN, J. Sequence of operations analysis for dynamic data structures. *J. of Algorithms.* (1981), 1, 111–141.

[57] FÖLLMER, H. and SCHIED, A. Stochastic Finance: An Introduction in Discrete Time, de Gruyter Studies in Mathematics. 27, Walter de Gruyter & Co., Berlin (2002).

[58] FRANÇON, J., RANDRIANARIMANANA, B., and SCHOTT, R. Analysis of dynamic algorithms in D.E. Knuth's model. *Theoretical Computer Science* (1990), 72, 147–167.

[59] FRANZ, U., and SCHOTT, R. Stochastic Processes and Operator Calculus on Quantum Groups. Kluwer Academic Publishers (1999).

[60] FREIDLIN, M., The Dirichlet problem for an equation with periodic coefficients depending on a small parameter. *Teor. Veroyatnost* (1964), I. Primenen 9, 133-139.

[61] FRIEDMAN, A. Stochastic differential equations and applications Vol. 1. Probability and Mathematical Statistics. 28 (1975), Academic Press.

[62] GALLARDO, L. *Une transformation de Cramer sur le dual de* SU(2). Ann. Sci. Univ. Clermont-Ferrand II Math. (1982), 20, 102–106.

[63] GALLARDO, L. and RIES V. La loi des grands nombres pour les marches aléatoires sur le dual de SU(2). *Studia Mathematica* (1979), 66, 133-148.

[64] GAMET, C. and SCHNEIDER, D. Théorèmes ergodiques multidimensionnels et suites aléatoires universellement représentatives en moyenne. *Annales de l'institut Henri Poincaré* (1997), 33, 2, 269-282.

[65] GILBARG, D., TRUDINGER, N.S. Elliptic partial differential equations of second order Second edition, Grundlehren der Mathematischen Wissenschaften, 224, Springer-Verlag (1983) .

[66] GREVEN, A. and DEN HOLLANDER, F. Large deviations for a random walk in random environment. *Ann. Probab.* (1994), 22 ,3, 1381-1428.

[67] GUILLOTIN, N. Asymptotics of a dynamic random walk in a random scenery. *C.-R.Acad.-Sci.-Paris* (1997), 324, Série I, 231-234.

[68] GUILLOTIN, N. Asymptotics of a dynamic random walk in a random scenery I. A law of large numbers. *Annales de l'Institut Henri Poincaré - Probabilités et Statistiques* (2000), 36, 2, 127-151.

[69] GUILLOTIN, N. Asymptotics of a dynamic random walk in a random scenery II. A functional limit theorem. *Markov Processes and Related Fields* (1999), 5, 2, 201-218

[70] GUILLOTIN-PLANTARD, N. Dynamic Z^d-random walks in a random scenery: A strong law of large numbers. *Journal of Theoretical Probability* (2001), 14, 241-260 .

[71] GUILLOTIN-PLANTARD, N. and SCHNEIDER, D. Ergodic theorems for dynamic random walks. *Mathematical Inequalities and Applications* (2003), 6, 1, 1-30.

[72] GUILLOTIN-PLANTARD, N. and SCHOTT, R. Distributed algorithms with dynamical random transitions. *Random Structures and Algorithms* (2002), 21, 3-4, 371-396.

[73] GUILLOTIN-PLANTARD, N. and SCHOTT, R. Dynamic random walks on Heisenberg groups. *Journal of Theoretical Probability* (in press).

[74] GUILLOTIN-PLANTARD, N. and SCHOTT, R. Dynamic quantum Bernoulli random walks. Prépublication Institut Elie Cartan, 45 (2005)(to appear in Infinite Dimensional Analysis, Quantum Probability and Related Topics).

[75] GUILLOTIN-PLANTARD, N., SCHOTT, R. and SCHURZ, H. Asset Pricing in Dynamic (B, S)-Markets. Prepublication Institut Elie Cartan, 17 (2004).

[76] GUILLOTIN-PLANTARD, N. and LADRET, V. Functional limit theorems for U-statistics indexed by a one-dimensional random walk. ESAIM: Probability and statistics, (2005), 9, 98-115.

[77] GUILLOTIN-PLANTARD, N. and LE NY, A. Random walks in dynamically oriented lattices. Prépublication, Université, Paris 11 (2003).

[78] GUIVARC'H, Y. *Sur la loi des grands nombres et le rayon spectral d'une marche aléatoire.* Astérisque (1980), 74, 47-98.

[79] GUIVARC'H, Y., KEANE, M., and ROYNETTE, B. Marches aléatoires sur les groupes de Lie *Lecture Notes in Mathematics 624*, Springer Verlag (1977).

[80] HABERMAN A. N. System Deadlocks *K. M. Chandy and R. T. Yeh, ed.* (1978), 256–297, Prentice-Hall.

[81] HEYER, H. Probability Measures on Locally compact Groups. Springer Verlag (1977).

[82] HIDA, T. Brownian Motion. Springer Verlag (1980).

[83] HOF, A. Quasicrystals, Aperiodicity and Lattice Systems, *Doctoral dissertation of Rijksuniversiteit.* Groningen (1992).

[84] ITÔ, K. Stochastic integral. *Proc. Imp. Acad. Tokyo.* (1944), 20, 519-524.

[85] HU, H. Decay of correlations for piecewise smooth maps with indifferent fixed points. *Ergodic Theory Dyn. Syst.* (2004), 24, 2, 495-524.

[86] ITÔ, K. On a formula concerning stochastic differential equations. *Nagoya Math. J.* (1951), 3, 55-65.

[87] JACOD, J. and PROTTER, P. Probability Essentials. Universitext, Springer Verlag (2000).

[88] JACOD, J., SHIRYAEV A.N. Limit theorems for stochatic processes. Grundlehren der mathematischen Wissenschaften, Springer Verlag (1987).

[89] JIANG, M. Sinai-Ruelle-Bowen measures for lattice dynamical systems. *J. Stat. Phys.* (2003), 111, 3-4, 863-902.

[90] JIKOV, V.V., KOZLOV, S.M., OLEINIK, O.A. Homogenization of differential operators and integral functionals. Springer Verlag (1994).

[91] KAKUTANI, S. Random ergodic theorems and Markoff processes with a stable distribution. *Proceedings of the Second Berkeley Symposium on Mathematical Statistics and Probability, 1950, 247-261.* University of California Press, Berkeley and Los Angeles (1951).

[92] KARATZAS, I. and SHREVE, S.E.Brownian Motion and Stochastic calculus. Springer Verlag (1988).

[93] KELLER, G. Equilibrium states in ergodic theory. *London Mathematical society*, students texts 42, Cambridge University Press (1998).

[94] KESTEN, H. and SPITZER, F. A limit theorem related to a new class of self-similar processes. *Z. Wahrsch. Verw. Gebiete* (1979), 50, 5–25.

[95] KHINCHIN, A. Continued Fractions. Chicago University Press (1964).

[96] KHOSHNEVISAN, D. and LEWIS, T. M. A law of the iterated logarithm for stable processes in random scenery. *Stochastic Process. Appl.* (1998), 74, 89-121.

[97] KLOEDEN, E., PLATEN, E. and SCHURZ, H. Numerical solution of SDEs through computer experiments. (1st edition), Springer, Berlin, 1994 (2nd edition, 1997, 3rd corrected printing, 2003).

[98] KNESSL C., MATKOWSKY B. J., SCHUSS Z, and TIER C., An Asymptotic Theory of Large Deviations for Markov Jump Processes., *SIAM J. Appl. Math.* (1985), 46, 1006-1028.

[99] KNUTH D. E. The Art of Computer Programming. vol 1, Addison-Wesley (1973).

[100] KNUTH, D.E. Deletions that preserve randomness. *Trans. Software Eng.* (1977) 3,351-359.

[101] KOLMOGOROV, A. Das Gesetz des iterierten Logarithmus. *Math. Ann.* (1929), 101, 126-135.

[102] KORÁNYI, A. Geometric properties of Heisenberg-type groups. *Adv. Math.* (1985), 56, 1, 28-38.

[103] KORÁNYI, A. Geometric aspects of analysis on the Heisenberg group. In: De Michele, L. Ricci, F. (ed.). *Topics in modern harmonic analysis* vol.1, Istituto Nazionale di Alta Matematica Francesco Severi, Roma (1983), 209-258.

[104] KOUKIOU, F., PETRITIS, D. and ZAHRADNIK, M. Extension of the Pirogov-Sinai theory to a class of quasiperiodic interactions. *Commun.Math.Phys.* (1988), 118, 365-383.

[105] KRENGEL, U. Ergodic Theorems. W. de Gruyter (1985).

[106] KREPS, D.M. Arbitrage and equilibrium in economies with infinitely many commodities. *J. Math. Econom.* (1981), 8, 1, 15-35.

[107] KRYLOV, N.V. Controlled diffusion processes. Springer Verlag (1979).

[108] KUIPERS, L. and NIEDERREITER, H. Uniform distribution of sequences. Wiley and Sons (1974).

[109] LACEY, M., PETERSEN, K., WIERDL, M. and RUDOLPH, D. Random ergodic theorems with universally representative sequences. *Annales de l'institut Henri Poincaré* (1994), 30, 3, 353-395.

[110] LAPEYRE, B. and PAGÈS, G. Familles de suites à discrépance faible obtenues par itérations de transformations de $[0,1]$. *C.-R.Acad.-Sci.-Paris* (1989), 308, Série I, 507-509.

[111] LEDRAPPIER, F. Systèmes dynamiques. Presses de l'école Polytechnique (1994).

[112] LEROUX, P. Coassociate grammars, periodic orbits and quantum random walks over $\mathbb{Z}^d$ Prépublication IRMAR, Université de Rennes 1 (2003).

[113] LEWIS, T. M. A self-normalized law of the iterated logarithm for random walk in random scenery. *J. Theor. Prob.* (1992), 5, 4, 629-659.

[114] LEWIS, T. M. A law of the iterated logarithm for random walk in random scenery with deterministic normalizers. *J. Theor. Prob.* (1993), 6, 2, 209-230.

[115] LIPTSER, R.SH. and SHIRYAEV, A.N. Theory of Martingales. Kluwer Academic Publishers, Dordrecht (1989).

[116] LIN, M., RUBSHTEIN, B. and WITTMANN, R. Limit theorems for random walks with dynamical random transitions. *Probab. Theory Relat. Fields* (1994), 100, 285-300.

[117] LIONS, P.-L., SZNITMAN, A.-S. Stochastic differential equations with reflecting boundary conditions. *Comm. Pure Appl. Math.* (1984), 37, 511-537.

[118] LIVERANI, C., SAUSSOL, B. and VAIENTI, S. Conformal measure and decay of correlation for covering weighted systems. *Ergodic Theory Dyn. Syst.* (1998), 18, 1399-1420.

[119] LOUCHARD, G. Random walks, Gaussian processes and list structures. *Theoret. Comput. Sci.* (1987), 53, 1, 99-124.

[120] LOUCHARD, G. Some Distributed Algorithms Revisited. *Commun. Statist. - Stochastic models* (1995), 11, 4, 563-586.

[121] LOUCHARD, G., and SCHOTT, R. Probabilistic Analysis of Some Distributed Algorithms. *Random Structures and Algorithms* (1991), 2, 151–186.

[122] LOUCHARD, G., KENYON, C., and SCHOTT, R. Data structures maxima. *SIAM J. Comput.* (1997), 4, 1006-1042.

[123] LOUCHARD, G., RANDRIANARIMANANA, B., and SCHOTT, R. Probabilistic analysis of dynamic algorithms in D.E. Knuth's model. *Theoretical Computer Science* (1992), 93, 201–225.

[124] LOUCHARD G., SCHOTT R., TOLLEY M. and ZIMMERMANN P. Random Walks, Heat Equations and Distributed Algorithms. *J. Comp. Appl. Math.* (1994), 53, 243-274.

[125] LOYNS, R.M. Products of independent random elements in a topological group *Zeitschrift für Warscheinlichkeitstheorie* (1963), 1, 446-455.

[126] MAES, C., REDIG, F., TAKENS, F., VAN MOFFAERT, A. and VERBISTKIY, E. Intermittency and weakly Gibbs states. *Nonlinearity* (2000), 13, 1681-1698.

[127] MAIER, R.S. A path integral approach to data structures evolution. *Journal of Complexity* (1991), 7, 3, 232-260.

[128] MAIER R. S. Colliding Stacks: A Large Deviations Analysis. *Random Structures and Algorithms* (1991), 2, 379–420.

[129] MAIER R. S. and SCHOTT R. Exhaustion of Shared Memory: Stochastic Results. *Proceedings of WADS'93.* (1993), LNCS no 709, 494–505, Springer Verlag.

[130] MANDELBROT, B. Fractals and Scaling in Finance, Discontinuity, Concentration, Risk. Springer Verlag, Berlin (1997).

[131] MANDELBROT, B. and VAN NESS, J.W. Fractional Brownian motions, Fractional Noises and Applications. *SIAM Rev.* (1968), 10, pp. 422-437.

[132] MARI J. F. and SCHOTT R. Probabilistic and Statistical Methods in Computer Science, Kluwer Academic Publishers (2001).

[133] MARTINEZ, S. and PÉTRITIS, D. Thermodynamics of a Brownian bridge polymer model in a random environment. *J. Phys. A* (1996), 29, 6, 1267-1279.

[134] MEL'NIKOV A.V. Financial Markets, Stochastic Analysis and the Pricing of Derivative Securities, *Translation of Mathematical Monographs.* 184, American Mathematical Society (1999).

[135] MEYER, P.A. Quantum Probability for Probabilists *Lect. Notes Math. 1538.* Springer Verlag (1993).

[136] MERTON, R. Option pricing when underlying stock returns are discontinuous. *J. Finan. Econom.* (1976), 3, 125-144.

[137] NAEH T., KLOSEK M. M., MATKOWSKY B. J., and SCHUSS Z. A Direct Approach to the Exit Problem., *SIAM J. Appl. Math.* (1990), 50, 595-627.

[138] NEUENSCHWANDER, D. Probabilities on the Heisenberg group. *Lect. Notes Math. 1630.* Springer Verlag (1996).

[139] NEUENSCHWANDER, D. and SCHOTT, R. On the local and asymptotic behavior of Brownian motion on simply connected Lie groups *J. Theoret. Probab.* (1995), 8, 4, 795-806.

[140] NEWMAN, C. M. and WRIGHT, A. L. An invariance principle for certain dependent sequences. *Annals of probability* (1981), 9, 671-675.

[141] NIEMINEN, A. Fractional Brownian motion and martingale-differences. Preprint (2003).

[142] OLLA, S. Central limit theorems for tagged particles and for diffusions in random environment. Comets, Francis (ed.) et al., Random media. Paris: Société Mathématique de France. Panor. Synth. (2003), 12, 75-100.

[143] OSGOOD F., C. Diophantine approximation and its applications. Academic Press (1973).

[144] PAGÈS, G. and XIAO, Y.J. Sequences with low discrepancy and pseudo-random numbers: theoretical remarks and numerical tests. *Journal of Statistical Computation and Simulation* (1997), 56, 163-183.

[145] PARDOUX, É. Homogenization of linear and semilinear second order parabolic PDEs with periodic coefficients: a probabilistic approach. *J. Funct. Anal.* (1999), 167 ,498-520.

[146] PARDOUX, É., VERETENNIKOV, A. YU. Averaging of backward stochastic differential equations, with application to semi-linear PDE's. *Stochastics Stochastics Rep.* (1997), 60, 255-270.

[147] PARTHASARATHY, K.R. A generalized Biane's process. *Lect. Notes Math. 1426.* (1990), 345-348, Springer Verlag.

[148] PARTHASARATHY, K.R. An Introduction to Quantum Stochastic Calculus. Birkhäuser (1992).

[149] PETERSEN, K. Ergodic theory. *Cambridge Studies in Advanced Mathematics.* Cambridge University Press, Cambridge (1983).

[150] PIANIGIANI, G. First return map and invariant measures. *Isr. J. Math.* (1980), 35, 32-48.

[151] PINSKY, R.G. Positive harmonic functions and diffusion. Cambridge Studies in Advanced Mathematics, 45, Cambridge University Press, Cambridge (1995).

[152] PROTTER, P. Stochastic Integration and Differential Equations. Springer Verlag (1990).

[153] REVUZ, D. and YOR, M. Continuous martingales and Brownian motion, Springer Verlag (1991).

[154] ROGERS, L.C.G., WILLIAMS D. Diffusions, Markov Proceses and Martingales. Volume 1, Foundations. John Wiley and Sons (1979).

[155] ROGERS, L.C.G., WILLIAMS, D. Diffusions, Markov Proceses and Martingales. Volume 2, Itô Calculs. John Wiley and Sons (1987) .

[156] ROZANOV, Y.A.. Probability theory: A concise course, Dover Pub, Inc (1969).

[157] SAISHO, Y. Stochastic differential equations for multidimensional domain with reflecting boundary. *Probab. Theory Related Fields.* (1987), 74, 455-477.

[158] SCHACHERMEYER, W. A Hilbert space proof of the fundamental theorem of asset pricing in finite discrete time, Insurance. *Math. Econom.* (1992), 11, 249-257.

[159] SCHACHERMAYER, W. Martingale measures for discrete-time processes with infinite horizon, *Math. Finance* 4 (1994), no. 1, pp. 25-55.

[160] SCHMIDT, W.M. Simultaneous approximation to algebraic numbers by rationals. *Acta Math.* (1970), 125, 189-201.

[161] SCHNEIDER, D. Convergence presque sûre de moyennes ergodiques perturbées. *C. R. Acad. Sci. Paris* (1994), 319, Série I, 1201-1206.

[162] SCHNEIDER, D. Convergence presque sûre de moyennes ergodiques perturbées. (*Thèse, I.R.M.A., Strasbourg* (1994).

[163] SCHNEIDER, D. Théorèmes ergodiques perturbés. *Israel Journal of Mathematics* (1997), 101, 157-178.

[164] SCHNEIDER, D. Polynômes trigonométriques et marches aléatoires multidimensionnelles : application à la théorie ergodique. *Ann. Inst. H. Poincaré Probab. Statist.* (2000), 36, 5, 617-646.

[165] SCHNEIDER, D. and WEBER, M. Weighted averages of contractions along subsequences. Proceedings of the Conference "On almost everywhere convergence in ergodic and probability theory". W. de Gruyter and Co., Columbus, Ohio State University USA (1996).

[166] SCHÜRMANN, M. White Noise on Bialgebras *Lect. Notes Math. 1544.* Springer Verlag (1991).

[167] SCHURZ, H. Stability, Stationarity, and Boundedness of Some Implicit Numerical Methods for Stochastic Differential Equations and Applications. Logos-Verlag, Berlin (1997).

[168] SCHURZ, H. Numerical analysis of SDE without tears. in *Handbook of Stochastic Analysis and Applications*, ed. by D. Kannan and V. Lakshmikantham, Marcel Dekker, Basel, 237-359 (2002).

[169] SCHURZ, H. Discrete Sub-, Super- and Martingale Theory and Related Stochastic Calculus. Manuscript, Southern Illinois University, 2004 (textbook in preparation).

[170] SHIRYAYEV, A.N. Probability, Springer Verlag (1984).

[171] SHIRYAYEV, A.N. Essentials of Stochastic Finance: Facts, Models, Theory, World Scientific (2000).

[172] SINAI, I. Dynamical systems, Vol 2, Springer Verlag (1987).

[173] SLOMIŃSKI, L. Euler's approximations of solutions of SDEs with reflecting boundary. *Stochastic Process. Appl.* (2001), 94, 317-337.

[174] SOLOMON, F. Random walks in a random environment. *Annals of Probability* (1975), 3, 1, 1-31.

[175] SOTTINEN, T. Fractional Brownian motion, random walks and binary market models. *Finance Stochast.* (2001) 5, 343-355.

[176] SOTTINEN, T. and VALKEILA, E. On arbitrage and replication in the fractional Black-Scholes pricing model. *Statistics & Decisions.* (2003), 21, 137-151.

[177] SPITZER, F. Principles of random walk. 2nd ed., New York, Springer-Verlag (1976).

[178] STRASSEN, V. An invariance principle for the law of the iterated logarithm. *Zeitung Wahrscheinlichkeitstheorie verw. Gebiete* (1964), 3, 211-226.

[179] STRATONOVICH, R.L. A new representation for stochastic integrals and equations. *SIAM J. Control.* (1966), 4, 362-371.

[180] STRICKER, C. Arbitrage et lois de martingale (French, English: Arbitrage and martingale laws *Ann. Inst. H. Poincaré Probab. Statist.* (1990), 26, 3, 451-460.

[181] STROOCK, D.W., VARADHAN S.R.S. Multidimensional diffusion processes. Springer-Verlag (1979).

[182] SZÜSZ, P. and VOLKMANN, B. On Strassen's Law of the iterated Logarithm. *Zeitung Wahrscheinlichkeitstheorie verw. Gebiete* (1982), 61, 453-458.

[183] TANAKA, H. Stochastic differential equations with reflecting boundary condition in convex regions. *Hiroshima Math. J.* (1979), 9, 163-177.

[184] TAQQU, M.S. and WILLINGER, W. The analysis of finite security markets using martingales. *Adv. Appl. Probab.* (1987), 19, 1-25.

[185] THALER, M. Estimates of the invariant densities of endomorphisms with indifferent fixed points. *Israel. J. Math.* (1980), 37, 303-314.

[186] VARADHAN, S.R.S. Asymptotic probabilities and differential equations. *Comm. on Pure and Applied Math.* (1966), 19, 261-286.

[187] YAO, A. C. An Analysis of a Memory Allocation Scheme for Implementing Stacks. *SIAM J. Comput.* (1981), 10, 398-403.

[188] YOCCOZ, J-C. Sur la disparition de la propriété de Denjoy-Koksma en dimension 2. *Astérisque.* 231 (1995).

[189] YOUNG, L. Recurrence times and rates of mixing. *Isr. J. Math.* (1998), 110, 153-188.

Index

Skorohod topology, 137

absorbing barrier, 4, 119
absorbing boundary, 4, 128
allocation algorithm, 119
annealed case, 169, 174
array, 144
asset pricing, 191
associated, 41

banker algorithm, 4, 128
Bernoulli shift, 186
Bernstein-Kolmogorov Inequality, 21
Birkhoff's theorem, 232
Black-Scholes model, 222
Boltzmann, 231
Borel subset, 233
boundary, 120
broken corner, 4, 128
Brownian motion, 12, 120, 122, 129

central limit theorem, 12
characteristic function, 12, 36, 89, 180
Clebsch-Gordan's formula, 103
colliding stacks problem, 119
Cox-Ross-Rubinstein's model, 226
critical phenomena, 167

data structures, 143
data type, 143
Denjoy-Koksma's inequality, 236
dictionary, 144
diophantine approximations, 235
Dirac distribution, 123
Dirac measure, 103
distributed algorithms, 119
Doeblin condition, 135
dual of $SU(2)$, 102, 105
dynamic random walk, 3
dynamical system, 5

dynamically oriented lattice, 167, 168, 170

embedding, 184
entropy, 181
ergodic, 134
ergodic dynamical system, 168
Ergodic theory, 231

financial market, 191
Fourier series, 243
free monoid, 144
functional limit theorem, 47, 171, 185

Gaussian random variable, 179
generalized convolution, 103
Gibbs measure, 186
gradient, 136

Hardy and Krause, 75
heap, 144
Heisenberg group, 83
Hilbert spaces, 245
history, 145
hitting time, 4, 128
homeomorphism, 152
homogeneous norm, 84
homogenization property, 137
hyperplane, 133

irrational angle, 126, 235
irreducible, 135
iterated logarithm, 20

kernel, 133
Killing inner produc, 105
Knuth's model, 145
Kolmogorov's criteria, 9
Korányi ball, 84

large deviation principle, 156

large deviation principle, 23
lattice , 169
Lie group, 83
limit theorems, 11
linear list, 144
linked list, 144
local limit theorem, 14, 106
long-term average, 231

Manneville-Pomeau map, 171
Markov chain, 137
Markov shift, 187
markovian model, 145
measure-preserving transformation, 231

nilpotent group, 84
non-random environment, 167

orbit, 231

pagoda, 144
Pauli matrices, 101
possibility function, 145
priority queue, 144
probability space, 5

quadratic form, 15, 36
quantum Bernoulli random walk, 105
Quantum random walk, 99
quenched case, 168

random orientation, 168
random scenery, 39
random walk, 119
rectangle, 128
recurrence, 33, 34
reflecting barrier, 4, 119
rotation on the torus, 31, 190, 233

schema, 144
shared storage, 119
Skorohod metric, 241
spectral lemma, 76
SRB measure, 189
stack, 143
Stationary stochastic process, 233
Strassen's Functional Law (of the Iterated Logarithm), 20
strong law of large numbers, 11
symbol table, 144

transience, 33, 35, 174
transient random walk, 167
translation-invariance, 169
translation-invariant Gibbs measure, 188
triangle, 119
two stacks problem, 4, 120

Van der Corput's Inequality, 75
vertical and horizontal embeddings, 172
Von Neumann's theorem, 232